INDUSTRIAL ABRASIVE MATERIALS AND COMPOSITIONS

INDUSTRIAL ABRASIVE MATERIALS AND COMPOSITIONS

Edited by M.J. Collie

NOYES DATA CORPORATION
Park Ridge, New Jersey, U.S.A.
1981

Library of Congress Catalog Card Number: 81-38326
ISBN: 0-8155-0851-4
Printed in the United States

Published in the United States of America by
Noyes Data Corporation
Noyes Building, Park Ridge, New Jersey 07656

Library of Congress Cataloging in Publication Data

Main entry under title:

Industrial abrasive materials and compositions.

(Chemical technology review ; no. 190)
Includes index.
1. Abrasives--Patents. 2. Patents--United States.
I. Collie, M. J. II. Series.
TJ1296.I53 621.9'2 81-38326
ISBN 0-8155-0851-4 AACR2

Foreword

The detailed, descriptive information in this book deals with industrial abrading processes, abrasive materials and compositions based on recent U.S. patents issued since January 1975.

This book is a data-based publication, providing information retrieved and made available from the U.S. patent literature. It thus serves a double purpose in that it supplies detailed technical information and can be used as a guide to the patent literature in this field. By indicating all the information that is significant, and eliminating legal jargon and juristic phraseology, this book presents an advanced commercially oriented review of recent developments in the field of abrasives.

The U.S. patent literature is the largest and most comprehensive collection of technical information in the world. There is more practical, commercial, timely process information assembled here than is available from any other source. The technical information obtained from a patent is extremely reliable and comprehensive; sufficient information must be included to avoid rejection for "insufficient disclosure." These patents include practically all of those issued on the subject in the United States during the period under review; there has been no bias in the selection of patents for inclusion.

The patent literature covers a substantial amount of information not available in the journal literature. The patent literature is a prime source of basic commercially useful information. This information is overlooked by those who rely primarily on the periodical journal literature. It is realized that there is a lag between a patent application on a new process development and the granting of a patent, but it is felt that this may roughly parallel or even anticipate the lag in putting that development into commercial practice.

Many of these patents are being utilized commercially. Whether used or not, they offer opportunities for technological transfer. Also, a major purpose of this book is to describe the number of technical possibilities available, which may open up profitable areas of research and development. The information contained in this book will allow you to establish a sound background before launching into research in this field.

Advanced composition and production methods developed by Noyes Data are employed to bring these durably bound books to you in a minimum of time. Special techniques are used to close the gap between "manuscript" and "completed book." Industrial technology is progressing so rapidly that time-honored, conventional typesetting, binding and shipping methods are no longer suitable. We have bypassed the delays in the conventional book publishing cycle and provide the user with an effective and convenient means of reviewing up-to-date information in depth.

The table of contents is organized in such a way as to serve as a subject index. Other indexes by company, inventor and patent number help in providing easy access to the information contained in this book.

16 Reasons Why the U.S. Patent Office Literature Is Important to You

1. The U.S. patent literature is the largest and most comprehensive collection of technical information in the world. There is more practical commercial process information assembled here than is available from any other source. Most important technological advances are described in the patent literature.
2. The technical information obtained from the patent literature is extremely comprehensive; sufficient information must be included to avoid rejection for "insufficient disclosure."
3. The patent literature is a prime source of basic commercially utilizable information. This information is overlooked by those who rely primarily on the periodical journal literature.
4. An important feature of the patent literature is that it can serve to avoid duplication of research and development.
5. Patents, unlike periodical literature, are bound by definition to contain new information, data and ideas.
6. It can serve as a source of new ideas in a different but related field, and may be outside the patent protection offered the original invention.
7. Since claims are narrowly defined, much valuable information is included that may be outside the legal protection afforded by the claims.
8. Patents discuss the difficulties associated with previous research, development or production techniques, and offer a specific method of overcoming problems. This gives clues to current process information that has not been published in periodicals or books.
9. Can aid in process design by providing a selection of alternate techniques. A powerful research and engineering tool.
10. Obtain licenses–many U.S. chemical patents have not been developed commercially.
11. Patents provide an excellent starting point for the next investigator.
12. Frequently, innovations derived from research are first disclosed in the patent literature, prior to coverage in the periodical literature.
13. Patents offer a most valuable method of keeping abreast of latest technologies, serving an individual's own "current awareness" program.
14. Identifying potential new competitors.
15. It is a creative source of ideas for those with imagination.
16. Scrutiny of the patent literature has important profit-making potential.

Contents and Subject Index

Introduction

Abrasives are rough materials which may be used to grind, wear away, clean, smooth or polish other materials. Among the earliest tools were those involving use of an abrasive, such as grinding stones for grain and cutting tools made by chipping the surface of one rock with another, harder one.

Abrasives are usually divided into two classes, bonded abrasives and coated abrasives. Bonded abrasives are in the form of rigid, shaped material, such as a grinding wheel. In bonded abrasives the grains of abrasive material are dispersed in a binder of resin or vitrified ceramic material. As abrasive grains, there may be mentioned diamond, natural and synthetic; cubic boron nitride; alumina, alone or in combination with other oxides; and carbides.

This book describes many processes related to formation of these and other abrasive grains, such as methods for control of crystal size, uniformity and porosity; rapid cooling of alumina-zirconia melts; formation of diamond of desired crystal structure, synthetic diamond manufacture and coating of diamonds. Processes for forming compacts and composites from these grains; binders and matrices for such composites; sintering, cold and hot press forming methods are included. Brazing alloys, fluxes and plating techniques for improved methods of manufacturing grinding wheels and cutting tools are described. Abrasive grains are also used in a loose form as in impact and tumbling cleaning. A section on these uses of abrasive grains is included.

In coated abrasives the grains are attached to a flexible backing material, usually cloth or paper, by a resin or glue. Ordinary sandpaper is the first example of a coated abrasive that comes to mind. Coated abrasive belts and discs are widely used in industry.

This book provides coatings and bonding agents, manufacturing processes and backings for coated abrasives. Also included are several patents for abrasive grains intended for use on coated abrasives.

Abrasives may also be classified as grinding abrasives or polishing abrasives. One chapter in this book discloses compositions and processes for polishing and finishing metals, lenses, semiconductors, photoreceptors and bearing surfaces. Another provides grinding aids and lapping compounds.

Abrasive-Surfaced Tools

GRINDING WHEELS

Interchangeable Composite Abrading Segments

C.V. Rue; U.S. Patent 4,212,137; July 15, 1980; assigned to Norton Company describes an annular segmental grinding wheel having a plurality of interchangeable composite abrading segments preferably with interfitting tongue-and-groove-like abutting side portions clamped to a rotatable chuck. Each composite segment comprises a preformed bonded abrasive segment of simple arcuate shape encased in a relatively nonabrasive precision molded shell of more complex shape covering at least a portion of each side except the working face of the abrasive segment of less consistent dimensions. On one end, the shell has a tongue or projection and on its opposite end a tongue or projection receiving depression, groove, slot, or recess molded to precise dimensions and shape for interfitting and abutting engagement with identical tongue-and-groove portions of adjoining composite segments in the segmental grinding wheel.

The shell is also provided with accurately molded convex, concave and top surfaces for precise mating, clamping engagement with the inner locating and retaining walls of the chuck and clamp means for securing the composite segments to the chuck.

Depending on the type of chuck and clamping means thereon, the composite segment may or may not be provided with a clearance slot for allowing the passage of a clamping screw for retaining the segment. The clearance slot, may be molded solely in a top portion of a plastic shell or in both the abrasive segment and the shell.

Various thermoplastic and thermosetting compositions, plastic materials and resins may be used in conjunction with suitable conventional injection or transfer compression molding or casting apparatus to mold or cast the complex shells about the abrasive segments of simpler shape.

Preferably, the material or composition of which the shell is molded or cast is itself a grinding aid or contains as much of any known grinding aid type filler as is consistent to maintain flowability and strength thereof.

The materials or composition of the shell may also contain modifiers such as colorizers, fire retardants, reinforcing fibers, active or inactive fillers and other additives to produce alloys thereof with exceptional physical and chemical qualities.

During grinding the jacket or shell material simultaneously wears to enhance the abrading process. Also, the remaining side portions prevent absorption of coolant swarf into the abrasive segments and reinforce the abrasive segments against breakage and chipping at the edges thereof.

Producing a typical composite abrading segment of the process similar to the Sterling type includes conventionally molding an abrasive composition containing abrasive particles such as aluminum oxide of the desired grit size, e.g., 30 grit, and a conventional vitrifiable bond composition containing glass into a self-supporting green abrasive segment of smaller size and of simpler arcuate shape than the desired Sterling type segment. The green segment is then fired in a kiln, cooled and removed therefrom in the well known manner. Each vitrified bonded segment is then cleaned of foreign matter, placed into and held centered in a molded cavity of the desired finished size and shape of the Sterling type segment. A thermoplastic grinding aid material such as polyvinyl chloride on sulfur reinforced with glass fibers and wollastonite fillers, heated to maintain it in a flowable state, is then injected in the usual manner into the mold cavity until the space between the abrasive segment and mold is filled. The mold is then cooled, to solidify the material, opened and the composite segment with a shell of specified thickness molded to the precise interfitting size and shape is removed therefrom.

When the shell is made of a thermosetting material, the material preferably contains a grinding aid filler and is usually preheated to a more flowable uncured state before it is rammed into the mold cavity between the abrasive segment and a heated mold. Thus, the thermosetting material forced into the mold is heated thereby to a permanently hardened shell of precise complex shape about the abrasive segment of simpler shape. Each composite segment as molded is ready for immediate interfitting supporting engagement with other identically molded composite segments of a segmental wheel and the mating, supporting and clamping surfaces of the chuck and clamping means.

Long Axis of Needle-Shaped Particles Normal to Working Face

C. Phaal; U.S. Patent 4,203,732; May 20, 1980 provides a method of making an abrasive product including the steps of providing a mixture of needle-shaped abrasive particles and starter ingredients for a bonding matrix, causing the mixture to flow so causing the particles to orientate with their long axes substantially in the direction of flow, and causing the starter ingredients to set, at least partially.

Preferably, the mixture is caused to flow along a passage which has a constriction. The flow of mixture accelerates through the constriction and this assists in the orientation of the particles. The constriction is preferably provided by a narrowing of the passage and, in particular, by a narrowing of the cross-sectional area thereof.

The passage may also be provided with a bend which, when provided, may be a right angled bend.

Flow along the passage may conveniently be achieved by means of pressure, e.g., in the form of a plunger acting in the passage. The mixture is usually in powdered form with the starter ingredients being of finer particle size than the abrasive particle.

The method may be used to form a grinding wheel as a whole, i.e., the rim is bonded in situ to the hub portion, or used to make a rim or segment which is subsequently bonded to a hub portion.

In the case of rim production using flow through a passage, the passage preferably leads into a cavity defining the rim, the direction of mixture flow being substantially normal to the working face of the rim. In this way, the particles will be orientated with their long axes substantially normal to the working face of the rim which is advantageous particularly in the case of resin bond grinding wheels.

In the case of segments, these are preferably extruded through an orifice of predetermined shape and then cut to a desired size and shape. The method can be used with particular advantage in making segments for cup grinding wheel rims. Such segments can be produced with the long axes of the particles orientated substantially normal to the working face of the segment.

In related work by *C. Phaal; U.S. Patent 4,008,055; February 5, 1977* the needle-shaped abrasive is cubic boron nitride. Cubic boron nitride particles include particles known in the art as "flats" which have a long axis and a short axis transverse to the long axis. The ratio of long axis to short axis is preferably at least 3:1.

The preferred bonding matrix is a resin matrix such as phenol-formaldehyde resin or a polyimide resin. In this case the abrasive product will be for use in, or form part of, a resin bond grinding wheel.

The abrasive particles are preferably metal coated, particularly when the matrix is a resin. The preferred metal is nickel. The coating preferably comprises between 30 and 80 wt % based on the coated particle.

The particles may be aligned by means of any suitable impressed field of force. Thus, pressure or electrostatic forces may be used. The preferred method is, however, the use of an impressed magnetic field. In this case, the particles will be provided with a magnetic coating and the bonding matrix will be of a nonmagnetic material. Resins are suitable nonmagnetic bonding matrices and nickel, cobalt and iron, preferably nickel, are suitable magnetic coating materials for the particles.

Example: Cubic boron nitride particles of particle size 60/80 U.S. mesh and containing about 80% needle-shaped particles having a ratio of long axis to transverse short axis of at least 2:1 were provided with a 60 wt %, based on the coated particle, nickel coating. The nickel-coated particles were then mixed with powdered phenol-formaldehyde resin starter ingredients and usual inorganic fillers such as silicon carbide. The quantities of the various components in the mixture was 12.5 vol % cubic boron nitride, based on the uncoated particle, 25 vol % resin starter ingredients and 62.5 vol % filler.

The mixture was used for making a resin bond grinding wheel and in particular the rim thereof. The mold for this wheel was conventional in the sense that it was provided with a bakelite hub portion around which was a rim-defining groove portion. The mold was unconventional in the sense that it was made of stainless steel and was provided with a first permanent magnet inside the hub portion and a second permanent magnet, of opposite pole, outside the rim-defining groove.

The mixture was poured into the rim-defining groove and in so doing passed through the magnetic field. The lines of magnetic force being radial, i.e., directed from the center to the periphery of the mold, caused a substantial portion of the nickel-coated needle-shaped cubic boron nitride particles to align radially. In other words, the long axes of these aligned particles were substantially normal to the periphery of the rim portion which defines the working face of the wheel. The magnetic field was maintained while heat and pressure was applied to the mixture to cause the resin to set.

The G ratio of a number of wheels made in the abovedescribed manner and tested on M-2 high speed steel was determined and averaged. The average of six wheels was found to be 78. By way of comparison similar wheels were made using nickel-coated cubic boron nitride which was not of needle shape. The average G ratio of six such wheels again tested on M-2 high speed steel was found to be 65.

Ceramic Bonded Tools with Graphite in Bond

The grinding tool of *P.P. Keat; U.S. Patent 4,157,897; June 12, 1979; assigned to Norton Company* makes practical the use of ceramic bonded diamond wheels in the dry grinding of cemented carbide tools. The grinding tool includes a grinding element consisting of diamond or cubic boron nitride abrasive grits bonded by a ceramic matrix. The ceramic matrix bond includes graphite in an amount of from 23 to 53% of the total bond volume for grinding wheels, and from 10 to

40% for hones (including the graphite). The porosity of the element is below 15%, and the grinding wheels is below 10% (by volume), and is as low as the fabrication technique permits. The graphite should be of high purity, preferably "flake" graphite and should be less than 200 μ, numerical average particle size. Particle size means the average of the longest and shortest dimensions of the flakes as they normally lie on a flat plane (that is, the thickness is disregarded).

The grinding elements are mounted on supports to form grinding tools, such as grinding wheels of standard shape. A suitable material for making the support member is aluminum-filled resin as disclosed in U.S. Patent 2,150,886. Epoxy resins of good thermal conductivity are convenient elements for joining the grinding element to the support or preform to produce the grinding tool.

This process relates solely to ceramic bonded elements and any of the known glass bonds may be used. Typically such bonds are made up of a glass containing silica, boron oxide, sodium oxide, aluminum oxide and alkali and alkaline earth metal oxides. Such glasses may include a microcrystalline inorganic filler which is added prior to forming the tool or which is formed in situ by partial devitrification of the glass bond composition.

Examples 1 through 3: In the tests noted below, variations of the rotary grinding tools of the process were compared with a standard resinoid bonded grinding wheel of the prior art. No vitrified wheels without graphite were employed as standards since such wheels are, as a practical matter, unusable in the dry grinding of carbides or high hardness steels.

Three wheels were prepared using nickel-coated diamond (150 grit), a devitrified glass bond containing varying proportions of flake graphite (finer than 325 mesh). The standard wheel was a resin bonded wheel containing 35 vol % of silicon carbide filler in the bond but no graphite. The grinding elements contained 18¼ vol % of diamond except that of Example 3 which contained only 17¾ vol % of diamond, and the resinoid standard which contained 16% diamond. The wheels were cup-wheels (type D6A9) 4 x 1¾ x 1¼ inches.

The grinding machine was a Norton S-4, 6 x 18 inches surface grinder with automated cross-feed, index and shutdown modifications, operating as follows: wheel speed, 3,800 surface feet per minute; table traverse, 100 inches per minute; unit in-feed, 2.0 mils; material ground, 44A Carboloy (cemented tungsten carbide); and operation, dry grinding. The results were as follows:

	% Graphite in Bond Vol %	Porosity Vol %	G Ratio	Power Drawn
Example 1	27	3.5	101	1,325
Example 2	41	5.1	303	795
Example 3	48½	7.8	47	500
Standard	—	—	33	1,130

The G ratio represents the total volume of material ground divided by the total volume of wheel wear. The power drawn is measured by a wattmeter in the power line on the motor driving the wheel.

Thermosetting Resin Bond for Abrasive and Base Member

I. Suzuki, T. Imai, M. Kitajima and K. Hota; U.S. Patent 4,099,934; July 11, 1978; assigned to Toyoda Koki KK, Japan provide a method for manufacturing a resinoid-bonded grinding tool in which metal-coated abrasive grains are adhesively attached upon the surface of a metallic base member with a thermosetting resin, comprising the steps of:

(a) washing the surface of the base member so as to expose a pure metallic surface thereon by washing the surface of the base member with a grease removing solvent, washing the washed surface with an alkali solution, and then washing the alkali washed surface with an acid solution;

(b) applying a liquid thermosetting resin to the pure metallic surface of the base member so as to prevent the formation of an oxide film thereon or the adherence of gas thereto during subsequent heating;

(c) washing the abrasive grains;

(d) coating the surface of the abrasive grains with a liquid thermosetting resin;

(e) mixing a powdered thermosetting resin with the abrasive grains covered with the liquid thermosetting resin and press-forming the mixture upon the resin-applied surface of the base member to form a press-formed article of a desired shape; and

(f) heating the press-formed article formed upon the base member so as to thermally set the formed article upon the base member.

Vitreous Bonded Cubic Boron Nitride

P.C. Balson; U.S. Patent 3,986,847; October 19, 1976; assigned to Cincinnati Millacron, Inc. describes an abrasive article, such as a grinding wheel, having an abrasive section comprising:

(1) as the abrasive phase cubic boron nitride abrasive and from 0 to 90% of the total volume of the abrasive phase of a second abrasive grain having a coefficient of thermal expansion substantially the same as the coefficient of thermal expansion of cubic boron nitride, and

(2) as a vitreous bond bonding the abrasive phase, a vitreous bond having a coefficient of thermal expansion substantially the same as the coefficient of thermal expansion of cubic boron nitride and being substantially nonreactive with the cubic boron nitride abrasive grain. The second abrasive grain is characterized in that on the surface of the cubic boron nitride abrasive grain interfacing the vitreous bond there shall have been chemically formed thereon a thin, adherent boric oxide layer.

In the grinding of metals the grinding wheels of this process exhibit high resistance to thermal and mechanical shock during grinding, good grinding efficiency and high retention of shape during grinding.

Example 1: The following formulation was used to form a grinding wheel: 37.88 wt % cubic boron nitride (Borazon I) 170/200 grit, 34.85 wt % black silicon carbide 180 grit, 19.44 wt % bond and 7.83 wt % 50% by weight solids aqueous paraffin wax emulsion.

The bond used in the above formulation had essentially the following composition expressed as the oxides: 1.0 part by weight K_2O, 1.0 part by weight Na_2O, 2.0 parts by weight Al_2O_3, 26.0 parts by weight B_2O_3 and 70.0 parts by weight SiO_2. The volume ratio of cubic boron nitride to silicon carbide was 50:50.

The cubic boron nitride and silicon carbide were blended and then the 50% by weight solids aqueous paraffin wax emulsion added and mixed into the blend of cubic boron nitride and silicon carbide. The bond was then added and thoroughly mixed into the blend and the resultant mixture was then dried in an oven at 105°F to essentially remove the water from the mixture. Upon being removed from the oven, the dried mixture was broken up by passing it through a 45 mesh sieve. The required amount of the dried, particulate mixture was then placed in a suitable mold, the completely assembled mold placed in a press and the wheel pressed to the desired size and shape. The wheel, upon being removed from the mold, was then fired in an oxidizing atmosphere (i.e., air) by heating at the rate of 300°F per hour until a temperature of 1525°F was reached and then maintaining the 1525°F temperature for 3 hours.

Example 2: In this example a 1.675 x 0.875 x 0.675 inch grinding wheel of this process, made in accordance with Example 1, was compared to a comparable commercially available vitrified bonded cubic boron nitride grinding wheel in the internal bore grinding of 52100 steel bearings. When using a commercial aqueous based coolant, a wheel speed of 12,000 surface feet per minute (SFM) and a work speed of 300 to 500 SFM the aforementioned grinding wheel of this process required a force to cut of 60 lb whereas the commercial wheel required a force of 110 lb to produce a cutting action.

In the grinding of metals the higher the force required to cut the greater are the power requirements, power consumption, the temperature of wheel and the temperature of the metal workpiece. The higher the temperature of the grinding wheel the greater is the susceptibility of the wheel to break down. The higher the temperature in metal workpiece during grinding the greater the distortion of the metal workpiece and the greater is the incidence of out-of-tolerance ground parts. Thus, lower cutting forces are desired and indeed advantageous in the grinding of metal, particularly in the precision grinding of metal, such as internal grinding.

Metal Bonded Wheel with Continuously Bonded Grains and Rim

T. Asaeda, I. Suzuki and T. Imai; U.S. Patent 3,973,925; August 10, 1976; assigned with part interest to Toyoda Koki KK, Japan provide an improvement in the prior art manufacturing process for a metal bonded grinding tool comprising the steps of (1) preliminary pretreating in which abrasive grains have their surfaces cleaned and prepared for subsequent metal coating, (2) press-forming in which the abrasive grains pretreated in step (1) are press-formed to specified structural dimensions, (3) bond metal coating in which the press-formed abrasive grains obtained in step (2) are placed in a chemical plating liquid and the chemical plating liquid is forcibly passed through the spaces between the abrasive grains, whereby the metal is precipitated on the surfaces of the abrasive grains and the abrasive grains are bonded with the precipitated metal, and (4) heat treating for crystallizing the precipitated metal. The improvement comprises the steps of:

(a) preparing a rim of porous material formed from a sintered alloy;

(b) press-forming abrasive grains in contact with the rim of porous material to a specified structural dimension; and

(c) forcibly passing a nickel chemical plating liquid through and radially out of a contacting body of the abrasive grains and the rim of porous material obtained in step (b) for continuously forming the metal precipitate on the surfaces of the abrasive grains, but not completely filling the interstices defined therebetween, and in the pores of the rim of porous material so as to bond the abrasive grains uniformly in an axial direction of the rim and continuously bond the abrasive grains and the rim.

Among others, the following features are obtained in the process. A uniform metal bonding characteristic can be obtained for all dimensional grinding wheels even if the width is larger than the thickness, because the chemical plating liquid can pass through the rim and the press-formed abrasive grains especially when the plating liquid is supplied in a radial direction of the rim. The abrasive grains can be strongly bonded on the rim because the surface area of the rim is sufficient due to the unevenness effected by the etching operation and the porosity thereof. Moreover, there is no boundary portion between the rim and the abrasive grains because the plating liquid is supplied through the rim and the abrasive grains in a radial direction thereof.

In addition, it is possible to get a rim which is fully durable against the normal severe force at the time of a heavy and high speed grinding operation since the pores of the rim can be filled by the chemical plating metal when the plating liquid is passed therethrough in the radial direction and outwardly. Furthermore, as the bonding layer is all metallic, high electric conductivity can be obtained at the portion of the rim and the abrasive grains and therefore, a grinding wheel which is suitable for electrolytic grinding can be obtained.

Blow Molding Process

Mixtures for grinding wheel manufacture are obtainable that include particles covering a substantially narrower range of grain sizes than in the past for all the raw materials in the mix. This is more particularly the case for so-called "granulated mixtures."

Research work has shown that mixtures of such type could be used in conjunction with grinding wheel manufacture by blowing, thus providing quality products with otherwise repeatable properties.

Though applicable to all kinds of grinding wheels, the method described by *D.R. Vinard; U.S. Patent Reissue 29,592; March 28, 1978; assigned to Norton Company* is best suited for organic bonded grinding wheels.

Studies performed in conjunction with the process have shown that satisfactory results may be achieved provided that the granulated mixture used corresponds, for instance, to one of the grain sizes that may vary in accordance with the relationship wherein D is the mean diameter of particles or granules of the mixture and 90% at least of the particles have a diameter between 0.5 and 2 D, 2% at most of the particles should have a diameter less than 0.1 D and the remainder may be larger than 0.1 D and less than 0.5 D. The process will be further described with reference to Figure 1.1.

Figure 1.1: Blow Molding a Grinding Wheel

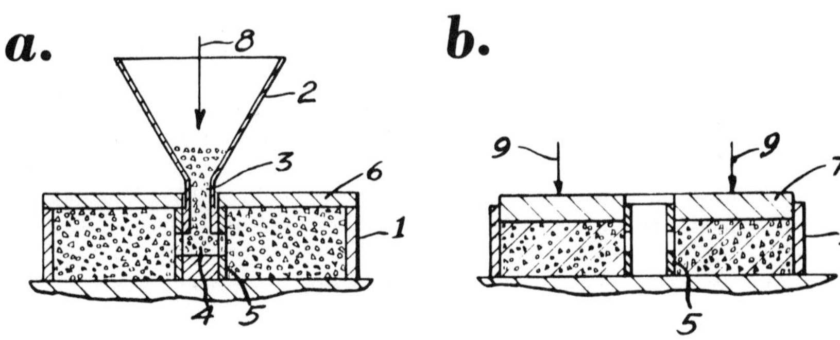

Source: U.S. Patent Reissue 29,592

Figure 1.1a represents a temporarily closed mold having a cylindrical body **1** over which is supported a hopper **2** connected through a passage **3** with cross passage **4** that passes through the center hole of the wheel that is defined by a cylindrical center wall **5** of the mold. Passage **3** opens into the inner cavity of mold **1** through the passage **4**. It should also be noted that mold **1** is provided with a cover **6** during filling and includes two or more parts **1** and **7** that are movable with respect to each other after being filled so that, during compression, its inner volume can be reduced as described below.

The mixture for manufacturing the grinding wheel is poured into hopper **2**. Then a compressed air or another gas under pressure is suddenly applied to the mixture in hopper **2**, as schematically represented by arrow **8**, so as to pneumatically expel the mixture, through passage **3**, into cross passage **4** and then into the inner cavity of mold **1** to fill the cavity.

Figure 1.1b represents the last step of the manufacturing process. In this case, mold **1** was separated from the hopper, a movable disc **7** is fitted into mold **1** and submitted to a pressure, as schematically shown by arrows **9**, to compress the mix to the desired grinding wheel size.

Resinoid Wheel Containing Trimanganese Tetroxide and Ferric Oxide Fillers

J. Ukita; U.S. Patent 3,960,517; June 1, 1976; assigned to The New Registon Company, Limited, Japan provides a resinoid wheel composed of abrasive grains, a binder resin and a filler and characterized in that based on weight one-third to two-thirds of the amount of the filler is a mixture of 20 to 55 wt % of trimanganese tetroxide and 80 to 45 wt % of ferric oxide, the remaining one-third to two-thirds thereof being a conventional filler material.

The mixture of trimanganese tetroxide and ferric oxide in the specified ratio is highly compatible with the thermosetting resin used as a binder, permitting the binder resin to effectively cover the abrasive grains, and moreover serves to efficiently dissipate the heat to be generated during grinding operation.

Example: 10 kg of brown fused alumina abrasive grain (fused alumina abrasive grain A 24, Nihon Kenmazai Kogyo Co., Ltd., Japan), 250 g of cryolite having a particle size of 250 to 325 mesh (Central Glass Co., Ltd., Japan), 250 g of a mixture of trimanganese tetroxide and ferric oxide in a weight ratio of 35:65, and 750 g of liquid phenolic resin were mixed together, and 750 g of powdered phenolic resin was then added to the resulting mixture with stirring. The mixture was then packed into a disclike die measuring 180 mm in diameter and 6 mm in thickness, with two sheets of glass cloth (Glass Cloth No. 350, Unitika Co., Ltd., Japan) placed in layers in the mass of mixture as spaced apart by 2 mm from its upper and lower surfaces respectively. The mass was then compressed at pressure of 100 kg/cm^2 at room temperature into a molded product, which was dried at 110°C for 10 hours and then heated at 170° to 220°C for 5 hours to obtain a resinoid wheel.

The grinding wheel obtained was used for grinding a workpiece of mild steel (JIS G 3141, Class 1). The wheel was driven at 7,000 rpm to conduct a grinding operation at a pressure of 7.5 kg/cm^2 for 5 minutes. The amount of the material removed from the workpiece and the amount of wear on the wheel were measured to calculate grinding ratio and operation efficiency by the following equations. The results are given in the table on the following page.

$$\text{Grinding Ratio} = \frac{\text{Amount of material removed}}{\text{Amount of wear on wheel}}$$

$$\text{Operation efficiency} = \frac{(\text{Amount of material removed})^2}{\text{Amount of wear on wheel}}$$

Comparison Example: A resinoid wheel was produced in the same manner as in the above example except that 500 g of cryolite was used in place of 250 g of the same, without using any amount of the mixture of trimanganese tetroxide and ferric oxide. Using the resinoid wheel, a performance test was conducted in the same manner as above. The results are given below.

	Process	Comparison Example
Amount of wear on wheel, g	3	4
Amount of material removed from workpiece, g	60	44
Grinding ratio	20	11
Operation efficiency	1,200	484

Grindstone Element Portion Embedded in Support

In the process of *G. Biardi; U.S. Patent 3,960,516; June 1, 1976; assigned to Norton Company* a grinding wheel is made by locating a grindstone element in a molding tool and injection-molding a supporting part so that a portion of the grindstone element is embedded in the supporting part and securely bonded thereto by the molding operation. Recesses in the embedded portion of the grindstone element ensure good torque transmission between the molded supporting part and the grindstone element.

Figure 1.2 illustrates the manufacture of a grinding wheel having a hollow cylindrical cup-shaped grindstone element **1** of abrasive material. The grindstone element is positioned in a molding tool comprising a punch suitably prepared. In the punch the grindstone element **1** is located in a space defined between an outer cylindrical element **2** and an inner cylindrical element **3** the space being closed at one end by a baseplate **4**.

An annular cavity **4a** is defined between the baseplate **4** and the cylindrical elements **2** and **3** of the punch. A supporting part **5** for the mounting of the grinding wheel on a rotary spindle of an operating machine is formed by injection-molding within the cavity **4a**. The baseplate **4** has, in manner known per se, internal passages **6** for injection of material, for example plastics material, into the cavity **4a**.

For carrying out this process, it is necessary that the grindstone element **1** of the grinding wheel projects for a certain distance into the molding cavity **4a** so that the material injected into the cavity **4a** to form the supporting part **5** adheres directly to the part of the element **1** embedded therein.

For the purpose of obtaining secure keying connection between the grindstone element **1** and the supporting part **5** which is capable of transmitting high torques, the end of the element **1** which is embedded in the part **5** is provided with recesses **1a**.

The baseplate **4** is provided with projecting pins **7** which bridge the cavity **4a** and which form through holes **8** in the part **5** molded therein for the passage of fixing bolts (not shown) for securing the supporting part **5** to the spindle of an operating machine.

Figure 1.2: Grindstone Element Portion Embedded in Support

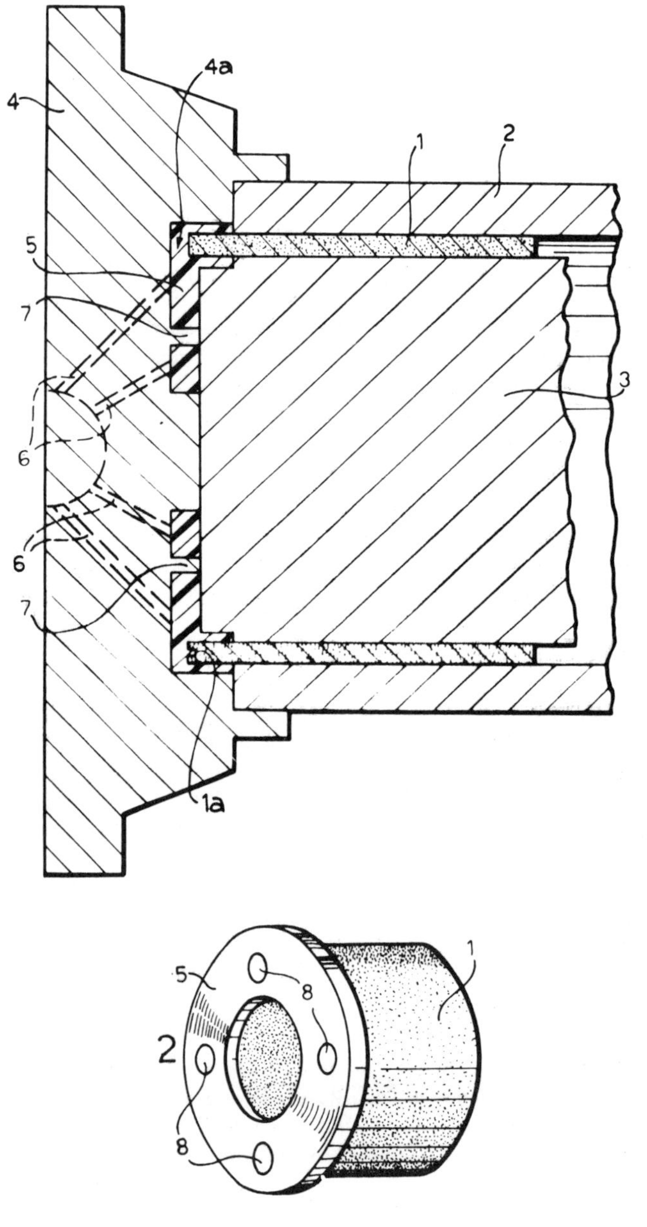

a.

b.

Source: U.S. Patent 3,960,516

After insertion of the grindstone element **1** into the punch and closure of the same, material is injected into the cavity **4a** so as to form the supporting part **5**. Direct bonding between the element **1** and the part **5** takes place during the stamping operation without the need to use any kind of adhesive or other specific fixing means.

The connection thus formed is very effective by virtue of the adhesion of the plastics or other material constituting the part **5** to the element **1**. In the embodiment shown, only a short portion of the grindstone element **1** is embedded in the supporting part **5**. To prevent any infiltration of the material from which the part **5** is formed beyond the molding cavity **4a**, there are provided in the punch suitable means, known per se, in proximity to the cavity **4a**. For example, the cylindrical elements **2** and **3** of the punch may be provided with metal expansion packings or passages for a cooling fluid, so that rapid solidification of the material injected into the cavity occurs in those regions where there is danger of infiltration. After cooling, the punch is opened and the complete grinding wheel extracted.

The process described above is suitable for making grinding wheels of any type, by ensuring in each case that a portion of the grindstone element projects into the cavity in which the supporting part is molded.

Arched Grinding Sections Separated by Grooves

The process of *J.G. Frangipane; U.S. Patent 3,951,619; April 20, 1976* is directed to a method of making a grinding wheel. The wheel has a plurality of abrasive grinding sections having convexly arched grinding surfaces and deep circumferential grooves between the sections, the grooves being filled with a relatively nonabrasive material. The convexly arched grinding surfaces of the sections are preferably formed to a radius half the diameter of the width of the sections.

The grooves are preferably cut to a width in the range of about five-thousandths to fifteen-thousandths inch and a depth of about three-eighths inch, with the grooves spaced apart to provide sections having a width of about ninety-five thousandths inch in a wheel having a diameter of about 6 inches.

Figure 1.3 illustrates a preferred embodiment of the grinding wheel.

Referring to Figure 1.3a, a grinding wheel **25** is illustrated in the form of a generally cylindrical, integral abrasive body **27** having opposed generally cylindrical hollows **29** with a web **31** between the hollows. A central hole in the web **31** snugly receives a suitable metal bushing **33** which in turn is snugly received on a reduced portion **35** on the spindle shaft **37** and is seated against a rigid metal washer **39** telescoped on the reduced portion **35** of the shaft and seated against a shoulder **41** of the shaft. A similar washer **43** is positioned on the reduced portion **35** of the shaft and the assembly is firmly clamped on the shaft by a nut **45** threaded onto a reduced end portion **47** of the shaft **37**.

Figure 1.3: Arched Grinding Sections Separated by Grooves

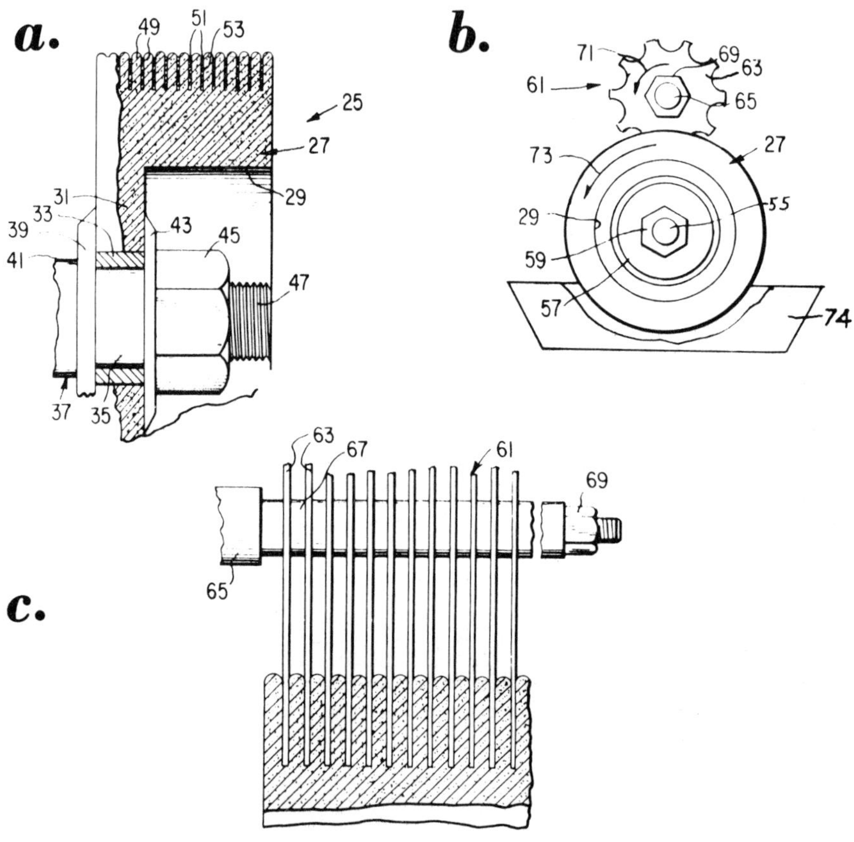

Source: U.S. Patent 3,951,619

The one piece abrasive grinding body **27** has a generally cylindrical grinding surface formed into a series of grinding segments or sections **49**, each section separated from the other by grooves **51**. The outer cylindrical faces of the grinding segments **49** are formed with a convexly arched, preferably radial configuration and the grooves **51** are filled with a relatively nonabrasive filler **53** to the point at which the radial configuration of the grinding segments **49** begins. The abrasive sections **49** and the grooves **51** may be proportioned as desired for the intended function of the grinding wheel, as for serrating knife edges, for example.

The abrasive portion of the wheels may be of any desired abrasive composition. A preferred vitrified composition is as follows: 1,362 g pure white aluminum oxide (100 grit size), 153 g feldspar, 175 g ball clay, 55 g borax, 30 g low melting frit, 75 g tragacanth solution and 18 g coloring agent.

Composition, by weight, of the filler material is preferably as follows: 35% ball clay and 65% fine alumina.

The method of manufacturing the vitrified grinding wheel preferably includes the following steps. First, the above composition is molded and pressed as is well known in the art, to form the grinding wheel body **27**. The wheel is heat set in typical manner, first to a relatively low temperature of about 1500°F, so that the body melts and upon setting is soft enough for cutting the grooves **51** and dressing the outer circumferential grinding surfaces of the sections **49**, while avoiding crumbling of the body **27**.

With reference to Figures 1.3b and 1.3c, upon cooling of the body **27** it is mounted on a spindle shaft **55** and is suitably clamped in place between washers **57** by a nut **59**, as previously described. A cutter **61** is mounted above the spindle shaft **55** and includes a plurality of circular notched blades **63** suitably mounted on a spindle shaft **65** with spacers **67** of a thickness exactly that of the desired thickness of the abrasive sections **49**, the blade assembly being suitably clamped in place by a nut **69** on a threaded end of the shaft **65**.

As indicated by the arrow **71** in Figure 1.3b, the cutter assembly is rotated counter-clockwise at about 17 rpm and is moved downwardly into engagement with the body **27** to cut the grooves. The blades **63** are preferably 2½ inches in diameter and 0.005 inch thick, and of any suitable blade material. The spacers **67** are approximately 1 inch in diameter. The grinding wheel body **27** is rotated counter-clockwise as indicated by the arrow **73** and at about 7 rpm so that the cutting blades **63** and the grinding wheel body **27** move in opposite directions at their points of contact and at any suitable speeds to avoid excessive wear of the blades. The grinding wheel body is rotated in a trough **74** of coolant to facilitate cutting and to wash out off-fall.

After the grooves **51** are cut into the grinding wheel body **27**, the arched configuration of the grinding surfaces of the abrasive segments **49** are formed in typical manner known in the art.

Next, the grinding wheel **25** is cleaned of loose material, which may be accomplished in any suitable manner as by blowing. The wheel is placed in a tank and under vacuum, and the filler material is inserted into the tank to fill the grooves **51**. The wheel is then removed from the tank and the excess filler is washed off the wheel leaving the grooves filled to approximately 0.047 inch inwardly of the outside diameter of the abrasive segments **49**.

The grinding wheel is then heated a second time to a higher temperature of approximately 1915°F for 8 hours, whereupon the wheel is at the desired stage of hardness for serrating typical knife edges.

Alternatively, the filler can be a rubberlike compound in which event the grooves are filled with powdered alumina to keep the thin sections **49** from warping when fired the second time. After the second firing the alumina is blown out of the grooves and the rubber compound is then inserted into the grooves to the depth previously described.

Wire-Reinforced Wheel

In order to provide a more efficient cutting process the cutting speeds of grinding wheels have been increased in recent years. These increases in speed in some cases create a centrifugal force field on the grinding wheel material that exceeds its inherent capacity to resist. This problem is particularly evident in those types of grinding wheels having a relatively large inner diameter. In this type of wheel a substantial stress concentration is built up near this inner diameter.

The process of *G.N. Peterson; U.S. Patent 3,939,612; February 24, 1976; assigned to Dresser Industries, Inc.* provides a reinforced grinding wheel which has a length of wire positioned on at least one of the side faces of the wheel in a spiral pattern beginning adjacent the inner diameter. Successive turns of the wire substantially abut one another and are positioned progressively radially outward with respect to the side face of the wheel. The wires are bonded to the side wall of the wheel so that the wheel is reinforced.

The wire is bonded to the wheel by a suitable adhesive. Preferably the adhesive is a reinforced epoxy adhesive that cures at room temperature. An example of such an adhesive would be a resin known as Epotuf 37-140 with a hardener known as Epotuf 37-164 and the resin would be reinforced with wollastonite. Epotuf compounds are available from Reichhold Chemicals, Inc., Azusa, CA.

Preferably the wire is formed from steel wire having a circular cross section that is rolled to produce an elongated generally oval section. This type of arrangement enables substantial contact between the adjacent turns of the spiral and, in addition, produces a greatly increased bonding area between the surface of the wheel and the wire.

The wire may be formed from a low carbon steel wire that is rolled to produce the generally oval shape. An example of a typical forming operation would be 16 gauge round wire with a diameter of 0.0625 inch rolled to a thickness of 0.040 inch. The cold working during the rolling operation produces an increase in the tensile strength of wire thereby enhancing its ability to reinforce the wheel.

From some applications the thickness of the grinding wheel cannot be increased because of limitation on mounting apparatus. For this application the embodiment of Figure 1.4a is utilized. A grinding wheel **30** having a peripheral grinding surface **32** had side faces **34** and **36**. Side faces **34** and **36** have generally annular recesses **38** and **40** extending from an inner hole **42** in the wheel **30**. Lengths of wires **44** are positioned in the recesses **38** and **40** in a spiral pattern. The wires are bonded to the wheel by adhesive **48** to reinforce the inner diameter.

Figure 1.4: Wire-Reinforced Wheel

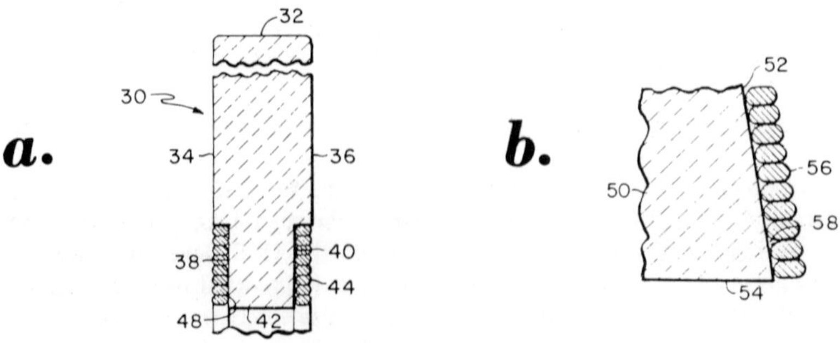

Source: U.S. Patent 3,939,612

The flattened wire can be used to reinforce grinding wheels with tapered side faces, as shown in Figure 1.4b. A grinding wheel **50** has tapered side faces **52** extending from an inner diameter **54.** A length of flattened wire **56** is positioned on the side face **52** in a spiral pattern. Adhesive **58** bonds wire **56** to side face **52.** Turns of flattened wire can be displaced laterally relative to one another and still provide an effective support. The degree of taper of the side face is a function of the wire thickness and width.

The process provides a highly effective reinforcement of the inner diameter of a grinding wheel. Since the turns of the wire generally simply abut one another during assembly, they can more readily conform to surface irregularities of the grinding wheel, or to a tapered side face, than a flat steel plate. However, once the turns of the wire have been bonded to each other and to the side face of the wheel, they provide a very effective reinforcement of the wheel. It should be noted that the spiral pattern of wire is formed by a given length of wire and that there is substantially no waste compared to the stamping out of a steel ring. This greatly minimizes the cost of manufacturing the spiral wire.

For Dry Grinding Cemented Carbides

P.P. Keat; U.S. Patent 3,925,035; December 9, 1975; assigned to Norton Company has found that relatively large amounts of the fine graphite in closely controlled sizes and amounts, result in an order of magnitude improvement in the efficiency of metal-bonded diamond cup-wheels in the dry grinding of cemented carbides, when weak or friable type diamond is employed. Specifically it has been discovered that, depending upon the particular metal bond employed, the graphite content may range from 15% of the total bond volume to 50%, and the particle size may range from under 10 to 200 μ, the coarser graphite being employed at the higher concentrations and vice versa. The graphite particles normally being flaky or platelike, when so shaped, are measured in terms of the average diameter of their face. Other fillers, specifically polytetrafluoroethylene, hexagonal boron nitride, and molybdenum disulfide, may be substituted in whole or in part for the graphite. The grinding element of the wheel is fabricated by conventional hot pressing techniques by placing the prepared mixture of graphite, metal powders, and diamond grit in a suitable mold and pressing while simultaneously heating. Diamond which is useful in the process is that known in the art as "resin bond type" diamond.

The diamond abrasive grit to be used in the process is defined by one of the following lists (a) and (b) of characteristics: (a) synthetic multicrystalline diamond having an irregular surface and a friability index F, for grit sizes of 130 and coarser $\leqslant 0.0052s^2$, and $\leqslant 1.07s^{0.9}$, for grit sizes finer than 130, wherein s is the arithmetic average of the number designations of the USS sieve series screens used to obtain a sized sample of the diamond, and (b) natural monocrystalline diamond with a friability index of $< 11.3^{0.43}$ and is selected from a population having a bulk density of 1.7 or less in the 140/170 grit size and 1.6 or less in the 200/230 grit size as measured by the American National Standards Institute method.

Example 1: The following materials were mixed to form a homogeneous blend: graphite powder, natural, 65 μ particle size (through 400 on 500 mesh), 3.13 g; bronze powder, 70 Cu (15 to 20 μ), 20 Sn (2-3 μ) 10 Ag (2-3 μ), 22.30 g; diamond (150 grit, copper clad, 50 wt % of Cu) friatest friability 50 to 60, 5.66 g.

The mixture of the above powders was cold pressed at 10 tons/in^2, pressure released, heated to 500°C, soaked for 5 minutes at 500°C, hot pressed at 6 tons/in^2 at 500°C, for 10 minutes, and removed from the mold to give a grinding element containing 15.25% by volume of diamond. The graphite represented 35% of the bond volume (metal plus graphite), ignoring in this case, a pore volume of 4% by volume. The pore volume is not critical and may typically range from 0 to 8%.

Example 2: A similar grinding element was made up of 15.5 vol % diamond, of the same type, the same metal bond, and 40% of the total bond volume was through 325 mesh on 400 mesh natural graphite (average size 80 μ). The grinding element had a porosity of 2.8%.

Wheels were made by cementing the grinding element with a metal filled epoxy resin on an aluminum filled resin wheel center to produce a standard 6A9 straight cup-wheel having a 4 inch outside diameter, 1.75 inches high with a 1.25 inch mounting hole.

In the dry grinding of carbide cutter material at a relatively heavy infeed rate of 2.4 mils per pass, and a total rate of cut of 0.037 cubic inches per minute, the wheel made from the grinding element of Example 1 drew the same power as a conventional standard phenolic bonded commercial wheel containing the same kind and amount of diamond (nickel clad diamond was employed for the phenolic wheels) but had a volumetric ratio of material removed to wheel wear (G ratio) of over 16 times that of the standard wheel. The wheel of Example 2, in the same test, drew 25% less power and had a G ratio of over 14 times that of the standard. Thus the wheel made from the element of Example 1, under these conditions, could do the work of more than 16 standard wheels, while the wheel of Example 2 could do the work of 14 of the standard wheels at 25% less power draw. The diamond was 150 grit size, before cladding.

The peculiar effect of graphite when employed with resin bond type diamonds in producing the order-of-magnitude, or greater, efficiency of performance of the grinding elements of the process is shown by the fact that many other materials also sometimes considered to be solid lubricants are ineffective in producing improved grinding results. Such materials which have been tested and found ineffective in this process are: mica, CaF_2, LiF, WS_2, WSe_2, $NbSe_2$, lamp black, and cryolite. Aluminum oxide was also tried and found ineffective.

Resin Bonded Wheel Containing Gas-Filled Thermoplastic Beads

L.A. Anna and K.C. Dixon; U.S. Patent 3,925,034; December 9, 1975; assigned to Research Abrasive Products, Inc. have found that certain types of flat surface grinding operations can be drastically improved, particularly the grinding of disc brake rotors and the like.

The process involves use of a refractory composition comprising at least 40% by volume of abrasive grains of aluminum oxide, silicon carbide or other suitable refractory material with a particle size preferably from 30 to 100 mesh, a suitable synthetic resin binder, and a substantial amount of volume of polystyrene beads or beads of other thermoplastic resin that are capable of forming voids in the cured grinding wheel and that enable the worn grains to break away at the proper rate during the grinding operation. The amount by weight of refractory is at least several times the total weight of the resin binder and preferably in the range of from 4 to 8 times the total weight. The thermoplastic resin beads are substantially larger than the abrasive grains and have a particle size or diameter preferably from 0.05 to 0.2 inch, a density preferably from 1.5 to 5 pcf, and a melting point preferably in the range of from 350° to 600°F.

Thermoplastic beads function in a unique manner in a grinding wheel because of their ability to melt, soften or shrivel when heated. They are selected to provide proper resistance to melting during curing of the wheel so that they provide large voids in the wheel. If some of the beads have not shrunk or shriveled during the final curing of the wheel sufficiently to loosen the bond between the beads and the surrounding portions of the wheel, then they can soften and shrivel when heated during grinding. The result is that they usually break away and expose the void in the grinding surface well before the wheel wears down to half the depth of such void.

Example: An epoxy resin composition is prepared according to the following recipe, in parts by weight. 76 parts silicon carbide (54 mesh), 8.3 parts aluminum oxide (54 mesh), 11.7 parts Epon 828 (diglycidyl ether of a bisphenol A), 2.4 parts curing agent Z (eutectic mixture of metaphenylene diamine and methylenedianiline), and 1.6 parts mica (water ground, 325-mesh).

The Epon 828 and curing agent Z are premixed and then mixed with the silicon carbide and aluminum oxide particles for at least 5 minutes. The mica is added toward the end of the mixing.

To this mixture is added preexpanded polystyrene beads with a diameter of approximately one-eighth inch (much greater than that of the abrasive grains) and a density of 4 pcf in an amount such that there is formed a mixture containing about 18 to 20% by volume of the beads. After this mixture is thoroughly mixed to provide uniform distribution, it is deposited in a suitable wheel mold and cured for 4 to 6 hours at a temperature of 300°F to produce a grinding wheel with a specific gravity of about 1.8.

A formulation of the above type, for example, could be employed to produce a large diameter straight-cup inserted-nut surface grinding wheel (or disc wheel) with a thickness of 1 to 1.5 inches, an axial length at the periphery of about 4 inches, and a flat annular grinding face with an internal diameter of 27 to 29 inches and an external diameter of 32 inches. Such a wheel would be suitable for surface grinding at speeds of 4,000 to 9,000 surface feet per minute.

Pairs of straight-cup epoxy-bonded grinding wheels with an external diameter of about 32 inches made according to this process have been used in conventional double-disc grinders used to grind flat brake rotors for automobile disc brakes and have been found to be far superior to the phenolic grinding wheels which were previously considered best suited for this grinding operation.

The typical procedure for grinding such brake rotors (or discs) is to rotate the brake rotor at a substantial speed (perhaps more than 100 rpm) while rotating the two grinding wheels in opposite directions to effect simultaneous grinding of both faces of the brake rotor in a short period of time, such as 12 to 20 seconds. The amount of grinding on each face may be, for example, about 0.005 inch so that the thickness of the rotor is reduced by 0.01 inch to the desired thickness.

An epoxy grinding wheel made according to the process and containing polystyrene beads has a useful life at least three to four times that of the previously used phenolic grinding wheels in such a double-disc grinding operation, and such epoxy wheel operates at a lower temperature with a more effective cutting action and produces a higher percentage of premium parts because of more uniform grinding.

Metal Coated MD Diamonds for Grinding Soft Metals

F.H. Hughes; U.S. Patent 3,902,873; September 2, 1975; assigned to Industrial Distributors (1946) Limited, South Africa provides a general purpose diamond abrasive grinding device which can grind a wide range of materials with practically acceptable efficiency and economy, and which is particularly suitable for grinding soft metals.

It has been found that contrary to the accepted view that diamond particles in general are not suitable for the grinding of soft metals, a substantial advance in diamond abrasive grinding of soft metals is achieved by substituting metal clad MD (metal diamond) particles for RD (resin bond diamond) particles in a bonding matrix of resinous material.

Any suitable metallic materials may be used for cladding the MD particles. Apart from the preferred metal nickel, cobalt, platinum group metals, gold, silver, copper, molybdenum, titanium, aluminum, manganese, cadmium, tin, zinc, chromium, tungsten, iron, zirconium and alloys containing at least one of these metals may be used.

The metal cladding may comprise 10 to 70% by weight of the total weight of the composite metal clad MD particles. More particularly, the metal cladding may comprise 50 to 60% by weight, preferably 55%, of the composite metal clad MD particles.

Any suitable resin matrix may be used, such as phenolic, epoxy, polyimide, alkyd, unsaturated polyester, silicone, polybenzimidazole and polyamideimide.

A flat grinding wheel according to the process comprises an aluminum phenolic hub carrying an abrasive peripheral skin of resin-bonded metal clad MD diamond particles. The peripheral skin is bonded to the hub under heat and pressure in well known manner.

Example: Tests were conducted on similar mild steel workpieces with grinding wheels carrying different diamond abrasive grits to determine the performance of the various grits under identical conditions.

Test Conditions

Workpieces	150 x 100 mm EN 9 (55 carbon) mild steel
Wheel specification	175 mm x 10 mm x 1¼" D1A1 peripheral wheel on a fiber-filled phenolic hub
Diamond types	100/120 mesh metal clad RD grit (designated RDA-MC)
	100/120 mesh metal clad MD grit comprising a blend of block shaped and irregularly shaped MD crystals (designated MDA-MC-Blend)
	100 mesh metal clad MD grit comprising block shaped MD crystals blended with a proportion of specially selected irregularly shaped MD crystals (designated MDA-S-MC Blend)

(continued)

Metal cladding	55% by weight nickel
Diamond concentration	100 (4.4 ct/cm^3)
Wheel speed	35.6 m/sec
Grinding machine	Jones & Shipman 540
Driving motor	2 hp
Cross-feed	1 mm per table traverse reversal
Down-feed	0.025 mm/pass
Table traverse rate	12 m/min
Coolant	Bryto 5 (1% solution)
Length of tests	3 hours
Volumetric removal rate	300 mm^3/min

D1A1 is the U.S. industry standard from USAS B 74.1-1963 and identifies a particular resinoid bond wheel made by practically all wheel makers. RDA means resinoid diamond abrasive, while MDA means metal-bond diamond abrasive. MC means metal clad, while S refers to the high strength of the specially selected MD crystals.

The grinding efficiencies obtained with the various types of diamond grit, expressed as the G ratio which is the ratio of the volume of workpiece removed to the volume of wheel worn away, are shown in the table below.

Diamond Type	G Ratio
RDA-MC	430
MDA-MC-Blend	660
MDA-S-MC-Blend	1,250

It will be seen that for the grinding of soft EN9 mild steels, considerably higher G ratios are obtained with metal clad MD particles than with metal clad RD particles.

Stress-Absorbing Particulate Spacer Material

A process is described by *V.K. Charvat; U.S. Patent 3,864,101; February 4, 1975; assigned to The Sherwin-Williams Company* for manufacturing resin bonded grinding articles such as grinding wheels and segments. The process comprises preparing a mixture of abrasive grain particles, particulate spacer material and a curable liquid resin, filling a mold with the mixture and curing the mixture to a solid article in the absence of external pressure. Grinding articles prepared in this manner and containing high concentrations of abrasive grain particles exhibit superior grinding wheel performance in rate of stock removal with no metallurgical damage to the steel being worked upon while maintaining precision of cut.

Examples of abrasives which may be used in the process include fused alumina, sintered alumina, silicon carbide, emery, garnet, talc, and pumice depending upon the abrading action and the resultant surface finish desired. While grit sizes of from about 600 to 10 mesh may be utilized, the range will be most frequently from 60 to 36 mesh.

The resin-bonded rigid grinding articles of this process are characterized as containing a high concentration, e.g., from 70 to 80% by weight of the mixture of abrasive grains which are nevertheless slightly spaced apart. This combination of requirements means that while a large amount of abrasive is used, the actual volume put in is slightly less than would fill the mold if such an amount was simply poured into it. The technique involves arranging and uniformly spacing this slightly smaller volume of abrasive so it entirely fills the mold exactly. Such an exact volume of abrasive affords a large number of cutting points, as is needed to do efficient grinding work, while yet providing the space and the capacity for the individual abrasive grains exposed at the working face to micro compress from their normal position to absorb impact energy.

This minimal micro adjustment or yielding action occurs on an individual grain basis and does not affect the essentially rigid character of the article. The amount of abrasive grains required in the articles of this process in order to obtain the desired properties should be above about 75% by weight and preferably from 30 to 50% by volume of the article.

In order to obtain the desired grinding articles of this process, it is preferred that an amount of abrasive equal to from 75 to 100% of the bulk or pack density of the particular abrasive be employed in the article.

In addition to the abrasive particles, the abrasive mixtures utilized in the preparation of the grinding articles of this process contain a stress-absorbing particulate spacer material of dual function. First it must serve to maintain the abrasive grains slightly spaced apart during the formation of the grinding article and second to act as part of the stress-absorbing bond material during wheel operation. The particulate spacer material can be either an organic resinous powder or an inorganic particulate material. Examples of organic resinous powders which can be utilized include polyurethane, polyethylene, polystyrene, epoxy, phenolic and silicone resins.

The third material required is the liquid phase bond plastic, preferably an epoxy. Ordinarily this is the primary bond substance since it not only fixes the abrasive mass to cake the structure as placed in the mold so that complete curing can take place without change in volume but it may be the major or only bond material present. Additionally the liquid phase plastic is capable of being used in varying amounts to permit total bond volume content to be adjusted to suit the work of grinding to be done. Hard strong wheels may require a total bond content of 45% whereas soft acting wheels need an amount in the range of 35%.

The spacer material should be so finely ground that it may be uniformly dispersed throughout the grinding article even when present in a small amount. The particle size of the spacer material will also influence the arrangement and concentration of the abrasive grains and should be taken into consideration when preparing the abrasive mixture. Generally, the spacer material utilized is rather fine particulate material such as 325 mesh mica or a similar size resinous powder.

The amount of spacer material included in the abrasive mixture will vary and may depend on the grain concentration desired in the final article.

Example: *Step 1* – Compute the wheel volume for a wheel having a 7 inch outer diameter and a ½ inch face (neglecting the small arbor hole)

$$V = 3.14 \times 3.5^2 \times \tfrac{1}{2} = 19.24 \text{ in}^3 \times 16.3872 = 315 \text{ cm}^3$$

Step 2 – Select the abrasive grain to be utilized and determine its bulk or pack density. For pure aluminum oxide, 60 grit size, the bulk or tap density is 1.98 g/cm^3.

Step 3 – Decide the abrasive grain density desired in the wheel to suit the work to be accomplished, the workpiece geometry, and operating conditions. In this example, a 95% tap density will be assumed. Therefore, the amount of abrasive on a weight basis will be

$$315 \text{ cm}^3 \times 1.98 \text{ g/cm}^3 \times 0.95 = 593 \text{ g abrasive}$$

Step 4 – The amount of particulate spacer material required to be mixed with 593 g of abrasive to obtain a volume of 315 cm^3 is determined by slowly adding and mixing the spacer material to the abrasive grain until the desired volume is reached. In this example it is found that approximately 41.5 g of powdered Epibond 100A is required to produce the desired volume. This represents approximately 7% by weight epoxy powder.

Step 5 – The amount of liquid epoxy utilized in this example is 10% by weight based on the weight of the abrasive. Thus, 59.3 g of liquid epoxy resin are required. This amount is, in fact, a mixture of resin and hardener which are utilized in the recommended proportions of 100 parts of resin to 27 parts of hardener which is equal to 46.7 g of the liquid Maraset and 12.6 g of the hardener. The liquid resin is "Maraset Resin 622C" and the hardener is identified as "Hardener No. 75."

Step 6 – A mixture of 593 g of the abrasive grains, 59.3 g of the epoxy liquid mixture and 41.5 g of the epoxy powder is prepared and placed in a mold constructed to provide the wheel having a 7 inch outer diameter, a ½ inch thickness and a 1¼ inch arbor hole. After placing the mixture in the mold, the mixture is leveled without packing and the mold is closed without asserting any significant pressure against the mixture except that required to obtain a flat surface. The filled mold is baked at a low temperature (200°F) for 2 hours and at a higher temperature (380°F) for 3 hours. The initial heating period cures the liquid epoxy whereas the elevated temperature completes the cure of the liquid epoxy and effects curing of the powder epoxy in the mixture.

The wheel prepared in accordance with the above procedure may be characterized as follows.

Abrasive volume = 150.2 cm^3
Epoxy liquid volume = 49.4 cm^3

Epoxy powder volume = 34.6 cm^3
Total solids volume = 234.2 cm^3
Void volume = 80.8 cm^3
Volume percent abrasive = 47.7%
Volume percent epoxy liquid = 15.7%
Volume percent epoxy powder = 11.0%
Volume percent voids = 25.6%

GRINDING WHEEL CORES

Magnetic Core for Diamond or CBN Wheels

In the process of *R.H. Sioui and E.B. Carver; U.S. Patent 4,184,854; January 22, 1980; assigned to Norton Company* iron or other magnetic metal powder is incorporated with aluminum in the core of a diamond or boron nitride grinding wheel. In addition to providing such characteristics as satisfactory bonding ability, strength, thermal expansion, thermal conductivity and turning characteristics, such cores are better held on a magnetic chuck for the performance of finishing operations.

The core member and rim are molded simultaneously at relatively high temperature. The core member includes a powdered bonding ingredient in the amount of 5 to 25% by volume, aluminum powder in the amount of 35 to 75% by volume, and powdered magnetic metal (iron, nickel, cobalt, or Alnico) in the amount of 20 to 40% by volume. Where the abrasive rim is resin bonded, it is preferred to use the same resin as the bonding ingredient in the core. A preferred resin is a high temperature addition cured resin such as Ciba Geigy P13N, Gulf Thermid 600, which are imide oligomers having unsaturated group terminated oligoimide groups. Upon cure by heat, a crosslinked polymer of high molecular weight is produced. Other polymers such as phenol-aldehyde resins, capable of being cured without objectionable decomposition above 200°C may also be used. Suitable resins which may or may not be crosslinked, but which do not melt upon heating to 250°C are polyimide resins, polyphenylene sulfides and polybenzimidazole resins. Zinc or tin metal may also be substituted for the resin.

The following is a typical mix composition for making the core members in wheels of the process:

Ingredient	Specific Gravity	Volume Percent	Weight Percent
Iron powder	7.9	30	57.4
Aluminum powder	2.7	60	39.4
Polyimide resin prepolymer (Thermid HR600)	1.3	10	3.2

The above resinous core is employed typically in making a wheel in which the same resin is employed as the bond in the diamond section, thus ensuring a good bond between the rim and the core.

Where the hydride bond of metal and resin is employed in the diamond (or boron nitride) grinding section, the composition of the core mix may be as follows:

Ingredient	Specific Gravity	Volume Percent	Weight Percent
Iron powder	7.9	30	50.4
Aluminum powder	2.7	60	34.5
Zinc powder	7.1	10	15.1

In making a grinding tool, such as a cup-type wheel, a cold pressed preform of the core material is desirably first produced, placed in a mold, and the diamond or cubic boron nitride grinding section mix is then put in place and the wheel is hot pressed to essentially theoretical density, at a temperature above 250°C, the specific temperature depending on the composition.

A typical diamond section mix bonded with polyimide bond is as follows: 35 vol % 150 grit, copper or nickel clad (~56 wt % coating), 25 vol % silicon carbide filler (finer than 315 mesh) and 40 vol % Gulf Thermid HR600 (as described above).

The core preform and the diamond section are placed together in an appropriate mold and placed in a press which has been preheated to 360°C. After the mold has reached 200°C, a pressure of 3 to 5 tons/in^2 is applied. This pressure is maintained as the mold is heated to 300°C and then cooled to within 130°C of room temperature before releasing the pressure and removing the wheel from the mold.

Crack-Free Cores for Diamond or CBN Wheels

A diamond or cubic boron nitride grinding wheel in which an abrasive section comprising abrasive grits distributed through and held by a phenolic resin matrix is bonded to a nonabrasive core, is improved in terms of ease of manufacture and elimination of cracking or poor adhesion in the process of *R.H. Sioui and E.B. Carver; U.S. Patent 4,042,346; August 16, 1977; assigned to Norton Company.* The improvement is effected by means of a core composition which includes between 5 and 20% by volume of a thermoset resin selected from the group consisting of phenol aldehyde, urea formaldehyde, and melamine formaldehyde resins, particulate metal and graphite, the remainder being a resin selected from the group consisting of epoxy resins, addition polymers of unsaturated polyesters, and thermosetting polyimide resins, in the range of from 5 to 80% by volume of the core.

A preferred core composition according to the process, for manufacturing a grinding wheel is as follows: 10 vol % phenolic resin molding powder (Union Carbide 5980, a two-stage novolak containing 9% hexamethylene tetramine), 15 vol % Kinel 1050 polyimide molding powder, 60 vol % aluminum powder, and 15 vol % flake graphite.

The above ingredients are thoroughly blended, placed in a suitably shaped mold and subjected to a pressure of from 10 to 20 tons/in^2 to produce a preform of sufficient green strength for subsequent handling. When necessary, depending on the shape of the wheel, a suitable recess for the diamond section is then machined, as by turning on a lathe, into the cold pressed core.

The mixture of abrasive grits, resin, filler and any other abrasive section component powders is then placed in position on the preformed core and the core and grinding section subjected to a pressure of from 2 to 5 tons/in^2 and a temperature of 165°C for about 30 minutes. The integral core and a grinding section are then placed in an oven and cured for 24 hours at top temperature of 220°C.

The abrasive section is a phenolic resin bonded diamond or cubic boron nitride ring. The composition of the grinding section can be as disclosed in U.S. Patents 3,868,232, 3,779,727, and 3,664,819, or may employ a more conventional bond consisting simply of silicon carbide filled phenolic resin.

Foamed Plastic Hub

It is an object of *E. Geissler, W. Haidacher and H. Ofer; U.S. Patent 4,035,161; July 12, 1977; assigned to Tyrolit-Schleifmittelwerke Swarovski KG, Austria* to provide a grinding wheel or a rotary abrasive element, the hub of which is smooth and exactly centric, no expensive subsequent treatment of the wheel being required.

This object is achieved by the provision of a hub formed of a foamed plastic, for instance polyurethane plastic. The process will be described with reference to Figure 1.5.

Figure 1.5: Foamed Plastic Hub

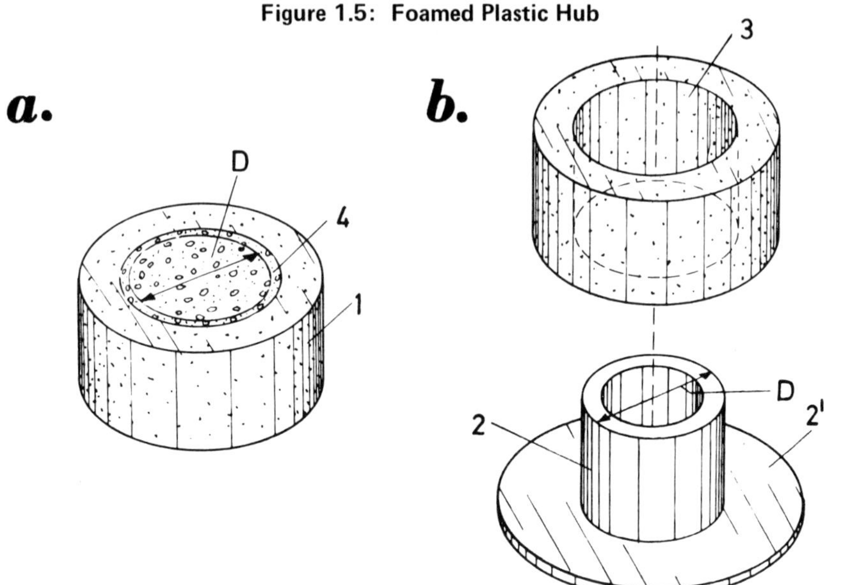

Source: U.S. Patent 4,035,161

The actual abrasive body **1** is mixed, molded and burnt in a furnace in a conventional manner.

After the binder has cooled, i.e., after the actual completion of the manufacture of abrasive body **1**, the latter is mounted onto a molding ring **2**.

The outside diameter **D** of molding ring **2** corresponds to the inside diameter of the hub to be formed by the supply of foamed plastic.

Whereas upon the previous manufacture of abrasive body **1** it was not necessary to pay particular attention to accurately center bore **3** of the wheel, this requirement is absolutely essential upon the mounting of abrasive body **1** onto molding ring **2**. In the example shown, molding ring **2** is provided with a lower core plate **2'**.

After abrasive body **1** has been mounted onto molding ring **2** and has been accurately centered, the gap or space formed between molding ring **2** and abrasive body **1** is filled in any known and conventional manner with foamed plastic, for instance polyurethane plastic. When the foamed plastic cures or sets, abrasive body **1** thereby is provided with a hub **4**.

The advantages of the method reside in that the supplying or injection of the foam can be carried out rapidly, and that losses of the granulated material due to heretofore necessary subsequent treatment of the bore of the abrasive body itself are avoided.

A further advantage is that upon the injection of the foam the hub area or inner surface of the abrasive wheel is impregnated with the two component liquid of the foamed plastic, and the bursting value of the wheel thereby is increased by 5 to 20% in comparison with a conventional wheel.

Example: In a specific example of carrying out the process, polyisocyanate and polyol were mixed in a proportion of 10/11 and a motive substance to produce gas was added thereto to produce a mixture having a viscosity of 1,700 cp. This mixture was poured into the molding space formed between molding ring **2** and the inner bore surface of abrasive body **1** to produce polyurethane which was cured at a pressure of from between three to four atmospheres and at a temperature between 130° to 140°C for seven minutes. A portion of the polyurethane foamed into the inner bore surface of the abrasive wheel **1**. After being cured and after a waiting period of 24 hours, the hub **4** was determined to have a volumetric weight of 600 kg/m^3, tensile strength of 195 kp/cm^2, a bending strength of 415 kp/cm^2, a crushing strength at 20°C of 195 kp/cm^2, a Shore hardness of 75, and a temperature stability of 90°C.

Diamond Wheel Core with Good Thermal Conductivity

The process of *E.B. Carver and R.H. Sioui; U.S. Patent 3,868,233; February 25, 1975; assigned to Norton Company* relates to the central core of a grinding wheel, and to such a core particularly useful in combination with a peripheral diamond abrasive section attached thereto for dry grinding operations at relatively high temperatures and stresses, since this core comprises a combination of a resin with particles of selected metals characterized by a high degree of thermal conductivity, by high strength at relatively high temperatures, and by ready dressability adjacent to the abrasive section.

The core may usefully be composed of from 49 to 90 vol % of particles of aluminum or copper or a mixture thereof, of from 2 to 35 vol % of particles of tin or a tin alloy, of 5 to 30 vol % of particles of resin, and of 0 to 15 vol % of particles of a solid lubricant consisting of graphite, hexagonal boron nitride, or polytetrafluoroethylene, or a mixture thereof.

According to the method of manufacturing such a grinding wheel, the core is formed first, and then the mixture of materials forming the abrasive section is introduced into a suitable peripheral recess on the core and then formed integrally with the core.

This method of manufacture may be carried out by filling a core mold assembly of the proper size and shape with a thoroughly blended dry mixture of the particulate metal components, the resin, and the solid lubricant forming the core. The core is cold pressed under 7½ tons per square inch pressure. The core is removed from the mold assembly and trued to provide a suitably shaped recess on the periphery of side wall **16** for an abrasive section **14** as shown in Figure 1.6.

Figure 1.6: Diamond Wheel Core

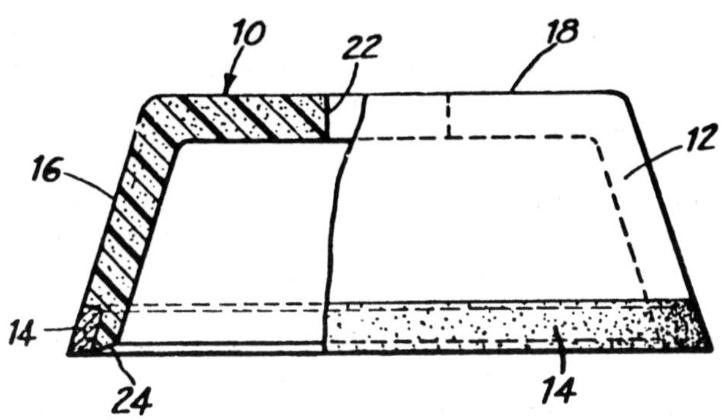

Source: U.S. Patent 3,868,233

The abrasive section **14** may, for example, consist of a mixture of diamond abrasive particles and a suitable resin serving as a binder therefor, used to fill the recess after the core is returned to the mold assembly. In order to form the abrasive section integrally with the core, the abrasive section is hot pressed at

160°C for 20 minutes under from 1 to 6 tons per square inch pressure, depending on the grade and composition of the abrasive section, the pressure used being determined in a manner well known in the art.

After the combined core and abrasive section is removed from the mold assembly, it is subjected to a postcuring operation at from 160° to 220°C, for from 12 to 38 hours depending upon the expected end use of the grinding wheel, the temperature and time used being determined in a manner well known in the art.

Alternatively, this method of manufacture may be carried out by first hot pressing the core material at 160°C and under 4 tons per square inch pressure for 20 minutes. The core is removed from the mold assembly and trued to form a suitable recess in the periphery of side wall **16** for the abrasive section **14.** The core is returned to the mold assembly and the recess is filled with the materials forming the abrasive section. Thereafter, the abrasive section is formed and cured as described above, assuming that the abrasive section in this case is also composed of a mixture of diamond abrasive particles and a suitable resin.

The range of compositions of the core material preferred for use in the first method of manufacture described above consists of from 58 to 80 vol % of aluminum or copper or a mixture thereof, from 8 to 16 vol % of tin or a tin alloy, from 8 to 16 vol % of resin, and from 2 to 10 vol % of the solid lubricant.

According to a second method of manufacturing such a grinding wheel, an annular abrasive section of resin bonded diamond particles is first formed and fully cured, then placed in the bottom of a core mold of the proper size and shape with a coating of phenolic or epoxy adhesive applied to its interface with the core, and then the mold is filled with a thoroughly blended dry mixture of the metal and resin particles and particles of solid lubricant forming the core material of the instant process.

Thereafter, in order to form the core integrally with the annular abrasive section, it is hot pressed at 160°C under 4 tons per square inch pressure for 20 minutes to complete the formation of the grinding wheel.

Alternatively, the second method of manufacturing such a grinding wheel may be employed with an annular abrasive section of resin bonded diamond particles first formed in a ring, but not fully cured, then formed integrally with the core material of the instant process as described above.

In this variation of the second method of manufacturing such a grinding wheel, after the annular abrasive section and the attached core are removed from the mold assembly, they are both subjected to a postcuring operation, as described above in connection with the first method of manufacturing such a grinding wheel.

The range of compositions of the core material preferred for use in the second method of manufacture described above consists of from 54 to 70 vol % of aluminum or copper or a mixture thereof, from 8 to 16 vol % of tin or a tin alloy, from 14 to 22 vol % of resin, and from 2 to 10 vol % of solid lubricant.

DRILLS AND DRILL BITS

Matrix with Lowered Hardness for Impregnated Diamond Drill Bit

L. Multakh; U.S. Patent 4,211,294; July 8, 1980; assigned to Acker Drill Company, Inc. provides a powdered metal matrix for forming the crown structure of an impregnated diamond drill bit which matrix has a lower degree of hardness but still securely retains the diamond particles.

In accordance with one preferred embodiment of the process, the metallic mixture contains up to 50% titanium carbide and 30% nickel-manganese along with 20% of iron powder. The resulting crown structure has a hardness of no greater than 10 Rockwell C. The resulting crown, however, provides a diamond retention ability of the same magnitude as that provided by the crowns formed in accordance with the prior art compositions that have a hardness within a range of 20 to 25 Rockwell C. Such a crown structure is capable of more easily wearing away so as to expose new diamond cutting edges while simultaneously preventing the diamonds from working themselves loose from the structure.

Although in accordance with the preferred embodiment, nickel-manganese is used in combination with the titanium carbide, it is possible for other alloys to be employed. The alloy selected must be capable of wetting the surface of the diamond particles. Such alloys typically have a relatively low melting temperature. When using such alloys within the powdered metal matrix, upon applying heat, the alloys prepares the surfaces of the diamond particles so that the surfaces have a greater affinity for the matrix material. Once the surfaces of the diamond particles have been so prepared, upon solidifying the metal matrix securely retains the diamond particles.

The amounts of titanium carbide and nickel-manganese alloy used in the mixture vary depending upon the particular use intended for the drill bit. Since the titanium carbide is more abrasive and provides for a higher degree of hardness, a greater quantity of that compound is used for more abrasive rock formations. On the other hand, where the drill is to be used in softer rock formations, a greater quantity of the nickel-manganese alloy can be employed.

In general, a powdered metal matrix includes between 15 to 50% by weight of a metal alloy having a relatively low melting temperature and capable of properly preparing the surface of the diamond particles (in the preferred embodiment the nickel-manganese alloy), 10 to 50% titanium carbide and 20 to 50% iron powder. In general, the iron powder is employed since it has good infiltration characteristics and serves to improve the soldering or fusing characteristics of the powdered metal matrix. Where a nickel-manganese alloy is used, the alloy should be formed from a mixture of approximately 50 to 60% nickel and 40 to 50% manganese.

An impregnated diamond drill bit made in accordance with the process, one that has a crown structure that is made from a powdered metal matrix containing an appropriate metal alloy such as nickel-manganese alloy, has vastly superior wearing characteristics than previously used drill bits. Since the crown structure has a lower degree of hardness, the structure is more easily worn away as the bit is used thereby exposing new diamond cutting surfaces. For this reason, it is possible to operate the drill bit at higher speeds, since new cutting surfaces are more rapidly exposed. Furthermore, the drill bit will last for a significantly longer period of time than prior drill bits. With respect to the expected lifetime, it is possible for the lifetime to be increased by a factor of two or three times the expected life span of prior drill bits.

Two examples of the content of the powdered metal matrix for two different types of drill bits are shown below:

Drilling Use	Titanium Carbide	Nickel-Manganese	Iron
Reinforced concrete, %	25	50	25
Granite, %	20	30	50

Small Diameter Drills with Filamentary Mandrel

An object of the process of *B.W. Dunnington and W.B. Gernert; U.S. Patent 4,203,262; May 20, 1980; assigned to The Glennel Corporation* is provision of a small diameter or micro abrasive drill with a filled cored portion that extends its usability to smaller dimensions than feasible in conventional core drills.

Another object is provision of a concentric solid cored abrasive drill that disintegrates the entire portion of material being removed from a workpiece by means of such drill.

These objects are accomplished via a cylindrical drill having an abrasive-containing metallic matrix portion surrounding a solid cored portion that is concentric therewith and with the longitudinal cylinder axis. The drill composition comprises abrasive particles distributed throughout a less abrasive matrix, and the cored portion comprises relatively nonabrasive material.

The drill is formed by electrical or chemical deposition of matrix material entraining abrasive particles onto a taut filamentary mandrel, which preferably comprises a round filament concentric with the longitudinal axis of the drill. The filament may be metallic, such as copper wire, or may be nonmetallic, such as nylon or other synthetic polymeric textile fiber composition. The deposition step forms the cylindrical drill outline to the desired diameter, which may be on the order of a millimeter or so or tenths thereof.

The drills are made possible by first selecting, arranging, and tautening the mandrel filament. The filament may be extended by heating, be gripped at its extended length (i.e., taut), and be cooled to tension it. Any filament may be heated externally, and metallic filaments may be resistance-heated by passing electrical current therethrough. Suitable filament diameter is from about one-twentieth to one-third the desired diameter of the resulting drill.

Then, the mandrel receives abrasive particles entrained in an electrically conductive matrix material to form the drill (to the desired diameter). Electroforming is readily carried out by deposition from a nickel solution in which the abrasive particles are suspended or at least circulating. As an alternative to cathodic deposition, electroless chemical reduction may be employed. When electrically nonconductive mandrel materials are employed they can be coated with graphite or conductive adhesive materials to assist in the deposition process. However, adhesive binding of the abrasive particles to one another, as often used in grinding wheels or the like, is not feasible in the drills of this process. With any cylindrical mandrel, uniform access of abrasive thereto is provided to locate the core concentrically as desired in a resulting simple cylindrical (round cross-section) drill.

The abrasive particles may be diamond, cubic boron nitride, alumina, tungsten carbide, or other recognized abrasives, such as oxides, carbides, or nitrides. Diamond is preferred because of its superior hardness and ready availability in desired size range. Cubic boron nitride is also very hard and, being more resistant to reaction with carbide-forming metals, is useful in drilling metals, such as tool steel.

The drills of this process are most useful in drilling nonmetals, usually hard brittle materials, such as ceramics and gemstones (both artificial and natural), which are difficult to drill by other means and methods and which tend to crumble when abraded. The smallness of the transverse cross-section of the concentric core in such drills ensures that such materials being drilled do not leave any central undrilled fragments, as lateral abrasion by the uniformly abrasive drill body throughout (including at the interface with the cored portion) crumbles the axial fragments.

A suitable range of particle size is from about a ten-thousandth to nearly half of the drill diameter. Thus, for a drill one or two millimeters in diameter, such particles could range from about one-tenth to one thousand microns in diameter and preferably from about 30 to 300 microns. Practical diameters for micro drills of this process are on the order of a millimeter or so or even tenths of a millimeter.

The matrix material may comprise metal, cobalt, chromium, or alloys thereof or other similarly suitable metallic material therein. It should effectively lock the entrained or occluded abrasive particles in place and be structurally strong enough to withstand the forces applied during drilling. Although hard, the matrix will be considerably softer than the abrasive particles themselves so as to wear away continually to expose them for drilling.

In related work *B.W. Dunnington and W.B. Gernert; U.S. Patent 4,128,971; December 12, 1978; assigned to Glennel Corporation* provide an abrasive drill of small diameter with a cored portion that is eccentric i.e., lacking a substantial concentric portion, relative to the longitudinal axis of the drill. The core, regardless of shape, is located so that during drill rotation all of the surface of a workpiece within the outline of the end of the drill is abraded by some part of the rotating end of the drill. The drill composition comprises abrasive particles in a less abrasive matrix, and the core is not hollow but comprises relatively nonabrasive material. The drill is formed by deposition of matrix material entraining abrasive particles onto a taut filamentary mandrel. The mandrel preferably comprises one or more round filaments, either straight or twisted together, alongside the longitudinal axis of the drill, and may be metallic, such as copper wire or may be nonmetallic, such as nylon or other synthetic polymeric textile fiber composition. The drill diameter may be on the order of a millimeter or so or smaller.

Drill Bit with Several Compact Cutters

W.H. Daniels and F.S. Knemeyer; U.S. Patent 4,156,329; May 29, 1979; assigned to General Electric Company describe a method for fabricating a drill bit comprising steps of (1) providing a plurality of composite compact cutters, (2) providing a drill bit comprising a drill crown with a plurality of recesses each for receiving a cutter, (3) positioning the cutters in the recesses and (4) bonding the cutters in the recesses.

In one embodiment, each cutter has a thin continuous layer of brazing filler metal coated on the exposed surfaces of the substrate which provides the filler metal to form a direct bond between the crown recess and the cutter. In a second embodiment, each cutter is comprised of a composite compact cutter blank bonded on one end of a cemented carbide pin. A thin layer of brazing filler metal is coated on the exposed surfaces of the substrate of the blank and on the surface of the pin. These layers provide the filler metal to bond the blank to the pin and to bond the pin in a recess of the drill crown. In both embodiments, the coating of brazing filler material is applied to the cutters by dipping in a molten bath of the filler material.

By way of example, a pilot bit **10** shown in Figure 1.7a illustrates a bit which can be constructed by this method. Bit **10** is comprised of an elongated, threaded shaft **13** and a drill crown **15**, e.g., of tungsten carbide, in which a plurality of peripheral composite compact cutters **17** and of central composite compact cutters **19** are mounted respectively in recesses. A plurality of waterways **21** are formed in the drill crown **13** for providing access of a cooling fluid to the interface between the drill crown and the earth during use of the drill. Fluid ports are provided longitudinally of the drill for transmission of a fluid to aid in mud and rock cutting removal.

Figure 1.7b illustrates, in fragmentary form, one of the compact cutters **17** such as shown in Figure 1.7a as provided prior to attachment to a drill crown. Cutter

17 is comprised of a (1) thin planar layer or mass **29**, preferably, of self-bonded polycrystalline diamond, (2) substrate **31**, preferably, of cemented carbide, bonded to layer **29** along an interface **30** and (3) a thin continuous layer of brazing filler metal **32** (shown in enlarged forms for purposes of illustration) coating the entire exposed surface of substrate **31** (i.e., except for the surface of substrate **31** along interface **30**). Although it is preferred that layer **32** cover, in their entirety, both base surface **33** remote from interface **30** and side surface **35** of substrate **31**, layer **32** may not and need not cover side surface **35** in its entirety.

Figure 1.7: Pilot Bit and Cutter

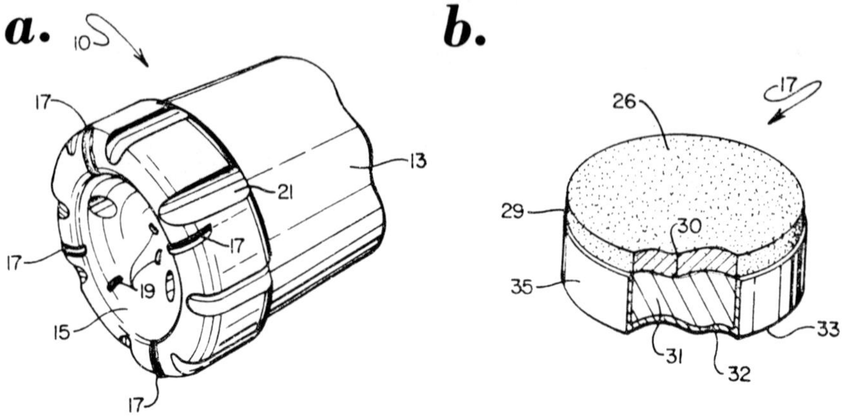

Source: U.S. Patent 4,156,329

Impregnated Diamond Drill Bit with Radial Spacers

R.E. Generoux; U.S. Patent 4,128,136; December 5, 1978; assigned to Lamage Limited, Canada provides an impregnated diamond drill core bit which, in tests which have been conducted, has been found capable of drilling at high penetration rates, with good tracking and flushing, through substantial footages before replacement is needed.

A diamond drill coring bit has an annular crown and inner and outer concentric side surfaces. The crown is formed from a number of radially extending composite segments spaced apart circumferentially by a circumferential spacer material, all integrally bonded together. Each composite segment consists of a number of diamond impregnated segments spaced radially from each other by a radial spacer material.

Reference is now made to Figure 1.8, which shows a drill bit **2** according to the process. The bit **2** comprises a shank **4** and a crown **6**. The shank is normally formed of steel and is commonly threaded as indicated at **8**, for attachment to a drill string. The crown has a cutting face **10**, inner and outer concentric side surfaces **12, 14** respectively, and a bottom **16** which is integrally secured to the shank, e.g., by brazing or by integral molding.

Figure 1.8: Impregnated Diamond Drill Bit with Radial Spacers

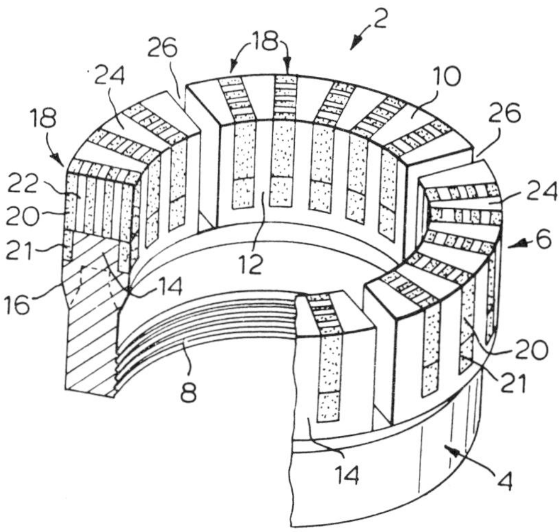

Source: U.S. Patent 4,128,136

The crown includes a number of circumferentially spaced composite segments **18**. Each composite segment is formed from a number of spaced parallel diamond impregnated segments **20** separated by a radial spacer material **22**. The composite segments extend generally radially between the inner and outer surfaces and also extend over a portion of the depth of the crown, from the cutting face **10** part way to the bottom **16**. A diamond impregnated segment **20** is located at each end of each composite segment, as shown, to help maintain the gauge of the hole to be drilled by the bit. Additional diamond impregnated segments **21** are located at the inner and outer surfaces below each composite segment, to further help to maintain the gauge of the hole.

The composite segments **18** are separated circumferentially on the crown by a circumferential spacer material **24** which extends the full depth and thickness of the crown, except for conventional radial waterways **26** which are cut to permit passage of drilling fluid and drilling debris. The spacer material **24** is integrally bonded to the composite segments **20** to hold and protect the composite segments. The spacer material **24** also separates the gauge holding segments **21** both circumferentially and radially and is integrally bonded thereto.

It is an essential feature of the process that the abrasion resistance of the diamond impregnated segments **20** be greater than that of the radial spacer material

22, and that the abrasion resistance of the radial spacer material **22** be in turn greater than that of the circumferential spacer material **24**. Specifically, the abrasion resistance of the radial spacer material must be less than that of the diamond impregnated segments, so that the radial spacer material will wear more rapidly than segments **20** and thus will cause circular ribs of rings to be formed in the material being drilled. This effect, which is known, assists tracking of the bit (since the ribs act as a guide for the bit) and also improves the penetration rate, since the ribs in the formation being drilled tend to break off and are flushed away by the drilling fluid, without the need for consuming diamonds to abrade the ribs. However, the abrasion resistance of the radial spacer material must not be too much less than that of the diamond impregnated segments, since otherwise the diamond impregnated segments will be prematurely exposed to an undue extent and will tend to break off, shortening the life of the bit.

The radial spacer material must be of greater abrasion resistance than that of the circumferential spacer material so that the material **24** will in turn wear away more rapidly than material **22**. This ensures that the radially oriented edges of the composite segments **18** are adequately exposed for cutting and ensures that the material **24** will not interfere with the cutting action. The more rapid wear of material **24** can also provide very small area additional waterways between the composite segments **18** to improve the flushing away of drilling debris, further reducing wear on the bit. Again, however, the circumferential spacer material must not wear too much more rapidly than the radial spacer material, to avoid premature exposure of the composite segments **18**. If the composite segments **18** are prematurely exposed, the radial spacer material **22** will wear away too rapidly and the diamond impregnated segments **20** will tend to break or chip away, shortening the life of the bit. The material **24** ideally acts to buttress the composite segments **18**, absorb shock, protect them against breakage, and conduct heat away from them, particularly in the region of the cutting face **10**.

As indicated, the materials from which the various portions of the crown are formed are important. Firstly, the matrix in which the diamonds are embedded to form the diamond impregnated segments must satisfy several requirements. The matrix must have sufficient hardness so that the diamonds exposed at the cutting face **10** are not pushed into the matrix material under the very high pressures used in drilling. In addition, the matrix must have sufficient abrasion resistance so that the diamond particles are not prematurely released. In addition, the sintering or hot-pressing temperature for the matrix material for segments **20** must be sufficiently low (below about 2500°F) that the diamonds are not graphitized during sintering or hot-pressing.

To satisfy these requirements, the following materials may be used for the matrix in which the diamonds are embedded: tungsten carbide, tungsten alloys such as tungsten/cobalt alloys, tungsten carbide or tungsten/cobalt alloys in combination with elemental tungsten (all with an appropriate binder phase to facilitate bonding of particles and diamonds), and the matrices produced by Wall Colmonoy Corporation of Detroit, Michigan under its numbers 11 and 50.

The number 50 material is an iron-bronze mixture containing 40 to 60% iron-bronze, and the remainder a self-fluxing alloy consisting essentially of 70 to 80% nickel and the remainder chromium, boron, silicon and iron. The number 11 material is fully described in Canadian Patent 781,677. One form of this matrix contains about 26% copper, 22% tungsten, 21.7% iron, 18% nickel, 4.4% chromium, 2.5% tin, 1.7% carbon, 1% boron, and 1.2% silicon. In general, the hardness on the Rockwell scale of the matrix in which the diamonds are embedded to form the segments **20** will be at least 100B, and probably at least 110B. The diamonds contained in the diamond impregnated segments **20** are present in a concentration of at least 75, and typically in a concentration of 100 or more. (A concentration of 100 is equivalent to 27 carats per cubic inch.)

The radial spacer material **22** will normally be the same material as the matrix used in the segments **20**, but of course without the diamonds. Material **14** may be a copper based material having a dispersion of secondary abrasives such as tungsten carbide, to improve its abrasion resistance. In very special applications, where the ground being drilled has very low abrasion, steel shims may if desired be used.

The circumferential spacer material **24** must have substantial thermal conductivity, so that it will act as a heat sink for the composite segments **18** (during drilling, the operating temperature of the crown can exceed 500°F). In addition, the circumferential spacer material should have a hot-pressing or sintering temperature low enough so that the diamonds in the composite segments are not degraded during the second heating required when the crown itself is fabricated. However, the hot-pressing or sintering temperature of material **24** should be high enough so that the material is not weakened at the operating temperatures of about 500°F. In addition, the circumferential spacer material **24** should have sufficient ductility to absorb shocks and impacts during drilling.

It is found that copper based materials are highly suitable for use as the circumferential spacer material. For example, simple bronze may be used (85% copper, 15% tin). Nickel or aluminum based materials may also be used, so long as the hot-pressing or sintering temperature required is less than the melting temperature of any phase of materials **20** and **22** to a maximum of 2200°F to prevent deterioration of the diamonds. Another suitable material for circumferential spacer material is the material sold by Wall Colmonoy Corporation under its material number 7, but with some of the abrasive particles removed. This material normally consists of 50 to 60% bronze, and the balance iron, except for 1 to 5% chromium boride, which acts as an abrasive. For use as the circumferential spacer material **24**, the proportion of chromium boride is held at a maximum of about 1%. Other secondary abrasives may be used in material **24**, so long as the abrasion resistance is less than that of material **22**, so that wear of material **24** is more rapid during drilling.

In the fabrication of the complete crown, a conventional mold may be used, as shown in Figure 1.8. The mold cavity is coated with a conventional release agent, then the composite segments **18** are placed on the mold cavity with the desired circumferential spacing between them. The gauge holding segments **21** are next put in place. The circumferential spacer material **24** in powder form is next poured into the mold cavity to fill the spaces between the segments **18** and to cover them to the desired depth. The top of the shank **4** is then inserted into the mold cavity with appropriate sealer blocks to seal the spaces between the top of the shank and the edges of the mold cavity, and the whole is then sintered. When material **24** is largely bronze, the sintering may be carried out at a temperature of between 1500° and 1700°F, which is well below the temperature at which significant damage to the diamonds will occur. Normally the circumferential spacer material will have a sintering or hot-pressing temperature of at least 1000°F.

Thin Film Treated Drilling Bit Cones

Features of the process of *G.W. White; U.S. Patent 4,054,426; October 18, 1977* include, in thin film treated bit cones, an outer dense hard smooth film surface with a phase zone of atoms extended into underlying base metal or intervening films of material all deposited by a high particulate energy level ion plating process. The drilling bit cones including the teeth thereof are made with the base metal tempered steel for ductility and shock resistance plated with ion process deposited hard film for minimizing chemical and physical failure. Internal bearing surfaces within the cones and bearing surfaces of bearing journals of the bit arm journal and bearing roller and ball components as well are given a dry lubricant thin film plating deposited by a high particulate energy level sputtering or ion plating process.

Figure 1.9a shows in enlarged partial section a single hard smooth material film layer **53** that is in some instances hafnium carbide deposited to as much as 0.1 to 5 mils thickness on the base material **54** over the outside of the cone including the cone teeth. A film layer **53** of hafnium carbide has been deposited over steel as the base material **54** by a high particulate energy level ion plating process providing a phase zone **55** of atom penetration into cone base material. This hard film layer along with the phase zone advantageously so improves hardened outer drill bit surfaces as to increase drill penetration rates and materially increase drill service life.

With the surface treatment of Figures 1.9b and 1.9c a plurality of film layers are deposited successively over drill bit cone outer base metal surfaces by high particulate energy level ion plating processes. In the instance of Figure 1.9b a titanium layer **56** is deposited first over the base metal **57** to a thickness of approximately 2,000 Å with a phase zone **58** of atom penetration into base metal again provided. The next successive layers deposited by high particulate energy level ion plating processes are a titanium carbide layer **59** to a substantial thickness as much as, for example, 4 mils thickness, and finally a titanium nitride layer **60** to a thickness, for example, of approximately 10,000 Å. Phase zones **61** and **62** are also attained in the upper surface region of both the titanium film and titanium carbide layers.

Figure 1.9: Thin Film Treatments

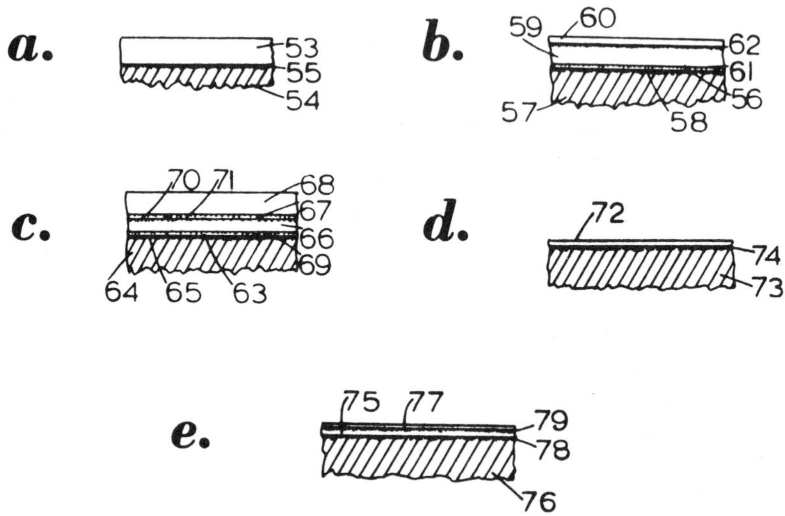

Source: U.S. Patent 4,054,426

With the film system of Figure 1.9c using high particulate energy level ion plating processes a titanium layer **63** is deposited first over the base metal **64** to a thickness of approximately 2,000 Å with a phase zone **65** of atom penetration into base metal provided. The next successive layers deposited are titanium carbide layer **66** approximately 1 mil thick, a hafnium layer **67** approximately 2,000 Å thick, and finally a hafnium carbide layer **68** as thick as, for example, from approximately 0.5 to 3 mils thick. Phase zones **69, 70** and **71** are also attained in the upper surface region of the titanium film, the titanium carbide film and the hafnium film, respectively.

In processing drill bit cones to provide an atomically dense coating of pure refractory material such as hafnium carbide with no binder as in Figure 1.9a, or graded shock resistant multilayer film system such as chromium on the base metal and then successively titanium carbide and hafnium carbide (not shown) or the systems of Figures 1.9b and 1.9c the substrate (i.e., drill cone) is held at a high negative minus 3,000 to minus 5,000 volt charge during deposition. Material to be deposited is thermally evaporated into plasma at a pressure of about 25 μ. Atoms of evaporant are ionized in the plasma and being negatively charged are accelerated directly to the substrate. When a pure binary carbide is desired as a deposited film for the outside of a drill bit cone approximately a 50-50 mixture of argon gas and methane gas is leaked into the vacuum system and in combination becomes the plasma sustaining gas. The methane gas cracks into elemental carbon and hydrogen as it enters the plasma leaving the carbon free to react with the evaporated metal. It is of interest to note that the process is so controlled as to maintain a sufficiently high temperature at the upper atomic layer of the surface deposition to prevent hydrides from forming in thin films being deposited and the hydrogen is pumped away by the vacuum pump of the system. While the temperatures of the upper atomic layer of a thin film material being deposited is difficult to determine with any high degree of accuracy it appears that the temperature at that location would be above at least 1000°C with atomic particle energy levels during deposition and is purposely maintained at a high level to prevent formation of undesired hydrides. Deposition rate is limited by the rate at which an evaporant may be evaporated that can be as high as 0.5 mil per minute and at times by the heat limit of the substrate object being ion plated. Thus, thick films of refractory materials can be grown as pure, uniform and atomically dense films several mils thick.

In order that the drill bits may be an optimally balanced operational design improved lubrication of bearing surfaces is critically important, in addition to improved hardened film coated outer cone surfaces, particularly in the extremely harsh operational environment imposed on such roller cone hard rock drill bits. Referring to Figure 1.9d, a single thin film layer in the form of a dry lubricant film plating **72** is deposited on bearing surfaces of base metal **73** by a high particulate energy level sputtering or ion plating process. Molybdenum disulfide (MoS_2) has been used as the dry lubricant film plating **72** to approximately 10,000 Å thickness and there is a phase zone **74** of atom penetration into base metal that helps anchor the dry lubricant in place and even with heavy friction loading still provides a high degree of lubricity. This thin film dry lubricant film plating has been applied to the internal bearing surfaces of the roller cones, bearing surfaces of drill bit arm journals and bearing roller and ball components as well and has been found in some testing to result in extended bearing surface life ranging from two to four times and more over bearings and bearing surfaces not so treated.

With the dry lubricant thin film approach of Figure 1.9e a plurality of film layers is deposited successively over bearing surfaces of base metal by a high particulate energy level ion plating process with at least the top most layer a thin film of dry lubricant material. These dry lubricant materials include metals such as gold, silver, lead and tin as well as layered lattice inorganic compounds such as molybdenum disulfide. One film system in accord with the showing of Figure 1.9e includes first a thin film deposit of chromium **75** to approximately 6,000 Å thickness on the base metal **76** and then a top layer of gold **77** approximately 4,000 Å thick. There is a phase zone **78** of atom penetration into the surface area of base metal **76** and a second phase zone **79** of atom penetration into the top surface area of the thin film of chromium. Should the thin film of gold be penetrated chromium oxide forms and continues to provide slippage or a thin film of chromium oxide could be initially provided between the thin films of chromium and gold or other dry lubricant material used in place thereof. These other dry lubricant thin film systems may be employed on any of the bearing surfaces such as described for the embodiment of Figure 1.9d.

Copper-Chromium Carbide-Metal Bond

A bond for abrasive tools described by *M.S. Drui, R.S. Ovseevich, G.A. Senkin; U.S. Patent 3,999,962; December 28, 1976* comprises 3 to 75 wt % of copper, 15 to 90 wt % of chromium carbide, 0.01 to 10 wt % of at least one metal selected from the group consisting of titanium, vanadium, chromium, zirconium, niobium, molybdenum, tungsten, 1 to 30 wt % of at least one low melting metal selected from the group consisting of Sn, Zn, Cd, Pb, and Al, 2 to 30 wt % of a metal belonging to the iron subgroup of Group VIII of the Periodic Table.

The operational characteristics of the drill bits of natural diamonds with the bond according to the process approach those of the bits made of natural diamonds with a tungsten carbide bond which makes it possible to substitute the scarce and expensive tungsten carbide in the bonds for drilling tools by a cheaper and easily obtainable chromium carbide.

Example 1: Drill bits of 59 mm diameter with the volumetric and cutting elements made from cubic boron nitride blanks 4 mm in diameter and 4 mm high employ a bond of the following composition: chromium carbide, 23.6 g; nickel, 4.30 g; copper, 99.8 g; tin, 4.8 g; lead, 0.3 g; titanium 0.2 g.

The bits are hot-pressed at a pressure of 150 kg/cm^2 and a temperature of 950°C. The bits made by this method are used for drilling iron quartzite of X-XI categories of drilling capacity with wash water. The drill speed averaged 2.5 m/hr at an average bit meterage of 1 m. The diamond drill bits with the bond based on tungsten carbide have produced an average mechanical drill speed of 0.8 m/hr at an average bit meterage of 0.8 linear meter.

Example 2: Drill bits of 59 mm diameter with the volumetric and cutting elements made from cubic boron nitride blanks 4 mm in diameter and 4 mm high employ the bond of the following composition: chromium carbide, 73 g; nickel, 11.7 g; copper, 26.1 g; tin, 8 g; cadmium, 2 g; and vanadium 4.5 g.

The bits are hot-pressed at a pressure of 300 kg/cm^2 and a temperature of 1000°C. The bits manufactured in this manner are used for air-blast drilling of monolithic quartz reefs of X category of drilling capacity. The average mechanical drill speed is 4.6 m/hr and the bit meterage, 0.80 linear meter.

Under the same conditions the diamond drill bits with a bond based on tungsten carbide display a drill speed of 1.35 m/hr at a bit meterage of 0.8 m.

CUTTING TOOLS

Rock Cutting Tip Inserts

P. Kenny and S.N. Johnson; U.S. Patent 4,194,790; March 25, 1980; assigned to Coal Industry (Patents) Ltd., England provide a rock cutting tip insert which comprises a cutting or working margin of one hardness level and a backing or base portion of a second hardness level, the first hardness level being at least one hundred units on the Vickers hardness scale greater than the second hardness level.

Referring to Figure 1.10, a rock cutter tool **1** comprises a body **2** having a recess or cutout **3** into which is mounted a tip insert **4**, the tip insert being secured to the body **2** by, for example, brazing. The tip insert comprises two separate layers of cemented carbide of different hardness. The cutting or working margin **5** of a tip insert is relatively hard grade carbide (having typically a hardness of 1,450 to 1,550 units on the Vickers hardness scale). Typically the thickness of the cutting or working margin **5** is 1 mm.

The layer of carbide forming the cutting or working margin **5** is secured (by, for example, brazing), onto a wider backing or base portion which is formed of a less hard grade carbide (having typically, a hardness of 1,000 to 1,100 units on the Vickers hardness scale).

Figure 1.10: Rock Cutting Tip Inserts

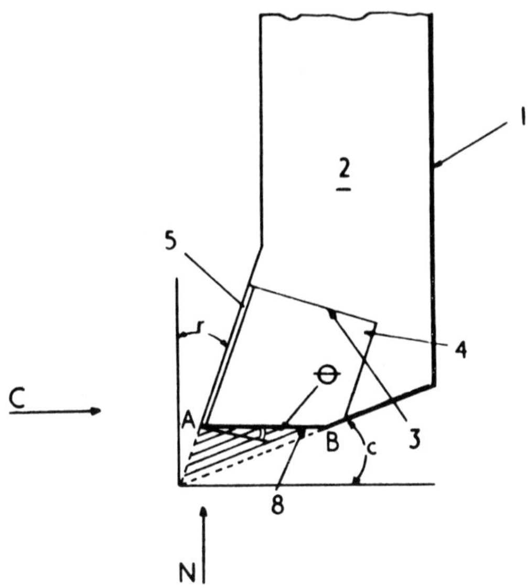

Source: U.S. Patent 4,194,790

Figure 1.10 indicates the direction of the cutting force on the cutting tool indicated by arrow **C** and the front rake angle **r** and back clearance angle **c** for a sharp tip insert. In addition the direction of a force normal to the cutting force is indicated by arrow **N**. This force is important since it represents the direction of force acting on a wear flat or worn surface **8** which is formed in use. The wear flat lies in one plane in the hard cutting or working margin 5 and in another plane in the less hard backing or base portion. The angle between the two planes as indicated in Figure 1.10 is called the wear angle. The so-called wear angle is formed because the less hard grade of carbide forming the backing or base portion is more readily worn away than the hard grade of carbide forming the cutting or working margin. Thus, during cutting as the pick wears and the wear flat is formed the cutting or working margin of the tip insert tends to cut clearance for the backing or base portion. Thereby the effect of rubbing the rock left uncut by the cutter tool is reduced.

In addition, as the moving tip insert presents a progressively increasing gap between a point on the uncut rock and the tip insert, particles of cut rock tend not to be crushed between the tip insert and the uncut rock. Consequently, the cutting force and the force in the direction normal to the cutting force are reduced. Thereby, the force exerted on the cutting tool during cutting is reduced and a more efficient cutting action is obtained throughout an extended operational life of the cutting tool.

In using a cutter tool having a tip insert constructed in accordance with the process the force exerted on the cutter tool remains constant at a relatively low value. The force **N** acting in the direction normal to the cutting force is affected in similar manner to that previously described with reference to the cutting force **C**. Thus, the cutting efficiency of the cutter tool remains relatively high throughout the extended operational life of the cutter tool.

Cutting Monofilament

F. Schneider and L. Bayersdorfer; U.S. Patent 4,172,440; October 30, 1979; assigned to Hoechst AG, Germany describe cutting filaments having a long service life which consist of a monofilament of a synthetic polymer into which from 0.3 to 10% by weight of an abrasive are incorporated. In the cutting process the abrasive is not ripped off, as happens with steel wires bearing the abrasive on their surface, but it is pressed into the monofilament and can be easily recovered therefrom.

Suitable synthetic high molecular weight polymers are preferably linear polyesters, especially polyethylene terephthalate without or with minor addition of modification agents and having a specific viscosity over 0.7 in the monofilament, the specific viscosity being measured in 1% by weight solution of the polyester in a phenol-tetrachloroethane mixture in a proportion by weight of 3:2 at 25°C.

As abrasive particles boron carbide and other carbides are preferred, although diamond can also be used for special applications. The preferred concentration of boron carbide in the monofilament is in the range of from 3 to 7% by weight, calculated on the polymer, the particles preferably having a size corresponding to 0.1 to 0.4 times the diameter of the monofilament. The monofilaments preferably have a diameter of from 0.1 to 3 mm.

It has surprisingly been found that the cutting effect of saws filled with abrasive particles does not increase with the concentration of the particles on the surface, but that an optimum cutting effect for different materials has to be determined by preliminary tests. It is quite obvious that the mechanical properties of the monofilament diminish with an increasing concentration of the abrasive in the polymer material.

In most cases, the abrasive particles will be as large as possible. Their size should be adapted, however, to the type of material to be cut and the diameter of the monofilament used.

Monofilaments having a larger diameter have, of course, a higher mechanical strength, but they require a higher sawing power because of the larger sawway. In the case of very valuable material to be sawn, the loss of material may recommend the use of very thin monofilaments.

Example 1: The properties of two cutting monofilaments according to the process were compared with those of a comparable commercial steel wire charged with diamond. In each case the wire had a diameter of from 0.5 to 0.35 mm, the sawing speed was 0.9 m/sec with a wire tension of about 350 g. For cutting potassium hydrogen phosphate monocrystals having a Mohs hardness of about 3 was used.

The cutting monofilaments of the process consisted of polyethylene terephthalate having a specific viscosity of 0.75 with a 3% by weight addition of boron carbide and of 0.70 with a 6% by weight addition of boron carbide. The average particle size of the carbide was 62 μ.

	Monofilament of Process with an Addition of		Diamond Charged Commercial Steel Wire
	3% B_4C	6% B_4C	
Cutting capacity	7 cm^2/hr	10 cm^2/hr	7 cm^2/hr
Service life	Up to 15 hr	Up to 15 hr	Up to 10 hr
Consumption of abrasive	About 10%	About 10%	100%

Example 2: The cutting effect of a monofilament of the process was compared with that of normal textile yarns containing 1.2% of TiO_2. Under the same cutting conditions and with the same cutting speed, the lengths of monofilament and of yarn were compared which were required to cut a standardized sample.

	Copper	Steel
Titanium dioxide yarn (1.2%), m	900	1,000,000
Boron carbide monofilament (6%), m	3.3	200

Abrasive Wire for Sawing Stone

The process of *N.R. Olson; U.S. Patent 4,097,246; June 27, 1978; assigned to Olson Manufacturing Company* is an improvement over the process of U.S.

Patent 3,847,569 which discloses a method of making an abrasive wire for sawing stone. The method of U.S. Patent 3,847,569 utilizes the following steps:

(a) forming a support element into a shape having two contiguous cylindrical portions, the first portion having a larger diameter, and the second portion having a smaller diameter;

(b) positioning the support element into the bore of a die;

(c) inserting abrasive powder containing a mixture of diamond grit and a binder into the bore of the die around the smaller diameter second portion;

(d) compressing the abrasive powder towards the larger diameter first portion so that a part of the smaller diameter second portion protrudes beyond the compressed abrasive powder;

(e) heating the abrasive powder to produce a solid abrasive body;

(f) forming a central axial bore through the support element;

(g) inserting a length of flexible wire rope through the support element bore;

(h) placing a collar concentrically over the smaller diameter second portion; and

(i) swaging the collar to the support element, and the support element to the wire rope.

The method modifies and improves the foregoing process by omitting the use of a collar.

The improvement resides in the step of affixing the solid abrasive body to the support element. This step is accomplished by deforming the axially protruding free end of the underlying smaller diameter portion of the support element so that it enlarges and assumes the same diameter as the larger diameter other end of the support element, thereby sandwiching and locking the abrasive body solidly therebetween.

Ultrasharp Diamond Edges from Ultrafine Diamond Powder

Ultrasharp diamond edges and points which are usable as cutting instruments and as high intensity point sources for the emission of electrons, ions, x-rays, coherent and incoherent light and high frequency electromagnetic radiation are produced in the process of *H.F.-M. Villalobos; U.S. Patent 4,084,942; April 18, 1978* by preparing and classifying ultrafine diamond powder having a particle size of 10 to 100 Å, placing the powder in a diamond mold defining the ultrasharp edge or point to be produced and applying a pressure of the order of 80 to 90 kbar while heating the powder to a temperature of the order of 2440° to 2500°K in an ultrahigh vacuum or inert atmosphere after degasing to avoid oxidation of the diamond powder.

The ultrafine diamond powder can be obtained by crushing, grinding, explosion, electric arc or thermal shock followed by ultrahigh speed centrifuging to separate out the finer particles whereupon oversized particles are further processed.

The bonding of the ultrafine diamond particles to form a molded article having an ultrasharp edge or point is achieved either with or without a binder. For example, ultrafine nickel, chromium, titanium, niobium, zirconium, molybdenum, tungsten, copper, beryllium metal particles or their oxides, may be mixed with the graded ultrafine diamond powder in appropriate amounts (e.g., 10 to 40% by weight) to serve as bonding and catalytic solvent agents. Alternatively, individual particles of the diamond powder can be coated with a suitable metal such as chromium or nickel which acts as a matrix for the diamond particles when the article is molded. Other suitable bonding agents are LaB_4 and LaB_6. Alternatively, it is found that by using suitable pressure and temperature in an ultrahigh vacuum or in an inert atmosphere such as helium, bonding of the ultrafine diamond particles can be achieved without the use of additives.

A diamond has a 1-1-1 or tetrahedral face, a 1-0-1 face as viewed edgewise and a 1-0-0 face as viewed from above. The 1-1-1 face is the hardest. Hence, to obtain an ultrasharp edge or point it is desirable to orient the diamond particles in the mold so as to present the 1-1-1 faces at the edge or point of the implement. Such orientation can be effected with a strong magnetic field, a strong electric field or a combination of magnetic and electric fields. This orientation is considerably enhanced when the ultrafine diamond particles or platelets are first coated (by coating with a thin film evaporated in a high vacuum) with nickel, cobalt, iron and other ferromagnetic metals and alloys.

The mold used in the process must be formed of material having the strength, hardness and temperature resistance to withstand the stresses and temperature imposed on it during the molding operation. Moreover, the female mold must have the ability of defining on the molded implement an edge or point having the requisite sharpness, e.g., a radius of 10 to 100 Å. In order to meet these requirements, both portions of the mold are preferably formed of diamond which may be either natural diamond when size permits or a molded diamond structure produced in accordance with the process. To avoid the molded article sticking to the mold a suitable lubricant or separating material is used. For example, this can be sodium chloride or graphite.

An implement formed in accordance with the process has an ultrasharp edge or point which is defined by the mold in the molding operation. Subsequent finishing or polishing not only is unnecessary but is highly undesirable as it would disrupt the extremely fine edge achieved by the process. Such edge has a sharpness defined by a radius which measures only 10 to 100 Å and is straight and uniform as in the case of a natural diamond knife as described in U.S. Patent 3,646,841. However, in contrast with the natural diamond knife, a diamond knife made in accordance with the process can be of any desired length.

Such ultrasharp diamond edges, commonly referred to as "diamond knives," are used in surgery particularly in delicate operations such as operation on the eye.

Wire Saw with Cutting Elements Brazed to Metallic Substrate

G.H. Hall; U.S. Patent 3,960,518; June 1, 1976 discloses a cutting tool comprising a plurality of blocks of composite, comminuted abrasive in a bonding matrix, connected to a metallic substrate by a braze comprising a layer of an elastically pliant metal having high thermal conductivity sandwiched between a metallic material which has a melting point below the melting point of both the matrix and the metallic substrate. The tool may be formed into a band saw and is especially useful for cutting large blocks of minerals such as granite.

The substrate is preferably a steel wire exhibiting negligible elongation under extreme tension short of its elastic limit, the wire preferably having two spiral grooves opposite one another. The cutting elements are brazed into the grooves as a series of spaced teeth. The wire is arranged in an endless belt and held in tension between at least two sheaves having grooves lined with a comparatively soft, elastic material which the abrasive elements can engage without substantially tearing, cutting or gouging.

Referring to Figures 1.11a and 1.11b there is shown a preferred embodiment of cutting wire **20** in the form of an elongated body or substrate member **22** made of a metal such as steel wire or the like. Body **22** supports a plurality of abrasive segments or blocks **24**.

Figure 1.11: Cutting Wire

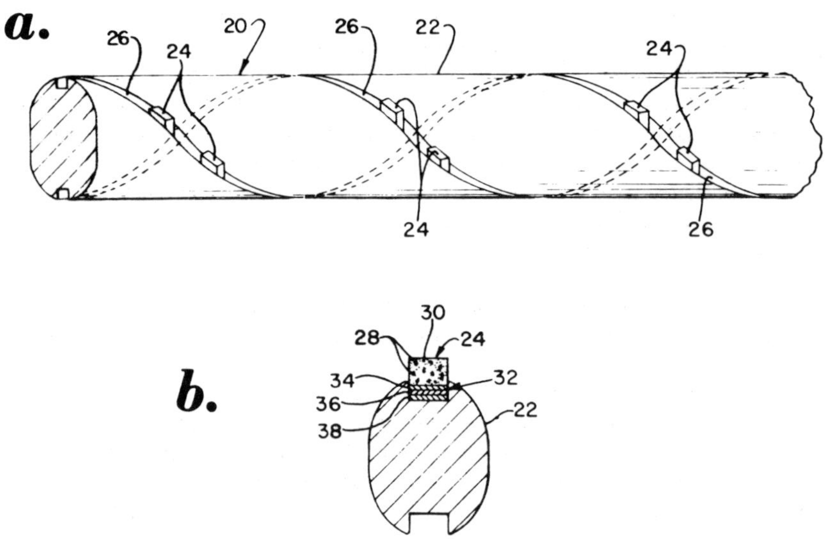

Source: U.S. Patent 3,960,518

Body **22** has one or more relatively thin, narrow notches or slots **26** formed on its surface, diagonal to its length. Preferably slots **26** are in the form of elongated grooves forming a helix around the long axis of body **22.**

A plurality of relatively short abrasive segment blocks **24** are mounted in spaced-apart relation in slots **26.** Blocks **24** typically comprise a mixture of diamond particles, e.g., diamond bits and grit **28** in a bonding matrix **30,** all as well known in the art. The size of the diamond particles and the proportion of particles to bonding matrix are not critical to the process, and may be varied over a wide range as is well known in the art.

It will be appreciated as above noted, that the wire is intended to flex. Hence, to avoid cracking matrix **30** by the forces arising during flexure of wire **20,** the matrix preferably is somewhat flexible. To this end, matrix **30** may comprise principally copper, e.g. up to about 70% copper, or some other malleable metal, and include lesser amounts of other metals such as iron, zinc, tin, titanium, chromium, zirconium and the like. The use of copper as the principal matrix material imparts the desired degree of flexibility to blocks **24.** Blocks **24** are formed as an unitary mass, e.g., by sintering the bonding matrix by techniques well known in the art.

Blocks **24** are sized relative to the depth of slots **26** so as to protrude beyond the surface of body **22** for a short distance, e.g., ⅛ inch when positioned in the slots. The diamond particles in the blocks form the abrasive or cutting elements of the blade.

Abrasive element blocks **24** are held in slots **26** by a bonding system comprising a layered metallic braze indicated generally at **32.** As noted, braze **32** is a sandwich of an elastically pliant metal between layers of a low melting point metal having a tensile strength of at least the order of the pliant metal and which bonds strongly to the pliant metal, to the substrate metal of wire **20** and to the matrix material of blocks **24.** Braze **32** typically then comprises a plurality of relatively thin metal-containing layers including an inner layer comprising metallic copper in major proportion, sandwiched between layers each comprising metallic silver in major proportion. By way of example, three metallic layers (**34, 36,** and **38**) are shown, although it will be understood that braze **32** may comprise four or more layers. Typically the metallic layers are as thin as about 0.005 inch and may be the same or different thickness. Layers **34** and **38,** the outside layers, each comprises a thin metallic film, comprising silver in major proportion, and may include in lesser amounts alloying metals with silver such as copper, zinc, cadmium and nickel.

The middle layer **36** comprises a thin film comprising metallic copper in major proportion. Thus, for example, layer **36** may comprise substantially pure copper, or one or more alloying metals may be included with the copper. Layer **36** may be substantially the same thickness as layers **34** and **38;** preferably, however, layer **36** is somewhat thicker than either of layers **34** and **38.**

The layers of braze **32** typically are laid in slots **26**, either one layer at a time or as a sandwich. Then abrasive blocks **24** are positioned in slots **26** on top of one outside layer. The blocks are then clamped, under pressure, to braze **32** and the blocks and body **22** are then brazed together by heating the sandwich until the silver containing layers melt and flow.

In a preferred embodiment of the process, brazing is effected by induction heating of the metal substrate body **22**, by passing an electric current through the body. This procedure has a number of important advantages. For one heating may be effected relatively rapidly (as compared with infrared heating procedures) with the result that the diamond particles may be subjected to high temperature for only a relatively short period of time, whereby high-temperature deterioration of the diamonds may be kept to a minimum. Thus typically, heating to brazing may be completed in about ten seconds or less. Another advantage is the elimination of substantial equipment often required by the prior art, e.g., for carrying out heating in inert or reducing atmospheres.

Diamond Abrasive Cut-Off Wheel

The basic techniques for manufacturing abrasive wheels of the general class typified by this process have been well and thoroughly documented in the prior art, best exemplified by U.S. Patents 2,818,850, 3,203,774, 3,298,806 and 3,443,343.

Briefly reviewing the prior manufacturing methods used, a circular sheet metal disc, having a discontinuous periphery, formed the central support and drive member, and arcuate segments containing abrasive diamond dust were welded or brazed to the periphery of the disc.

Cutting segments are generally produced separately, and are formed of diamond dust or chips embedded in a heat-settable, or bonded, bronze or nickel-copper alloy matrix.

The segments of prior cut-off wheels of this general class, have been formed with generally regular, substantially smooth cutting surfaces, for the reason that the actual cutting was accomplished by embedded diamond chips or dust and, as would be the normal expectation, little or nothing would be achieved by any predetermined cutting surface pattern.

It is therefore an object of the process of *H. Schwarzkopf and J.J. Bouvier; U.S. Patent 3,863,401; February 4, 1975; assigned to Ferro Corporation* to provide a cutting segment having greatly improved cutting efficiency by virtue of its uniformly serrated cutting surface.

Utilizing any of the techniques previously known, the segment **1** depicted in Figure 1.12c is manufactured, the upper, serrated portion **2** carrying diamond abrasive, the lower section **3** being of a nondiamond bearing, softer more readily machinable metal for brazing to the arbored, steel support and drive disc **4**.

Figure 1.12: Diamond Abrasive Cut-Off Wheel

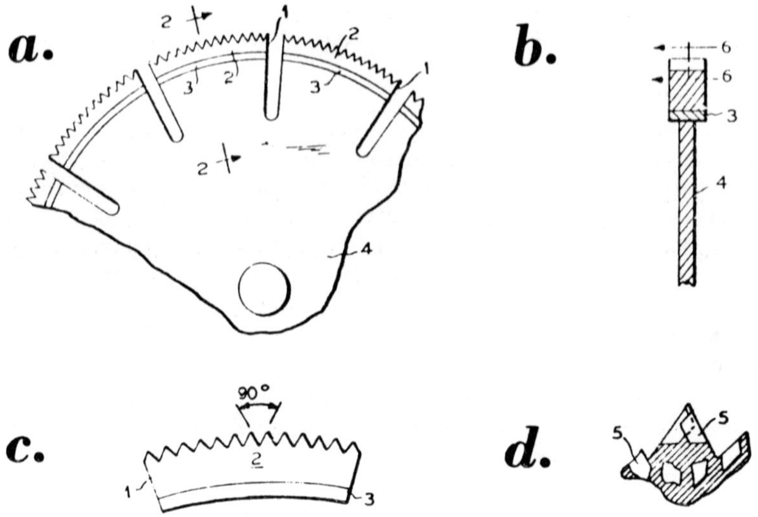

(a) Side view of a section of a cut-off wheel produced in accordance with this process.
(b) Cross section taken along the line **2–2** of Figure 1.12a.
(c) Enlarged view of one of the serrated segments of the wheel depicted in Figure 1.12a.
(d) Enlarged section taken along the line **6–6** for Figure 1.12b.

Source: U.S. Patent 3,863,401

The concave face of the segment die or mold may be readily serrated in a transverse direction, the serrations preferably being substantially, uniformly, approximately 90°, the serrations transferred to the convex face of the segment as a consequence of the forming operation.

It is thus to be understood that the transverse serrations as depicted in Figures 1.12a and 1.12c, particularly in diamond abrasive cut-off wheels, are not in any way to be confused with conventional, metallic saw teeth.

Rather, the serrations are designed, as best illustrated in Figure 1.12d, to cause certain flat faces of the diamond dust, or chips, **5**, to be oriented parallel to one of the inclined faces of the serrations during the pressing and sintering operations.

Using this technique, it will be readily apparent that a much greater percentage of relatively sharp angular points of the diamond dust or chips, will be oriented in such a manner that such points will be more quickly and readily available for immediate and improved cutting when the wheel is initially placed in operation, since the long axes of the diamond particles tend to orient more nearly radially, thus positioning any points outwardly from the center of the wheel, in a position to cut far more efficiently as the binder matrix is sacrificed during operation.

Utilizing prior, conventional, relatively flat arcuate cutting surfaces, it will be seen that the natural tendency is for surface diamond particles, or dust, to be oriented, both at the surface and interiorly, in such a manner that their longest, flatest surface is parallel to the convex surface of the segment; and, in operation, it is therefore likely that a flat surface of a diamond particle will be repeatedly whirled past the article to be cut, providing a much lower degree of cutting efficiency as compared to diamond particles positioned by the serrated segments of this process.

From a more quantitative standpoint, the serrated portion of the segment of Figure 1.12c may be approximately one-fourth the total height of the diamond bearing portion of the segment.

Yet surprisingly, up to one-half the life of the segment has been found to reside in the serrated 25% of the segment, based on the total height of the diamond bearing portion. Once the serrations are abraded away, the remaining three-fourths of the segment then cuts with diamonds in the conventional attitude.

Optionally, the cutting life of a segment may be further extended by forming serrations on both the convex and concave surfaces of the diamond bearing portion of a segment.

In this manner, once the surface serrations are worn away, inverted subsurface serrations become exposed, with their serration-oriented diamond particles providing their increased cutting efficiency.

An angle of serration of approximately 90° provides the greatest efficiency, although an angle of anywhere from 60° to 110° will provide much improved results.

Cable-Type Saw with Sleeved Abrasive Elements

H.A. Snow; U.S. Patent Reissue 29,491; December 13, 1977 provides a saw of the type utilized in cutting stone or the like having an elongated flexible cable and a plurality of small diameter abrasive wheels mounted on the cable. Each of the wheels is provided with a centrally located passageway in which lies a metal sleeve having a bore substantially the diameter of the cable to provide a tight threading of the cable therethrough. Means is provided for holding each of such sleeves and abrasive wheels in fixed, nonrotatable, spaced relationship on the cable.

Referring to Figure 1.13a, it can be seen that the saw **10** consists of a cable **18** formed from strands of high strength steel on which are mounted in space relationship a plurality of abrasive elements **19**. Each abrasive element consists of a main body **21** formed of a malleable metal on which is mounted a cylindrical abrasive portion **22**. The diameter of the cylindrical abrasive portion is equal to the desired width of the cut to be made in the stone. The main body **21** is provided with a radial flange **23**, having an outer cylindrical surface **24** which constitutes an extension of the cylindrical surface of the abrasive portion **22**.

Figure 1.13: Cable-Type Saw with Sleeved Abrasive Elements

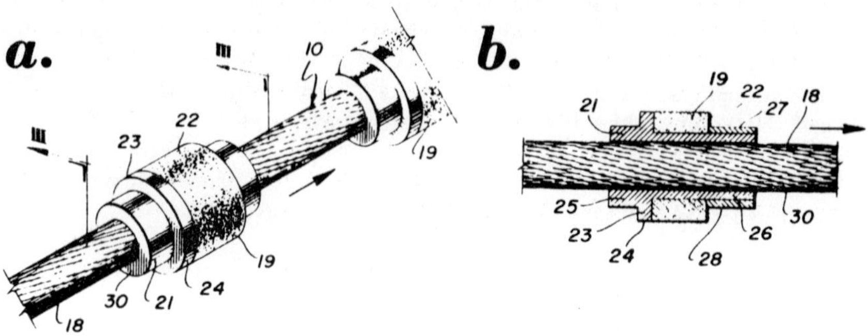

(a) Perspective view of a portion of the saw.
(b) Sectional view of the saw taken on the line **III–III** of Figure 1.13a.

Source: U.S. Patent Reissue 29,491

In Figure 1.13b it can be seen that the main body **21** consists of a first cylindrical portion **25** of large diameter and a second cylindrical portion **26** of small diameter. The abrasive portion **22** is in the form of a wheel which fits over the second portion **26**. The flange **23** extends outwardly of the first portion **25** at the shoulder between the first portion and the second portion. Also arranged on the second portion **26** is a sleeve **27** having an outer cylindrical surface **28** which is the same diameter as the first cylindrical portion **25** of the main body. The sleeve is tightly fixed on the second portion **26** of the main body and is pressed tightly against the radial side of the abrasive portion **22** which faces it.

Although it is not evident in the drawings, the main body **21** has a central bore **30** through which the cable extends rather tightly and the body is securely fastened to the cable **18** by swaging. The free end of the first portion **25** is swaged through the cable by use of a swaging tool operative on its cylindrical surface adjacent the end. The other end of the main body and the sleeve **27** is fastened to the cable by swaging downwardly on its cylindrical surface **28**, which has the effect of also swaging the outer end of the reduced or second cylindrical portion **26** of the main body. Furthermore, in the preferred embodiment, the abrasive portion **22** is formed on the main body **19** by compacting a mixture of abrasive particles and bonding material under the main body and centering the entire assemblage.

The arrows in Figures 1.13a and 1.13b show the direction that the saw is moved through the stone. It can be seen that the abrasive element **19** is advanced so that the abrasive portion **22** moves first through the cut and is followed by the flange **23**. Since most bonded abrasive materials are somewhat brittle as compared with the malleable metal of the main body (which in preferred embodiment is formed of copper), the strong flange lends support to the abrasive support. The net effect is an abrasive element which has a cutting quality of a more friable abrasive substance (which may be in the nature of a ceramic) and yet the strength and toughness of the material from which the main body is formed. Naturally, when the abrasive material wears, the surface **24** of the flange will be exposed to the stone. Since the material of which the main body is formed is much softer than the abrasive material, this flange will wear down and maintain at all times the same diameter as the abrasive portion **22**.

In the preferred embodiment, the abrasive portion **22** is formed as a mixture of diamond particles and powdered copper, which mixture is suitably compressed and sintered in the well-known manner to produce a tough yet friable abrasive product. The entire abrasive element, of course, is prevented from moving along or rotating about the cable **18** by the swaging, which has taken place between the ends of the main body, the sleeve **27**, and the cable itself. The fact that the cable is formed of a plurality of strands provides it with a configurated surface into which the malleable copper of the main body and the sleeve may be swaged or forged.

LENS GRINDING AND POLISHING

Lapping Tool with Accurate Curvature

It is the object of the process of *S. Taguchi; U.S. Patent 4,035,160; July 12, 1977; assigned to Canon KK, Japan* to provide a method for manufacturing a lapping tool having an accurate curvature, wherein the pellets, each having the same curvature as that of the curved surface of the lapping tool, are bonded thereon, and the shape of the pellets is controlled simultaneously with the manufacture of the completed tool.

The liquid casting resin which can be adopted for use in the method may be any kind of conventional pouring resin having appropriate adhesive and polishing properties with respect to a cast iron tool body. Particularly suitable liquid resins for the purpose of the process are, for example, liquid epoxy resin, liquid polyurethane resin, and liquid acrylic resin, liquid polyethylene terephthalate resin.

As shown in Figure 1.14, a master lapping tool **1**, on which pellets **2** are adhered in conformity to the surface curvature of a lens to be obtained is first manufactured. Next, this master lapping tool **1** is secured firmly by a holder and stand (not shown), and is set at the center part of a mold frame **4**.

Subsequently, silicone rubber and a hardening agent in measured quantities are uniformly mixed and poured into the mold frame **4**, and allowed to harden therein for about one hour at ambient temperature or an elevated temperature of about 70°C, for example. Upon verification of sufficient hardening, the master lapping tool is removed from the mold, whereby the silicone rubber mold **3** is obtained.

When the mold is thus formed, a liquid pouring resin is poured into the mold at the place corresponding to the pellets portion **2** of the mold **3**, and a blank body for the lapping dish is then placed in the mold in a manner to press against the mold surface to thereby harden the pouring liquid resin, thereafter the lapping tool is removed from the mold. In this way, the lapping tools identical with the master lapping tool can be produced easily, accurately, and in a large quantity.

Figure 1.14: Production of Lapping Tool

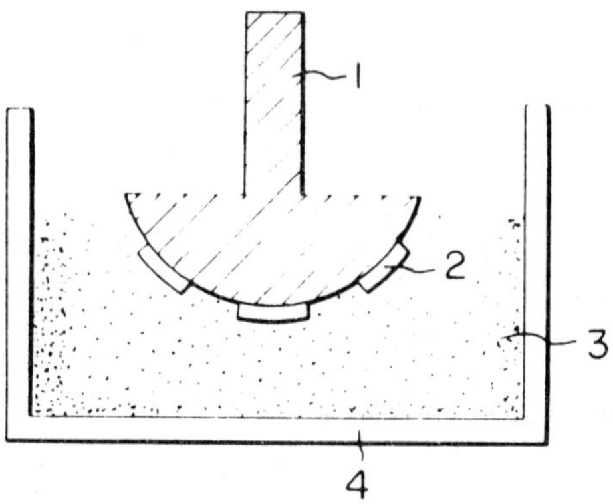

Source: U.S. Patent 4,035,160

Although a convex tool is shown in the figure, concave tools may also be manufactured by this process.

Example: A master lapping tool with pellets of circular contour and the same curvature as that of a lens to be obtained adhering onto the tool surface was first produced. Each circular pellet was 7 mm in diameter and 1.5 mm in thickness. The number of the pellets provided on the tool surface was thirteen pieces.

This master lapping tool was hung in a frame for a rubber mold. Then, a mixture of 20 g of silicone rubber and 2 g of a hardening agent was poured into the mold frame, and hardened at a temperature of about 70°C for one hour. Upon sufficient hardening of the rubber mold, the master lapping tool was removed from the tool, whereby the intended mold could be obtained.

Next, 13 g of liquid epoxy resin and 1.3 g of an amine type hardening agent were sufficiently mixed by agitation, after which the liquid resin was poured into the silicone rubber mold at places corresponding to the pellets portion thereof. While the liquid resin was still in the fluid state, a blank tool body of cast iron having the same curvature as that of the master but not having pellets adhering thereon was placed in the mold in a manner pressing against the mold surface from above. In that state, the mold was kept in an isothermal chamber maintained at a temperature of 60° to 70°C for 15 minutes to harden the epoxy resin.

After sufficient hardening of the resin, when the tool body was removed from the mold, a lapping tool identical with the master lapping tool and having the epoxy resin adhering onto the tool surface in the pellet form was obtained. Although it was almost unnecessary to correct the curvature of this lapping dish, it was still polished for about 15 seconds by a correcting tool of cast iron to correct its curvature to the most precise degree. In contrast to this, a lapping tool, on which pellets of epoxy resin of 7 mm in diameter and 1.5 mm in thickness were made to adhere by first applying an adhesive agent, had a curvature so different from that of the lens to be polished, that it had to be polished for about 3 minutes by a correcting tool to correct the curvature into conformity to that of the lens.

The abovedescribed lapping tool according to the process was subjected to lens polishing under a load of 4 kg/cm^2 at a varying speed ranging from low (10 rpm) to high (3,000 rpm). No peeling off of the pellets was observed at all, and satisfactory polishing could be achieved.

Fixed-Superabrasive Tool for Toric Lenses

The improved fixed-superabrasive tool for grinding and polishing toric and other curvilinear lenses of *C.W. Highberg; U.S. Patent 4,010,583; March 8, 1977; assigned to Engelhard Minerals & Chemicals Corporation* comprises a substantially rigid, superabrasive-free backing member and a malleable preformed superabrasive wafer conformed and releasably secured thereto. The substantially rigid, superabrasive-free backing member has a curved support surface determinative of the curvature of the curvilinear working surface. The substantially malleable preformed abrasive wafer has a predetermined thickness and one surface which is conformed to the curved support surface, either before or at the time of securing. The wafer comprises superabrasive particles having a Knoop hardness in excess of about 3,000 kg/mm^2 in a malleability-imparting powdered metal matrix, which facilitates the conformation step. A porous surface may be created by introduction of particles which are dissolved during use of the tool, leaving pores which assist the grinding and polishing processes.

Referring to Figures 1.15a, 1.15b and 1.15c, the preferred two-layer or dual-layer embodiment of the fixed-superabrasive wafer includes a first layer **10** comprising a substantially nonabrasive malleable buffer medium and a second layer **12** comprising superabrasive particles homogeneously distributed throughout and bonded by a powdered metal matrix. The two layers are integrally formed in the sense that they are fused or otherwise blended at the interface to form an integral structure.

Figure 1.15: Fixed-Superabrasive Wafer

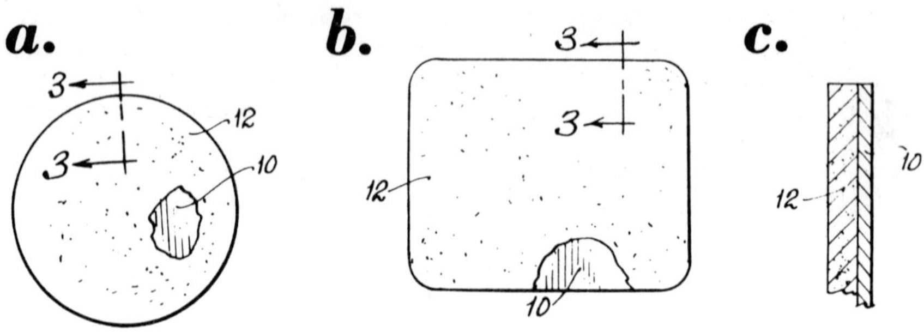

(a) Cutaway view of the preferred two-layer abrasive wafer as viewed from the abrasive side, prior to being conformed and releasably secured to a backing member.

(b) Cutaway view of another embodiment rectilinearly configured to conform to a similarly configured backing member.

(c) Magnified sectional view taken along the lines **3–3** of Figures 1.15a and 1.15b.

Source: U.S. Patent 4,010,583

For long tool life, abrasive layer **12** has a thickness of at least about $^1/_{32}$ inch, preferably in excess of $^1/_{32}$ inch. Since nonabrasive buffer layer **10** is not normally subjected to wear, it typically does not exceed a thickness of $^1/_{32}$ inch.

Since the wafer has a substantial predetermined thickness usually in excess of about $^1/_{32}$ inch, e.g., about $^1/_{16}$ inch, and must be conformed precisely to a curved backing member having a substantial curvature, e.g., radius or radii of curvature of less than about 36 inches, both layers **10** and **12** must have substantial malleability despite the presence of nonmalleable superabrasive materials. In preferred embodiments, layer **10** comprises a copper sheet of commercial purity, e.g., in excess of about 99% copper; and layer **12** comprises superabrasive particles in a powdered metal matrix comprising about 85 to 98% by weight of copper and about 15 to 2% by weight of tin. Alternatively, both layers **10** and **12** may comprise a powdered metal bond, layer **12** also having the superabrasive particles dispersed therein.

The superabrasive may comprise any suitable grinding or polishing medium having sufficient hardness to efficiently and economically grind or polish the grades of glass normally employed for optical and ophthalmic lenses. As already indicated, a Knoop hardness in excess of about 3,000 kg/mm^2 is required. While this requirement eliminates common abrasives such as quartz, aluminum oxide and silicon carbide, it does permit the use of cubic boron nitride, as well as natural or synthetic diamonds.

The size and concentration of the superabrasive particles in the powdered metal bond depend in part upon the type of operation being carried out and the particular material being worked. Other factors, representing tradeoffs, include quality of surface, speed of grinding and related cost considerations. As those skilled in the art will recognize, the requirements for a rough grinding operation are different from those of a finish polishing or lapping operation, In preferred embodiments, diamond powders conforming to Bureau of Standards (C.S. 261/63) Grade Numbers 9, 15 or 30, which correspond to a size range of about 8 to 36 μ, may be employed.

The concentration of diamonds in the bond may range from about 2 to 25% by volume of the total matrix. Thus, for example, a matrix may comprise about 10% by volume of superabrasives and about 90% by volume of bond, the latter comprising about 95% by weight copper and about 5% by weight tin.

In a preferred embodiment cooling and swarf removal during the grinding operation are achieved by introducing particles which are soluble in the cooling fluid used and thus create a porous surface in the matrix adjacent the superabrasive particles. Table salt, having 80-120 mesh size particles and between 10 and 50% by volume, has been found particularly suitable for introducing porosity.

While the superabrasive wafers depicted in Figure 1.15 have a constant thickness, the process is not necessarily limited thereto. The thickness may be varied, e.g., from the center to the outer edges, so as to achieve, when conformed to the curvature of the backing member, a desired configuration. Thus, backing members having the same configuration may be utilized to produce grinding tools having different curvatures at the grinding surface. In multilayer embodiments, the variations in wafer thickness may be achieved by varying all layers or less than all. Thus, for example, in the preferred dual-layer embodiment, the superabrasive layer may have a constant thickness to achieve the desired curvature when conformed to the backing member.

Radially Adjustable Grinding Wheel

A lens grinding apparatus for simultaneously grinding infinitely variable primary and secondary radii on the concave surfaces of a plurality of lens blanks is provided by *L. Coes, Jr.; U.S. Patent 3,986,303; October 19, 1976; assigned to Norton Company.*

The apparatus comprises a fixed wheel slide base or base on which is slideably mounted an adjustable wheel slide or wheel support for supporting a radially adjustable segmental grinding wheel for rotation in a single grinding plane and about an axis of rotation to produce the secondary radius. Feed means are provided for moving the wheel along the single grinding plane relative to and toward and away from the concave surfaces of a plurality of lens blanks fixed to a lens carrier mounted on a radially adjustable work holder movable about a pivot or center situated in the grinding plane at the intersection of another plane passing through the rotational axis of the grinding wheel and at right angles to the grinding plane. The work holder is either of arcuate or circular shape extending respectively partially or completely around the grinding wheel and comprises a plurality of radially adjustable and angularly spaced lens carrier support brackets for receiving and supporting lens carriers of various curvatures or radii and lens blanks thereon at various predetermined radial distances from the pivot or center of movement.

A radially contractable and expandable segmented grinding wheel of predetermined initial diameter is mounted on an adjustable wheel mounting which is removable from and extends axially from an end portion of a wheel spindle journaled in the wheel support or slide and rotatably driven about its axis. The grinding wheel has abrasive segments attached to angularly spaced peripheral portions of a slotted support center with angularly spaced deformable wall portions of identical shape about a hub portion. The configuration of the deformable wall portions can be varied by the wheel mounting to either increase or decrease the diameter and radius of the wheel. The abrasive segments are bonded with a heat softenable and alkali soluble metal bond which aids in heating, truing and dressing the wheel to the desired secondary radius of the lens.

A wheel truing device mounted on the wheel slide comprises a truing roll rotatably mounted on a truing slide slideably mounted on the wheel slide and movable by a feed means into forceful engagement with the periphery of the abrasive segments of the grinding wheel rotatably driven at a relatively slow truing speed by auxiliary drive means. During rotatable engagement of the grinding wheels by the truing roll, an electrical current is passed between the truing roll and the wheel to heat soften the metal bond which is displaced by the truing roll to produce the desired secondary radius on the wheel and hence the lens blanks ground thereby. Thereafter, the abrasive surface is dressed by applying an alkali solution which dissolves and removes a portion of the displaced bond to expose the abrasive cutting elements.

Simultaneous grinding of the primary and secondary radii on the lens blanks is accomplished by moving the carrier and lens blanks thereon at a constant radius about the pivot or center and at right angles across the peripheral abrasive surface of the grinding wheel whose radius is equal to the secondary radius and intermittently feeding the grinding wheel radially into the lens blanks to a final position equal to the primary radius from the pivot or center. The complete patent includes figures with detailed descriptions of the apparatus.

MASTERS AND BLANKS

Cutting Master for Total Form Abrasion Machining

The process of *T.J. O'Connor; U.S. Patents 3,990,868; November 9, 1976 and 3,948,620; April 6, 1976* is directed to the method of making a cutting master for total form abrasion machining.

Cutting masters for total form abrasion machining have a surface which is a mirror image of a form to be abrasion machined, which mirror image surface differs in dimension from the form to be machined in two of three mutually perpendicular directions. The mirror image surface of the cutting master is abrasive and includes a flat portion which is equal in size to the difference in size between the cutting master and the form to be machined in the two mutually perpendicular directions.

The cutting masters may be formed in a plurality of ways. Thus, for example, a model of the form to be machined which is different in size from the form to be machined by a predetermined amount may first be produced by conventional machining such as single point or line machining or by electro-erosion methods, and a mirror image master cutter may be cast over the model of a settable material having an abrasive mixed therein. The settable material may be set by chemical curing or change of temperature or the like after the settable material is cast on the model.

The cutting master may be formed by a plurality of small diameter rods having ends positioned on the machined different size model, which rods are then locked in position by a binder or poured thereover or placed thereon. With such method, the ends of the rods may be pointed as desired and the ends of the rods may be exposed by sandblasting or an electrical discharge machining process.

The cutting master may also be formed by spraying metal such as aluminum on a model of the shape to be total form abrasion machined to form a mirror image model of relatively soft metal, spraying electrode material such as copper on the soft metal mirror image model formed to produce an electrode which is the exact form to be abrasion machined, using the electrode so formed to sink a mirror image form of the shape produced by an electro-erosion process in conjunction with a rotary orbital movement applied to the electrode whereby the electrode is moved in two planes perpendicular to the direction of the movement of the electrode toward a workpiece during the electro-erosion process, casting a cutting master form in the electro-erosion machined form which also differs from the shape to be abrasion machined by a predetermined amount, and casting a settable material having abrasive material therein over the cast cutting master form.

Alternatively, the cutting master may be formed by an electro-erosion machining process directly either by first manufacturing an electrode which differs in size from the shape it is desired to abrasion machine by a predetermined amount and producing the cutting master by a direct plunge in electrical discharge machining. Alternatively, an exact size model of the shape to be produced may be produced and used as an electrode in a straight plunge electrode discharge process, after which the electrode is orbited in all directions in a plane perpendicular to the direction of plunge, to produce the different sized cutting master which has an abrasive surface due to the normal characteristics of electrical discharge machining.

In such processes, if it is desired to use the form to be machined as an electrode to produce a cutting master, and if it is not constructed of a conductive material, a conductive material may be sprayed over the nonconductive form to be machined. Either male or female master cutters may be produced by the above methods. Wherein the master cutter is female, it will be oversize, while a male master cutter will be undersize, whereby the rotary, orbital motion provided in abrasion machining will produce a machined form which is of the desired dimension.

Cutting Tool Blank from High Purity Alumina

The process of *T. Vasilos, W.H. Rhodes and J.E. Niesse; U.S. Patent 3,954,930; May 4, 1976; assigned to Avco Corporation* relates to a cutting tool blank formed from high purity alumina and to the procedure for forming the cutting tool blank.

The process involves vacuum hot-pressing a mixture of high purity alumina and from 0.05 to 0.35% by weight of magnesia. The particle size of the alumina-magnesia mixture should not exceed about 0.5 μ. The hot-pressing is effected in a temperature range of about 1300° to 1525°C with conventional hot-pressing equipment and conventional pressure conditions. The applied pressure may, for example, be in the range of about 1,000 to 5,000 psi. Important to the process is that care is taken during the heating and pressing cycle to achieve a controlled crystal growth. The average crystal size of the molded alumina tool blank is within the range of 1 to 2 μ. The fabricating temperature of the blank should not exceed about 1525°C; otherwise excessive crystal growth occurs.

The magnesia content of 0.05 to 0.35% by weight is present specifically to control crystal development during processing. A preferred magnesia content is within the range of 0.1 to 0.25% by weight. The presence of magnesia together with careful temperature control of the hot-pressing cycle provides the desired average crystal size range in the formed cutting tool blank of from 1 to 2 μ. Additives other than magnesia and nickel oxide, discussed later on, do not appear necessary. The presence of other additives may be detrimental to the cutting properties of the ultimate tool. Pure alumina seems to form the highest quality tools.

The formed tool blank has a density in excess of 3.97 g/cc and preferably in excess of 3.98 g/cc. The latter figure is very close to the intrinsic density of alumina, i.e., 3.986 g/cc.

In practice, it has been found that the tool blank and ultimately the tools formed are not uniform in color, being varying shades of grey, batch-to-batch. Color variations sometimes appear in the same tool. Whatever may cause the color differences has no effect on the performance of the tools. Unfortunately, some customers associate uniform optical appearance with product quality. To allay customer fears about off-color tools, and color variation on a tool, the tool blank may be tinted black by including up to 0.10% nickel oxide, preferably 0.03 to 0.10% as an additive during the formation thereof. The nickel oxide is included in partial substitution of the magnesia. Nickel oxide acts much like magnesia to control crystal development in the formed alumina tool blank. Here, however, the nickel oxide has a special advantage, cosmetic in nature. Its presence converts the tool blank into a black article and no noticeable optical differences show up batch-to-batch. An essentially uniform black tool blank can always be provided. Importantly, presence of the nickel oxide has not been found to have any effect on the quality of the ultimate cutting tool.

A significant process feature of the process is conducting the hot-pressing operation under evacuated conditions. Specifically contemplated is evacuation of the hot-pressing chamber to relatively high vacuum, e.g., the vacuum attainable by commercial diffusion pumps which may be as low as 0.1 μ of gas pressure.

Example: Powdered high purity alumina and magnesia (both less than 0.5 μ) are mixed thoroughly, the magnesia content being 0.25% by weight. A portion of the mixture then had 0.10% by weight nickel oxide added thereto.

Tool blanks (right discs about 6 inch diameter, 0.2 inch thick) were formed from each mixture in an evacuated hot press electric furnace. During each run, the applied pressure was 4,000 psi and the heating cycle was 3 hours from ambient to 1470°C, then a 3½ hour hold at 1470°C still at 4,000 psi after which the pressure was released and the furnace cooled to ambient (6 hours) before breaking the vacuum. The internal pressure was held to about 10 μ during the heating cycle.

The tool blanks containing nickel oxide were black, a uniform shade of black. The other specimens were various shades of grey, with some of the blanks being nonuniform in color.

Three tools of geometry SNG 434 were cut from blanks for comparison with like tools formed under nonevacuated hot-pressing conditions. Both sets of tools were employed to cut steel of Rockwell C-40 hardness (Carpenter Stentor Tool Steel) to a depth of 0.050 inch using an in-feed rate of about 0.004 inch. A turning rate of 850 to 1,200 surface feet per minute was employed with the actual rate being varied over this range in a random fashion during each test.

The best vacuum pressed tool cut through 15.8 miles of metal. In the air pressed tools, the best tool failed at 10.9 miles of metal cut.

Tools from the nickel oxide containing blanks and tools from blanks containing magnesia only proved to be of the same high quality.

TOOLS FOR SPECIFIC PURPOSES

Cleaning Aluminum Aircraft Skin

An abrasive disc prepared according to the process of *W.E. Early and M.G. Linn; U.S. Patent 4,133,144; January 9, 1979; assigned to the U.S. Secretary of the Navy* removes paint and sealant from aluminum aircraft skin without damaging the skin.

The soft abrasive disc contains a metallic wool matrix impregnated with an elastomeric polyurethane binder, both of which have been effectively mixed in such manner as to have a substantially uniform consistency and/or density.

As a means for removing paint, sealant, and other coatings from the aluminum skin of an aircraft, the metallic wool matrix would be an aluminum wool matrix, the aluminum of which is preferably of commercial medium grade with hard temper, which is thoroughly mixed with a polyether type elastomeric polyurethane binder which is a very abrasion resistant, tough, long-lived material, with the mixture thereof preferably containing (but not limited to) approximately 55% aluminum wool and 45% elastomeric polyurethane binder by weight.

The abrasive disc may be manufactured by using the following steps:

(1) Place a commercial medium grade, hard tempered aluminum wool matrix in the bottom of a suitable mold.

(2) Compress the aluminum wool matrix against the bottom of the mold at a pressure of 3,000 to 4,000 psi gauge.

(3) Bring the polyether elastomeric polyurethane binder material to room temperature of approximately 75°F.

(4) Mix the elastomeric polyurethane binder material with a setting catalyst of the amine catalyst type.

(5) Pour the elastomeric polyurethane binder and catalyst mixture into the mold on the aluminum wool matrix already deployed therein.

(6) Compress the elastomeric polyurethane and catalyst mixture at 300 to 400 psi gauge, so as to cause the former to impregnate the latter with a substantially uniform consistency and/or density.

(7) Cure the polyurethane binder and aluminum wool mixture at room temperature until it becomes a

solid disc (or other shape) having approximately a Shore hardness A of 90.

(8) Remove the cured elastomeric polyurethane and aluminum wool matrix disc from the mold.

When the disc is removed from its mold, it is ordinarily connected to the shaft of a drive motor. Then, while being rotatably driven by the motor, the abrasive disc is held (by hand or otherwise) against the material or materials desired to be cut, removed, and/or polished.

Because the cutting element of the disc is a matrix of aluminum wool, the substrate aluminum is polished without any significant damage thereto, as far as loss of strength, cracking, fatigue, or the like, are concerned.

Knife Sharpening Tool

P.K. Church and A.G. Phillips; U.S. Patent 4,102, 085; July 25, 1978; assigned to Kaman Sciences Corporation provides a knife sharpening tool having a cutting surface consisting of abrasive grains embedded and held in place in a ceramic-metal matrix, chemically bonded to a substrate such as a steel core.

A substrate is provided to which an abrasive coating can be adhered. The substrate may have a predetermined shape such as that of a sharpening steel. The abrasive coating is initially applied to the substrate in an aqueous slurry form by dipping, spraying, brushing and the like. The slurry consists basically of a suitable mixture of metal powder, abrasive grain, water and a water-soluble chromium compound. A small amount of a suspension agent such as kaolin, Cab-O-Sil and the like may be added to help keep the powders in suspension. When used in small amounts the suspension additive does not have a noticeable effect on the cutting properties of the sharpening steel.

The slurry coating, after being applied to the substrate is then dried and thermally cured in an initial heat cycle at a temperature high enough to chemically convert the chromium compound to a water-insoluble chromium oxide but low enough to prevent excessive oxidation of the metal powder constituent of the coating or the substrate.

During this initial heat cycle, a chromium oxide bond is established between the coating and the substrate.

Following the initial heat cycle, the applied coating is too soft for use at this point as a sharpening steel and must next be further bonded, densified and hardened. This is accomplished by impregnating the still porous coating with a solution of a soluble chromium metal compound capable of being converted to insoluble chromium oxide in situ. This impregnation-cure cycle procedure is repeated at least once and for a sufficient number of times to achieve the desired hardness and density. The additional chromium oxide conversions also establish an extremely strong bond between the particular coating constituents and also between the coating and the substrate.

A curing temperature of about 600°F is adequate for conversion of the chromium compound but a higher temperature is often employed, e.g., 800° to 1000°F to allow for faster cure times because of the thermal mass of the substrate, etc. Temperatures as high as about 1500°F may be considered in some instances but may also require the use of a nonoxidizing or inert atmosphere. Temperatures up to about 100°F may be used without the precaution of a nonoxidizing or inert atmosphere but temperatures above about 1000°F in a normal atmosphere produces excessive oxidation of the metal powder. Higher cure temperatures are normally avoided so as not to over-oxidize, deform or melt either the metal coating powder or the substrate and not to dull the sharp cutting edges of the abrasive grains. In addition, higher cure temperatures in general have been found to require a closer thermal expansion match between the coating and substrate.

Seed Defiberization

C.F. Harrington, J.M. Johnson and H.T. Prince; U.S. Patent 3,943,604; March 16, 1976; assigned to The Procter and Gamble Company provides an improved process and equipment for seed defiberization, for example, cottonseed delinting, wherein the seed is defiberized by contact with particular abrasive surfaces. In one preferred embodiment, a conventional cottonseed delinter is equipped with an abrasive faced cylinder instead of the conventional shaft mounted ginning saws.

The abrasive cylinder comprises a shaft mounted cylinder of approximate gin saw diameter which cylinder has bonded or adhered to its facial surface an open pattern of about 40 to 80 abrasive grit particles per square inch. The abrasive grit particles, preferably tungsten carbide grit, are sized to U.S. Sieve Series numbers of about 12 to 40. Replacement of the ginning saws with the abrasive face cylinder requires modification to the delinter gratefall rake and seed seals to adapt their shape and clearance to the comparatively flat abrasive cylinder face. The abrasive cylinder embodiment provides means for convenient adaptation of conventional gin saw delinters to achieve increased cottonseed handling capacity, less power requirement and decreased, safer labor requirement benefits.

Dry Grinding Hard Steel and Carbide Tools

The process of *R.H. Sioui and H.M. Cohen; U.S. Patent 3,868,232; February 25, 1975; assigned to Norton Company* relates to abrasive tools, in particular grinding wheels and coated abrasive belts, containing the hardest known abrasives, diamond and cubic boron nitride which, when metal coated diamond grit is the primary abrasive are particularly suitable for the dry grinding of cemented carbide tool material (e.g., cemented tungsten carbide), and when metal coated boron nitride grit is the primary abrasive are particularly suitable for grinding hard tool steels such as T15, M2, M3, and M4.

The process involves modifying the bond of resinoid bonded abrasive tools, containing metal clad diamond or cubic boron nitride, by inclusion of from 10 to 60% by volume, preferably 30 to 50% by volume of the total of bond and fillers (exclusive of the metal clad abrasive) of molybdenum, and from 5 to 30%, preferably 10 to 20% by volume of molybdenum disulfide.

Any of the known synthetic resins useful in making coated or bonded abrasives may be employed in the process.

One of the preferred features of the process is that the abrasive, diamond grit, or cubic boron nitride, have a metal coating encapsulating the abrasive grit such that the metal is present in the coated particle in an amount between 10 to 70% by volume.

Suitable metal coatings are copper, silver, nickel, cobalt, molybdenum and, in general, any metal melting above about 500°F which is chemically stable in the grinding tool. Although, for wet grinding the volume percent of metal coating can be higher, for dry grinding, to which the tool of the process is particularly directed, the volume percent of metal coating should be between 10 to 60% by volume.

The grit size of the abrasive is not relevant to the process, but grit sizes of 60 through 320 (based on the uncoated grit) are commonly used in diamond wheels.

Grinding elements according to the process may be formed by pressing the mixture in a mold of the desired shape. The mold may be heated and the resin may be completely or partially cured in the mold.

For the production of coated abrasive discs or belts, a liquid phenol-formaldehyde resin can be used. A size coat of liquid resin should be employed after the maker coat, and at least the size coat should contain the fillers of this process. The size should be high, that is, it should extend from the maker coat to close to the tips of the abrasive so that the fillers in the coat contact the workpiece during grinding.

The required fillers of the process are: (1) molybdenum present in the bond in the amount of from 10 to 60% by volume, preferably 30 to 50%, and (2) molybdenum disulfide present in the bond in the amount of from 5 to 30% by volume, and preferably from 10 to 20% by volume.

In addition to the metal and dry-film lubricant powder fillers, inert fillers such as silicon carbide may be added to improve the strength of the bond or otherwise control its physical properties. Where lower diamond concentrations are employed it may be desirable to add such fillers to reduce the overall bond content of the tool or grinding element.

Application of Abrasive Particles to Tools

BRAZING AND SOLDERING

Silver-Manganese-Zirconium Brazing Alloy

In accordance with the process of *G.A. Slack and W.S. Knapp; U.S. Patent 4,239,502; December 16, 1980; assigned to General Electric Company* a silver-manganese-zirconium brazing alloy used in an oxygen-free and nitrogen-free environment has been found to make strong, high thermal conductivity bonds between diamond or CBN and a molybdenum or tungsten support member. In the silver-base alloy, the active metal additives manganese and zirconium form carbides with diamond and form borides and nitrides with CBN to obtain a chemical attachment at the diamond or CBN alloy interface.

Molybdenum and tungsten both match the thermal expansion coefficient of diamond and CBN so that there is no cracking as the bond cools to room temperature, although other metals such as chromium, iridium, niobium, platinum, rhenium and tantalum with similar low expansion coefficients can also be used.

Manganese is added in the range of 0.1 to 15 atom percent, and zirconium in the range of 0.1 to 5 atom percent, although a high percentage of silver is preferred to produce a high thermal conductivity bond. In one process for fabricating such bonds, a diamond or CBN body is dipped into liquid silver-base alloy heated above its melting temperature (about 1000°C) and kept under an oxygen-free and nitrogen-free gaseous atmosphere, and removed and cooled to room temperature. The metal support member and coated diamond or CBN body are then clamped together and redipped in the silver-base alloy and subsequently cooled. Excessive solidified alloy is removed from surface areas of the bonded assembly as required.

Typical improved diamond and CBN products with a strong, high thermal conductivity bond between the diamond or CBN and a molybdenum, tungsten or other support member include a semiconductor device heat sink and a diamond grinding wheel. Other industrial products, for example, are a compact cutting tool for machining operations and a diamond compact wire drawing die. For many products, in view of its lower cost, molybdenum is preferred as the metal support member. The silver-base alloy bonding layer according to one composition that gives good results, consists essentially of about 94 atom percent silver, 3 atom percent manganese, and 3 atom percent zirconium.

Brazing Diamond Crystals to Metal Substrate

The process of *J.T. Lowder and E.M. Tausch; U.S. Patents 4,018,576; April 19, 1977; and 3,894,673; July 15, 1975; both assigned to Abrasive Technology, Inc.* relates to a method of manufacturing diamond abrasive tools by a direct brazing technique which requires none of the conventional techniques for preconditioning of the diamond. In general, the method revolves about the discovery that certain brazing alloys readily wet the surface of diamond crystals to achieve a very secure bond of the diamond to the metal substrate.

These alloys might be best generally described as those brazing and hard surfacing alloys of the cobalt and/or nickel based type which contain certain percentages of chromium. It has been found that this family of brazing alloys may be applied directly to the untreated diamond surface with superior results relative to any prior method. The configuration of the product so formed is unique in that the braze alloy clings to the diamond surface rather than flowing away from it during treatment and therefore greater surface contact between the diamond and the alloy is achieved to provide a more secure bond.

Alloy systems including preferably at least approximately 60% or more of nickel or cobalt or a combination of those metals and at least more than approximately 2%, but preferably above approximately 5% chromium, are essential to the method. Further, boron, silicon and phosphorus are very desirable at least in contributions up to about 12% but preferably less than 10%.

Relative to the melting point of the given alloy within the general composition limits disclosed herein, it should be pointed out that processing temperatures above 1300°C are very undesirable. Thermal degradation of the diamond crystals above this temperature is likely to be serious enough to substantially interfere with the usefulness of the tool. In fact, processing temperatures below 1200°C are greatly preferred.

It is convenient to use a temporary binder or other means to hold the diamond grains and braze alloy particles in place against the metal surface to which the diamonds are to be attached. This binder may be of a fluxing type or simply a resinous compound dissolved in a suitable organic solvent. It is important that the binder not be a source of oxygen because the atmosphere in the furnace should be as substantially free of oxygen as economically reasonable to assure the proper flow of the braze alloy. The presence of oxygen inhibits the wetting of the diamond by permitting oxides to form on the surface of the braze. Upon preparation of the substrate, braze alloy, and diamond assembly, the whole is submitted to a furnace cycle in a reducing atmosphere or a vacuum and heated until the braze flows and wets the diamond crystals and metal substrate. An entire furnace cycle need take no longer than is employed when following standard brazing practices for metals. The optimum brazing temperature is generally slightly higher than the liquidus temperature of the alloy because good fluidity of the braze is required.

The process may be performed in two steps rather than a single one and in some cases this may be desirable. In the two-step process, the brazing alloy would be applied in the same manner as described above; however, the diamond crystals would not be present. After the braze alloy has been fused, the diamond crystals would be attached to the layer of braze alloy on the metal substrate by use of a temporary binder and then the combination would proceed through the furnace cycle again. The same level of diamond bonding is achieved in this two-step method as in the one-step method; however, a somewhat greater control of surface uniformity of the abrasive tool may be achieved in the two-step method. If greater uniformity is desired, the braze coated surface may be machined before the step of applying the diamond.

Application of the described process to the manufacture of diamond abrasive tools encompasses a great variety of sizes, shapes, and types of tools from extremely thin abrasive discs to large diameter grinding wheels and saw blades. Acceptable substrate or mandrel materials are readily found among the plain carbon and alloy steels. Tool and stainless steels are particularly suitable because of their general resistance to warpage or dimensional instability during the necessary thermal processing.

Example: A small quantity of prealloyed braze metal in the form of minus 325 mesh particles of the composition, chromium (7.0%), boron (2.9%), silicon (4.5%), iron (3.0%), carbon (less than 0.1%), and nickel (balance) was mixed into a solution of 2-ethoxyethyl acetate saturated with a foamed urethane. The braze "paint" was applied by brush to a 1⁄16 diameter tool mandrel of M-2 steel. Without allowing the braze "paint" to dry, 100/120 mesh diamond crystals were applied uniformly over the braze by sprinkling from a sieve. The assembly was then dried at 225°F for 10 minutes to remove excessive volatiles. The assembly was then placed into a furnace (maintained at 2000°F) for 6 minutes in a dry hydrogen atmosphere and cooled under same.

The urethane binder successfully held braze and diamond particles to the mandrel but volatilized during the furnace treatment. The resultant tool consisted of a monolayer of diamond crystals firmly bonded to that portion of the tool to which they were applied with very good protrusion of each crystal above the bond metal surface.

Tools fabricated in this manner and similar in size, shape and diamond crystal size to commercially available tools were submitted to experienced practitioners in the dental field for evaluation in high speed air turbine grinders. The opinions of these practitioners were virtually unanimous; the tool manufactured by this process was superior in all respects to the best commercially available tools (all of which are electroplated tools).

Soldering Metals with Superhard Manmade Materials

J.V. Naidich, G.A. Kolesnichenko, N.S. Zjukin, B.D. Kostjuk, S. S. Shaikevich, Y.F. Motsak, V.P. Fedulaev, N.A. Kolchemanov, V.M. Ugarov, V.V. Losev, M.S. Drui, A.A. Lavrinovich, D.F. Shpotakovsky and S.V. Chizhov; U.S. Patent 4,117,968; October 3, 1978 provide a method for soldering metals to diamond- and boron nitride-base materials. A metalizing coating is applied to the soldering surface of the diamond- and boron nitride-base materials, with the materials to be joined being exposed to a temperature sufficient to achieve adequate adhesion of the metalizing coating to the materials, and the metal is joined with the metalized material by use of soldering agents exhibiting ductile properties and having melting points of up to 1000°C.

The metalizing coating is constructed by metals showing high adhesiveness towards the diamond- and boron nitride-base materials, viz chromium and/or molybdenum, tantalum, tungsten or titanium taken at the rate of from 3.0 to 65.0% by weight of the coating, the balance of the coating being accounted for by metals exhibiting improved oxidation resistance, viz copper and/or silver, lead, tin, nickel or cobalt.

The application of the coating is effected in two operational steps: first, a coat of a hard metal showing high adhesive activity towards diamond- and boron nitride-base materials is built up to a thickness between 0.0005 and 0.001 mm, and then a coat of a hard metal exhibiting improved oxidation resistance is built up to a thickness between 0.001 and 0.01 mm.

An alternative technique of application of the metalizing coating uses a metal melt made up of metals highly adhesive towards diamond- and boron nitride-base materials and oxidation-resistant metals, the coating thickness being from 0.01 to 0.5 mm.

The soldering procedure is carried out at the solder melting point for 0.5 to 2 minutes at a rate of heating and cooling of the surfaces being joined, of from 10 to 30 degrees per second in air under a layer of a liquid flux. The liquid flux belongs to the category of fluxes employed for soldering metals by means of metallic spelters. The melting points of such fluxes lie within the range from 500° to 600°C and they further exhibit stability of properties in the temperature range up to 1000°C.

An acceptable flux (Flux No. 1) may have the following composition in weight percent: 35% boric anhydride, 42% anhydrous potassium fluoride and 23% potassium fluoborate. Another flux (Flux No. 2) similar to Flux No. 1 in terms of properties has the following composition in weight percent: 70% boric acid, 9% calcium fluoride and 21% sodium tetraborate.

Example 1: A polycrystal of a cubic modification of boron nitride such as Elbor of diameter 3.9 mm and height 4.4 mm and a steel holder of diameter 5 mm and height 20 mm are to be soldered together.

The boron nitride polycrystal is premetalized with a chromium layer by electron-beam sputtering of the metal and its deposition to a layer thickness of 0.001 mm, the procedure being effected in a vacuum of 1-5 x 10^{-5} mm Hg. Then the polycrystal is given a protective coat of copper chemically deposited to a thickness of 0.005 mm. The coating components are in the weight ratio of 20 chromium to 80 copper. Then the semicrystal is subjected to baking under vacuum (2-5 x 10^{-5} mm Hg) at 950°C for 20 minutes. The metalized polycrystal is placed in a prepared cylindrical bore hole formed in the end face of the holder along the axis thereof, the solder slit being 0.15 mm on a side.

The soldering procedure is effected in air under a layer of Flux No. 1 by high-frequency heating with a soldering agent comprising 20% by weight of tin and 80% by weight of copper at a heating cooling rate of 20 degrees per second. The process is carried on at a temperature of 900°C for 1.5 minutes. The solder melts down and, by the action of capillary forces, flows into the solder slit.

The soldered joint is even, free from cavities and cracks in the cutting element as well as from other soldering flaws; the adhesion of the soldering agent to the polycrystal is satisfactory.

Example 2: An Elbor polycrystal of diameter 4.2 mm and height 4.6 mm and a steel of diameter 5.5 mm and height 20 mm are to be soldered together.

The polycrystal is premetalized with a 0.008 mm coating of chromium applied by chemical deposition. Then the polycrystal is given a protective coat of cobalt by metal sputtering under vacuum to a thickness of 0.005 mm. The weight ratio of the components is 15 chromium to 85 cobalt. The metalizing coating is subjected to vacuum baking at 1-5 x 10^{-5} mm Hg and 950°C for 30 minutes. The metalized polycrystal is placed in a prepared cylindrical bore hole formed in the end face of the steel holder along the axis thereof, the solder slit being 0.10 mm on a side.

The soldering procedure is effected in air under a layer of Flux No. 2 by high-frequency heating with a solder made up of 20% by weight of tin and 80% by weight of copper at a heating and cooling rate of 30 degrees per second for one minute. The soldering temperature is 900°C. The soldering agent melts down and, by the action of capillary forces, flows into the solder slit.

The resultant soldered joint is even and free from cavities and cracks in the cutting element as well as from other soldering flaws; the adhesion of the soldering agent to the polycrystal is adequate.

Nickel-Chromium Alloy Bonding Layer

G.R. Held; U.S. Patent 3,868,235; February 25, 1975 provides a method for applying hard carbide particles to the surface of a metal substrate which eliminates functional copper from the matrix and results in an improved bond of the carbide particles.

The substrate may consist of a conventional metal cutting or abrading tool, such as a saw blade, teeth of a hobbing wheel or milling cutter, a broaching tool, a surface smoothing abrading tool, and the like. Generally, the cutting edge or surface or abrading surface of the tool, which is normally made of a steel material, is to be coated to a predetermined thickness with particles of cemented hard carbides, such as tungsten carbide.

The first step in the process herein, as in the past, is to clean the surface to be coated by various conventional cleaning processes to remove foreign material, dirt, grease, previous coatings, etc. It is also conventional to rough the surface to some extent as by abrading it or bombarding it with a spray of hardened particles. This is to improve the mechanical interlock between the bonding matrix and the substrate surface.

Once the material has been suitably cleaned and roughened, a matrix or bonding layer is applied thereto, preferably by using a hot spray technique.

A convenient and inexpensive spray gun is a simple oxygen-acetylene torch into whose flame particles of the matrix are blown for melting and applying the particles in a uniform layer upon the substrate. This further oxidizes the matrix, adding to already present oxides.

Commercially available matrix materials are usually in the form of a nickel-chromium alloy. The important characteristics of the metal matrix are that it: (a) be copper-free, (b) have a melting point around 2000°F, plus or minus 100°F (i.e., nominally at least 200°F below the melting point of the substrate), and (c) exhibit a flow capability (capillary action). Flow capability is a characteristic of the molten matrix wherein the matrix forms a thin film or covering over the carbide during the "baking" or volatilization of the binder.

Once the matrix or bonding layer is applied, the material is cooled and then a layer of an organic adhesive or bonding material is applied thereto.

The particular type of organic adhesive or binder may vary considerably, but its principal function is to hold the later-applied carbide particles in place temporarily. A second function is to assist in creating porosity in the matrix. Thus, the adhesive or binder must be tacky after application and preferably will crystalize upon the application of heat. An organic binder or adhesive which is essentially volatilized at the furnace temperature is also preferred.

The preferred organic adhesive for this process is a mixture of ordinary glycerin and water, applied in a liquid form.

Following the application of the tacky adhesive layer, particles of hard carbide are simply physically sprinkled upon the surface thereof and are held in place adhesively. The thickness of the layer of particles may be varied by applying a thicker binder layer and by sprinkling on more particles.

The hard carbide particles contemplated here are the usual cemented or sintered carbides formed conventionally of tungsten.

Grit sizes for the carbide particles may range from 5 to 300 mesh depending on the utlimate use of the cutting tool.

The next step in the process is placing the layered substrate into a baking furnace which is maintained at a temperature high enough to melt the matrix and volatilize the temporary adhesive. The matrix material has a flow capability, i.e., it tends to surround and thinly coat the carbide particles which sink into the molten matrix. Simultaneously, the adhesive dries out or vaporizes and may form crystals which tend to float to the surface or recrystallize on the surface. The crystals on the surface, as well as those trapped, cause porosity in the solidified matrix. Those which form on the surface may be brushed after baking.

The temperature of the furnace must be maintained below the melting point of the substrate so as not to affect it, but since the baking time is short, e.g., roughly 15 to 20 minutes at roughly 1950° to 2050°F, the substrate is generally not affected by the baking.

The furnace may be provided with a reducing atmosphere, such as a primarily hydrogen-nitrogen atmosphere, which protects the carbide particles against oxidation and which reduces some oxidized portions of the matrix. By leaving some matrix oxides, hardness is maintained as it is some help towards creating porosity. The amount of oxides may vary considerably depending upon the desired results.

Alternatively, a vacuum-type furnace may be preferred in certain applications. The primary requirement is that the furnace be nonoxidizing and protect the carbide particles against oxidation. The hydrogen in a reducing atmosphere furnace provides a decarburizing atmosphere which tends to anneal the oxides. Further, humidity may be a problem in reducing furnaces, especially on humid days. These problems are eliminated in a vacuum furnace. A vacuum of 1 to 150 microns is suitable in the furnace, however, the preferred range is from about 0.1 to 10 microns.

After baking, the coated substrate is cooled, resulting in the finished product.

PLATING

Layer Having Single Particle Thickness

For many purposes electroplating, particularly electrodeposited nickel, has been the most satisfactory matrix for securing the abrasive particles to the working area of the tool blank. However, the abrasive tools heretofore available having electrically deposited matrices have had recognized disadvantages. Thus, problems were encountered with peeling of the plating from the tool blank. At times the tool life was unduly short due to overplating. Where the tool was underplated the abrasive particles would be stripped from the tool and the tool blank was frequently damaged. Also there was a tendency to obtain "grooving" in the abrasive coating and, in addition, lack of uniformity in the concentration of the abrasive particles, both of which produced undesirable results and lack of uniformity and departure from tolerances in the products produced by the grinding tool.

Difficulty was also encountered in bringing and holding the abrasive particles in contact with the working area surface during the plating operation and cumbersome and unsatisfactory expedients were resorted to in attempting to overcome this problem.

W.M. Haack; U.S. Patent 4,082,640; April 4, 1978; assigned to Keene Corporation has found that the problems heretofore encountered can be overcome and improved results can be obtained by utilizing abrasive particles of predetermined size, i.e., between 0.005 and 0.009 inch (preferably 0.007 inch) in cross-sectional dimension, then bringing a mass of the particles into contact with the working area surface of the tool blank and lightly securing or tacking the particles to the working surface in a layer of single particle thickness by electrodeposited nickel of a thickness no greater than approximately one-third of the height of the particles. The surplus abrasive particles are then removed from the area and further nickel plating is applied to a thickness of between one-half and two-thirds of the height of the abrasive particles.

During the initial plating or tacking the tool is held in place in a container having an impervious side wall surrounding and spaced from the working surface and having a base with a porous mesh portion having openings smaller than the abrasive particles beneath the space between the side walls and the working area surface. A nickel anode is supported inside the container spaced from the working surface. The abrasive particles are packed in the space between the side walls and working surface with particles in uniform contact with the working area surface, and nickel plating solution is then poured downwardly over the working area surface, through the abrasive particles and porous mesh until a layer of particles of single particle thickness is lightly tacked in place.

The preferred abrasive materials are borazon and natural and synthetic diamonds.

The nickel plating solution may be a standard Watts solution. A Watts solution of the following proportions gives satisfactory results: 2.75 lb/gal nickel sulfate, 0.40 lb/gal nickel chloride and 0.33 lb/gal boric acid.

Small quantities of the usual additives such as hydrogen peroxide, sodium lauryl sulfate and organic brightener compounds may also be included.

As previously explained, the initial or "tacking" plating is carried on until the plating reaches a height of no more than approximately one-third the height of the abrasive particles. Generally speaking, satisfactory results are obtained employing materials and particle sizes of the type explained above by carrying out the initial plating operation for approximately 50 minutes at 20 A/ft^2 of plating area. As previously explained the final overplating is carried on until the height of the plating is between one-half and two-thirds the height of the abrasive particles. Satisfactory results are obtained where the overplating is carried on for approximately 3½ hours at 10 A/ft^2 of plating area.

Interspersed Conductive Zones and Resist Zones

A diamond abrading disc is provided by *T.R. Green; U.S. Patent 4,078,906; March 14, 1978; assigned to Elgin Diamond Products Co., Inc.* in which the abrading surface has interspersed conductive metal and resist zones which are laid down in a predetermined pattern. When the abrading disc has a metal body or surface, the resist zones are applied through a screen which is commonly referred to as a "silk screen".

The resist material may be an acid resist ink which is brushed, or otherwise applied across the screen so it passes through the plurality of pattern openings in the screen onto the abrading surface. The screen is removed and the disc is then preferably deposited in a tank so an initial layer of plating is formed on the surface to "brighten" the abrading surface. Diamond particles are then sprinkled or distributed in a nonselective manner over the entire abrading surface, preferably while the discs remain in the tank, covering both the conductive metal and resist zones. The disc body is moved in level or horizontal orientation to ground surface to prevent spillage of the distributed diamond particles, and such movement is executed into an electrochemical bath wherein the disc body or its container is connected as a cathode. An anode of nickel is provided in the usual way, and such bath is conventionally prepared, such as the standard Watts bath.

The practitioner attends to the usual controls of pH and temperature, as well as attending to desired salt concentrations of the bath. Amperage is controlled according to the usual practice so that nickel may be electrodeposited around the cathode or abrading disc to initially bond the diamond particles resting on the conductive metal zones, but not bonding diamond particles resting on the resist zones. The abrading disc is maintained in horizontal orientation within the electrochemical bath so that the substantially uniform layer of diamond particles is not disturbed prior to initial bonding.

Thereafter, the orientation of the disc body is changed by tilting and shaking to dislodge the diamond particles not initially bonded. The disc body with the initially nickel bonded diamond particle zones is then repositioned in vertical orientation in the bath to successively and finally bond the diamond particles to the metal abrading surface by nickel electrodeposition.

The discontinuous diamond pattern on a selected abrading surface preferably covers a minor portion of the total abrading surface, from about 20 to 50%.

The amount of diamond will depend on the design of the abrading tool, but the discontinuous diamond pattern is used to advantage when a substantial portion of the abrading surface is made up to the free resist zones. Diamond particles may be used in a wide variety of sizes for laying down the discontinuous pattern, i.e., from 80 to 800 grit size, although this may vary.

The metal surface may be the abrading surface of a solid metal disc of various thicknesses and diameters, or it may be a metal layer bonded to an isoelectric body such as composition board. This form is the familiar circuit board type wherein a light sensitive metal layer, usually copper, is adhered to one side of the nonconductive board.

In another form, the light sensitive metal layer of the circuit board is used to receive light images of the conductive metal zones which are "printed" on the nonconductive body following removal of the metal not chemically reduced by the light pattern. The nonconductive portions exposed by the removal of metal are the resist zones. The resulting disc body with conductive metal and resist zones is then similarly positioned in the electrochemical bath of initial and final selective nickel bonding of the diamond particle zone.

The disc body with the discontinuous diamond pattern can then be utilized in many recognized ways in the abrading art. Several improved forms have found utility particularly in applications of the lapidary art. A metal abrading disc with a discontinuous diamond pattern which is sufficiently thin to be flexible is attached to an intermediate elastomeric backing and an outer rigid backing to provide a desired abrading tool. The circuit board type of abrading disc is substantially rigid and is used alone, or is attached to a rigid body to provide another lapidary tool of particular utility. Lapidary type processing includes the foregoing as well as areas such as the optical where ophthalmic lenses may be ground in spherical or toric form.

Pre-Etched Workpiece

The process of *J.L. Fletcher; U.S. Patent 4,079,552; March 21, 1978* comprises a process for more securely bonding polycrystalline diamond or other abrasive particles to the surface of a metallic workpiece. The process includes preceding the diamond/metal plating step with an etching step so as to suitably etch the workpiece prior to plating with the diamond/metal surface.

Referring to Figure 2.1, a workpiece **10** is shown in an etching bath **14** comprising a solution **15** of aqueous sulfuric acid. One suitable etching solution has a 60% sulfuric acid concentration. To assist in the etching of workpiece **10**, an electromotive force indicated at **12** is imposed between workpiece **10** and a cathode **16** or even a metal vessel **13** containing the acid solution **15**. A reverse dc current of about 4 A at 5 to 6 V for 6 or 7 minutes has been found adequate. To improve uniformity in the etching process, workpiece **10** is preferably rotated in the bath with a rotatable shaft **18**. Rotation of shaft **18** and workpiece **10** also agitates the solution and minimizes undesirable concentration of electrolytic action of any one portion of the surface of the workpiece thereby assuring more uniform etching.

Figure 2.1: Flow Diagram for Diamond/Metal Plating

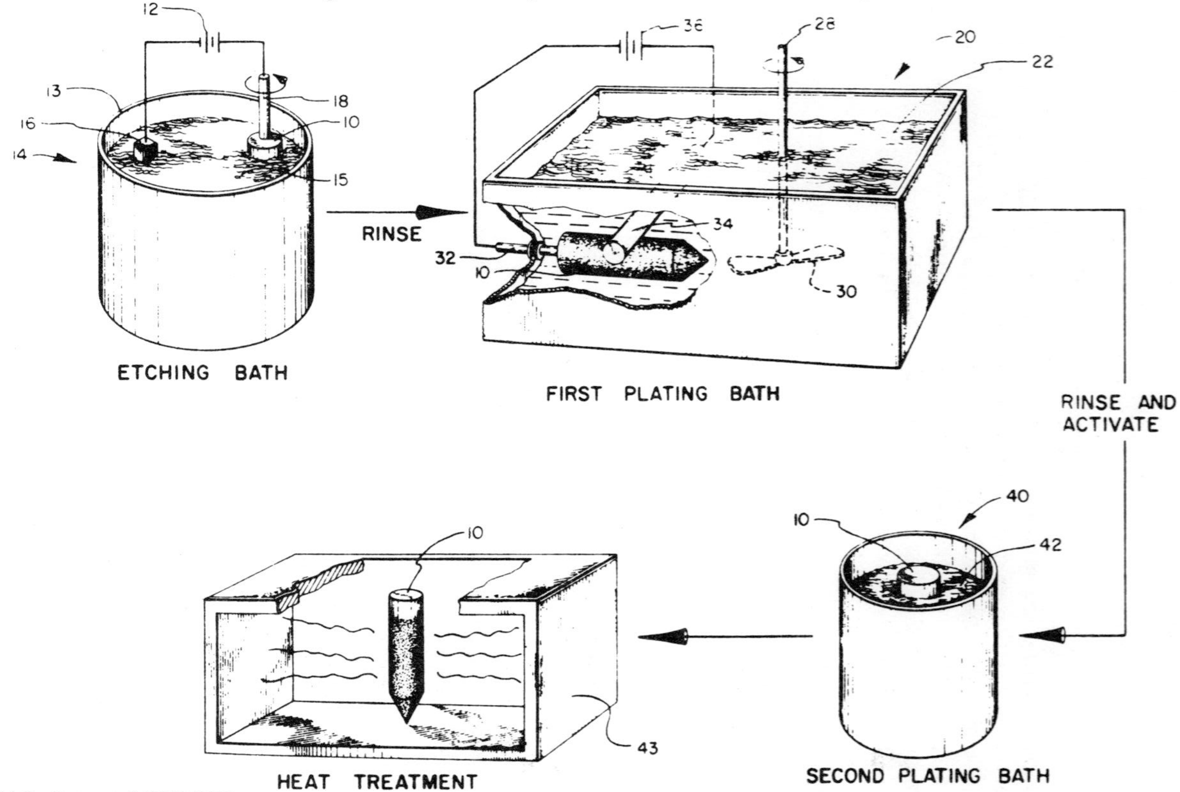

Source: U.S. Patent 4,079,552

After etching, any remaining sulfuric aicd is removed by rinsing workpiece **10** with water. The workpiece **10** is then placed in a first plating bath generally designated **20**.

Plating bath **20** contains a nickel plating solution **22** which may be a standard aqueous solution of nickel sulfate and nickel chloride heated to about 120°F. This plating solution is well-known in the art and is commonly referred to as a standard Watts bath.

Diamond particles are suspended in the aqueous solution **22** so as to facilitate uniform distribution of diamond particles on workpiece **10**. Diamond particles may be of any suitable size although the very fine particles (24 to 41 microns) are preferred for sharpening tools and the like. Grinding wheels and related tools may require particle sizes upwards of 100 mesh. Diamond concentration of about 10 to 50% by volume was found to be most efficient to effect essentially uniform diamond distribution over the workpiece. Suspension of diamond particles is maintained by vigorously agitating bath **22**.

Workpiece **10** is mounted upon a rotatable shaft **32** and suspended in plating bath **20**. Substantial improvement in diamond plating has been found where the surface to be plated is situated out of the vertical plane. For example, it is preferred that the workpiece **10** be oriented horizontally in the electroplating bath.

A nickel anode **34** is also suspended in plating bath **20**. An electromotive force is applied through circuit **36** between workpiece **10** and anode **34** with workpiece **10** connected to circuit **36** so as to act as a cathode. Workpiece **10** is plated with nickel metal ions. The plating action simultaneously entraps diamond particles on the surface of workpiece **10** and the plated nickel metal serves to mechanically bond diamond particles to the surface of workpiece **10**.

The cavities created in workpiece **10** during the etching step also greatly assist in forming a strong bond between workpiece and the diamond/nickel matrix. Moreover, many of the diamond particles are recessed into the cavities so as to limit their exposure to a small area in the shear plane formed along the diamond/metal surface. The diamonds thus secured have surprising resistance to shear and breakage away from the workpiece.

Rotation of workpiece **10** in bath **20** assures a more uniform plating of metal thereon and agitation by rotation of impeller **30** assures an even dispersion of diamond particles throughout electroplating bath **20** and, accordingly, on the surface of workpiece **10**.

It has also been discovered that the imposition of an electromotive force through circuit **36** appears to cause an attraction between diamond particles and workpiece **10** so as to more densely pack diamond particles on the surface of workpiece **10**. The thickness and concentration of the diamond particles can be determined by the duration of the electroplating step. Approximately six minutes has been found satisfactory.

After suitably electroplating the diamond particles to the surface of workpiece **10**, workpiece **10** is removed from the plating bath and rinsed with water to remove any unplated residue from bath **20**. While not essential, it has been found desirable to follow the rinsing step with an activation step wherein the diamond plated workpiece is treated by dipping or rinsing in a 50% hydrochloric acid solution prior to immersing the workpiece in a second plating bath **40**. Surface activation is primarily used where the workpiece surface has been oxidized. If care is taken to avoid drying of the workpiece **10** during the etching and electroplating process, activation can usually be avoided. Prior to treatment in the second plating bath **40**, the diamond adheres to the workpiece **10** as a soft pack.

The second plating bath **40** comprises an electroless nickel plating solution **42** of nickel ions. Any suitable electroless plating solution could be used such as solutions known as Product No. 794 A, B and HZ (Allied Kelite Div., Richardson Chemical Co.). The workpiece is held in this electroless plating bath for sufficient time to achieve a suitable coating, for example, approximately 70 to 80 minutes has been found adequate. The temperature in the electroless plating bath is elevated to about 195°F or such other elevated temperature as may be recommended by the manufacturer of the solution.

It is pointed out that while electroless plating is preferred electrolytic plating may be used. Nickel plating in this bath has been found to deposit about 0.0008 inch nickel per hour (0.02 mm per hour) in this bath and it is preferred to substantially interfill the surface area around the diamonds and/or cover the diamond adhering to the workpiece.

After removal from the second or electroless plating bath **40**, workpiece **10** is cleaned with water, dried and then subjected to heat treatment, e.g., in a furnace **43** wherein workpiece **10** is heated to approximately 600°F for approximately one hour. Heat treatment between 650° and 750°F for one hour yields a workpiece having a Rockwell C-Scale hardness of 72. Hardness of 46 to 72 has been found desirable. The actual hardness achieved is a function of both temperature and firing time.

Masking Technique

An abrasive product using abrasive particles, such as diamond particles, with the abrasive particles being positioned on the products by the use of a special masking technique and with abrasive particles being held in place by deposited metal is provided by *R.K. Wiand; U.S. Patent 4,047,902; September 13, 1977.*

The method of manufacturing is carried out by masking a platable conductive substrate or backing with a masking material of perforated nonconductive material. The perforations in the masking material can be of any suitable shape, spacing, design, or pattern and for example, can be of varying hole size to permit varying percent of surface exposure. Also varying shapes of perforations and/or arrangements of perforations can be used. The platable conductive substrate itself can be preperforated and then masked or it can be in essentially solid shape and then masked prior to the plating operation. The perforations in the conductive platable substrate can also be of varying shape and/or hole size as well as taking up varying percentages of the surface of the platable conductive substrate. The conductive substrate may also suitably be supported by a support member which may be of either a flexible or rigid nature. The support member can satisfactorily be formed of a number of different materials such as plastic, fiber glass, metal, wood, rubber or the like.

The platable substrate can be rigid or flexible and it is preferably a material which is metalized on one surface thereof to give a conductive surface for the subsequent plating operation which bonds the abrasive particles to the conductive substrate. Also the conductive platable substrate can preferably be made of metallic filled plastic materials, or, even from nonconductive materials whereby the abrasive particles are bonded by chemical reduction or vacuum metal deposition.

The masking material can be of a number of different materials, such as masking tape, silk screening materials, rubber materials, an unplatable metal (for example, titanium) a thin plastic, a painted material and various shapes of masking materials which are applied to the conductive substrate. Hot stamping to leave exposed areas in the conductive platable substrate is also possible for the purpose of carrying out the masking techniques, or a photoresist method can also be used to carry out the masking.

The plating operations to bond the abrasive particles to the substrate may suitably be electrolytic, chemical reduction or vacuum metal deposition. However, it is preferred to use an electroplating technique. The preferred electroplating technique to bond the abrasive particle to the conductive surface is generally carried out using the following steps.

First, the conductive platable substrate or tool on which the abrasive particles are to be positioned is properly mounted or maintained for subsequent application of the following manufacturing steps. The properly positioned tool is then treated such that the conductive surface of the substrate is prepared for adhesion of the metal plating. Temporary or initial adhesion of the abrasive particles is then carried out by a small application of metal plating. Lastly the balance of the metal plating is applied to finally bond the abrasive particles to the conductive substrate. Following this the masking material, which is removable can be pulled away, or left in place as desired.

The preferred abrasive particles or grit are generally of the nonferromagnetic type. Very good results are obtained with diamond abrasive particles.

The bonding of the abrasive particles with the metal plating is generally carried out with a static plating process, or technique as opposed to a dynamic or moving process. Furthermore the grit particles and the metal plating binder are applied generally simultaneously in carrying out the bonding operation. A nickel metal plating is the preferred plating material for bonding the abrasive particles to the conductive substrate.

POSITIONING OF PARTICLES

Parallel Fibers

M.H. Zoiss; U.S. Patent 3,972,161; August 3, 1976; assigned to Barnes Drill Co. provides an abrading tool which comprises a shaped and relatively rigid body of wearable matrix material having at least one working face for engagement with the workpiece to be abraded, and abrading elements in the form of elongated fibers of selected harder material disposed within the body and extending transversely of the working face with ends of the fibers exposed at the face for abrading engagement with the work, and with the remainder of each fiber extending inwardly through the body and securely anchored therein to brace and back the abrasive ends against bending and breaking off prematurely during the abrading operation.

For optimum abrading consistency and effectiveness, the fibers are closely spaced and arranged within the matrix to provide and maintain general uniformity in the pattern of the exposed ends, both initially and as the tool wears away, and all of the fibers are oriented to extend inwardly from the working face of the tool, preferably substantially parallel to each other.

As shown in Figure 2.2, the representative honing stone **10** has the usual shape and is made up of a large number of longitudinal rows or layers of parallel abrasive fibers **33**, each herein having a circular cross section and being generally perpendicular to the working face **17** (ignoring the transverse curvature) and parallel to the side **19** of the stone.

Figure 2.2: Honing Stone with Parallel Fibers

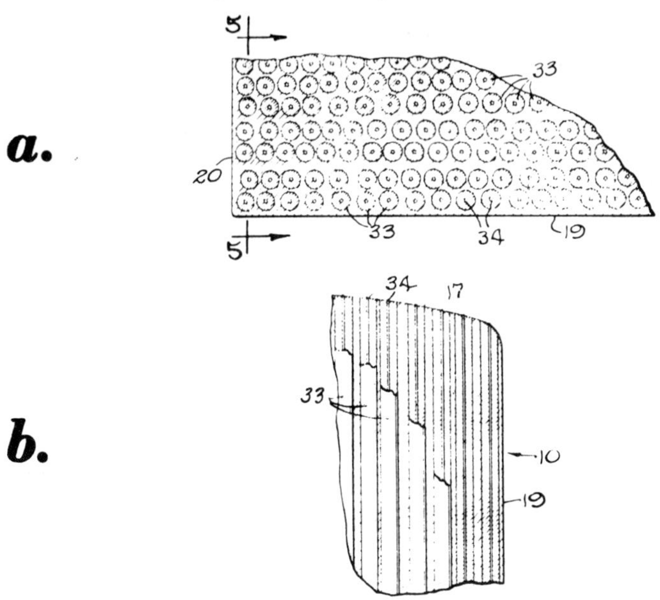

Source: U.S. Patent 3,972,161

The spacing of the fibers in each longitudinal row is generally uniform, and the spacing between adjacent rows is similar, but these spacings may vary in practice as a result of different manufacturing techniques used. Alignment of corresponding fibers transversely of the stone is to be avoided in order to eliminate transverse "blank" lines of matrix material that would produce inactive areas completely across the stone. Thus, the layers of fibers are longitudinally staggered in a random manner.

The particular material used for the fibers depends upon the material of the workpiece, that is, the fibers must be harder than the work and the difference in hardness should be at least 250 as measured on the Knoop scale. While different sizes of fibers **33** may be used, two representative sizes that are commercially available have diameters of 0.004 and 0.008 inch.

Where the work is iron or steel, a comparatively hard material should be used for the fibers and, while various hard abrasive materials may be, or become, available in fiber form, the preferred material is boron fiber which is supplied by manufacturers for use in structurally reinforced materials, particularly in the aircraft industry.

In one form, such fiber is produced by vapor deposition of boron on a 0.0005 inch tungsten filament which remains in the center of the fiber as a core **34**. Such boron fibers have a hardness approaching that of diamond (7,000 Knoop hardness with 100 gram loading), but the hardness value could not be precisely measured because of the closeness to the hardness of diamond.

Orienting Grit by Magnet

L.R. Oliver; U.S. Patent 3,918,217; November 11, 1975; assigned to Lloyd R. Oliver & Company describes cutting and abrading devices made with special predetermined protrusions armed with metal bonded refractory metal grit.

As presently preferred, the method of manufacturing the abrading devices is as follows. First the substrate material or metal is coated with a temporary binder means such as a film resin binder, similar plastic resin binder, or the like. Next a layer of the protrusion means such as the layer of steel shot is applied over the temporary binder and held in place thereby due to the adhesive nature of the temporary binder. Then a brazing metal or bonding material usually in finely powdered form (such as a nickel-chromium alloy brazing metal) is applied as the next layer of the device. Following this, additional temporary binder material may be added if desired; and, then a suitable magnet means (such as an elongated bar magnet) is contacted with a central base portion of the metal base member.

Subsequent to contacting the steel base member with the magnet as referred to, the grit material is applied to the device by contacting or dipping the device into a body of, for example, tungsten carbide grit, and as the grit is positioned on the temporary binder material of the device, the magnetic force field caused by the magnetic means causes the tungsten carbide grit to in effect stand up or orient itself on the protrusion means such that the grit is oriented outwardly from the protrusion means generally in each grit's long axis direction. Additional brazing metal may then be applied if desired over the top of the grit. The device is then subjected to heating to the braze metal melting range. In accordance with this process, the grit particles used herein, such as tungsten carbide contain a certain amount of cobalt alloyed into the tungsten carbide to render same magnetizable.

However, other grit particle means may be used herein so long as there is a magnetizable metal means associated therewith such as iron, cobalt, or nickel. In certain instances the grit means may have the magnetizable metal applied as a very thin coating over the outside of the individual grits in order to render same such that they may be magnetically oriented. The abrasive particle grit means used here may be selected from at least one material of the group consisting of tungsten carbide, silicon carbide, titanium carbide, aluminum oxide, refractory oxides, refractory borides, garnet, and diamond. The protrusion means herein or the spherical particles used for same may be selected from at least one material of the group consisting of steel, metal, or ceramic particles.

Oriented Crystalline Particles

It is known that diamond particles form low-wearing, high efficiency cutting members when held with the axes perpendicular to the direction of movement of the abrasive article and to the surface of the workpiece which is to be modified. When the particle is rotated so that its axis lies parallel to the surface to be modified, wear of the particle increases with little cutting effect. Furthermore, diamond and other particles of high crystallinity, for example, alumina, ruby, must be provided with a predetermined orientation of the cutting faces of the crystal to the workpiece, the cutting faces having a fixed relationship for each crystal, to the predominant crystal axes. Thus, if the proper orientation of the particle is not observed, the cutting effect is reduced or rendered negligible.

It is an object of the process of *K. Inoue; U.S. Patent 3,877,891; April 15, 1975* to provide an improved abrasive article with oriented abrasive particles, which can be made in a simple and automatic manner.

This object is attained by passing a succession of crystalline particles along a predetermined transport path, scanning the particles as they pass along the path, past a viewing station, detecting the orientation of the particles as they move along the path, and selecting particles of a predetermined orientation from a succession and transporting them, in turn, with fixed orientation, to a support in which the particle is planted.

The transport means consists of an endless conveyer provided at one end with means for delivering a supply of the crystalline particles to the upper stretch of the conveyer, preferably in a row of spaced-apart particles. Trained on this row, at a location downstream of the detector station, is a guide tube having its mouth closely juxtaposed with the conveyer belt and leading to the substrate. The tube constitutes the guide means and is activated to draw a selected crystal from the belt, retain its predetermined orientation with respect to a crystallographic axis and convey the crystal to the support into which it is to be implanted.

Advantageously, the detector station comprises at least one and preferably two sources of electromagnetic radiation and/or high-energy particles capable of projecting beams of subatomic particles at the successive crystals as they arrive at the detecting station. The detector may respond to the diffraction effect of the crystal upon the respective beams. Alternatively, the detector may respond to refraction or some other characteristic related to the crystallographic orientation. Preferably, the sources of electromagnetic radiation are x-ray beam sources arranged so that the respective beams intersect at the location traversed by the row of particles, the detector being an x-ray diffraction detector.

In this case, the process makes use of a principle long recognized in the crystallographic art, namely, that the crystal planes and axes can be established by sweeping the crystal with an x-ray beam and detecting the defraction pattern resulting from the x-ray scanning of the crystal. The system is based, in part, upon the fact that the respective crystal planes function as the lines of a diffraction grating and cause interference and reinforcement patterns. Either crystal or the beam is rotated in a conventional diffraction-pattern crystallographic analysis, but it has been found that no such rotation is required where only proximity to the correct orientation of a crystallographic axis is desired.

In other words, the beams may be fixed and may scan the individual particles as they move past the detection station, the sensor registering a predetermined output when the axis of the crystal is not in proper orientation vis-a-vis the desired orientation. At this point, the detector triggers the system for implanting the crystal in a support.

The control system of this process is preferably of the numerical type, for example, may be provided with a memory in which are stored coordinates of the sites at which the individual crystals are to be implanted. In general, the memory may consist of a band, tape or the like in which the information is digitally recorded and represents two coordinates capable of defining any point in the plane of the surface in which the particles are to be seated. The inputs to the memory may be calculated mathematically but also can be established by mutually setting a stylus or pointer to the desired position over the surface and then registering the two coordinates of the stylus in the recording medium.

This process is applicable also to the production of articles with discrete crystal tips, for example, scribers, glass cutters or the like. In this case, entire arrays of such elements may be provided in place of the disk or support wheel and the discharge end of the guide tube can be aligned with these tubes to receive individual particles also under numerical control.

To insure a random positioning of the particle upon the conveyer and to guarantee that the particles will remain in set positions during their movement toward the detection station, also to impart some relative movement of the particles and the detector station, it has been found to be advantageous to provide means for vibrating the conveyer. The patterns in which the crystal particles can be arrayed are, of course, variable within the capabilities of numerical control. In this respect, it must be mentioned that the coordinates may be of the Cartesian type, (i.e., orthogonal to one another) or of the polar type, depending upon the results desired. When polar coordinates are employed, it is a simple matter to implant the particles in a spiral or concentric circular arrangement while the use of Cartesian coordinates facilitates the positioning of the particles in rectangular arrays.

Compacts and Composites

DIAMOND

Fabricated with High Temperature Brazing Filler Metal

F.S. Knemeyer; U.S. Patent 4,225,322; September 30, 1980; assigned to General Electric Company describes a component comprised of a composite compact bonded to a substrate with a high temperature filler metal which, to form a bond, requires the exposure of the surfaces to be bonded to a temperature substantially greater than the degradation temperature of a particulate layer of the compact.

Figure 3.1 shows a component **11** in accordance with one embodiment of the process. The component is comprised of a composite compact **13**, a pin or substrate **15** and a thin, continuous layer **17** of filler metal (shown disproportionately large for purposes of illustration) disposed between and bonding the substrate to the compact along an interface **16**. The component configuration may, of course, be widely varied to suit the particular application.

The composite compact is comprised of a particulate mass or layer **19** of bonded abrasive particles of either diamond or hard phase boron nitride and a mass or base layer **21** of cemented carbide bonded to the compact along an interface **23**. Carbide mass **21** is also provided with an exposed surface **25** remotely spaced from interface **23**.

In a preferred embodiment of the compact, layer **19** is comprised by volume of at least 70% diamond and preferably in excess of 90% diamond. The remaining 30% or less is primarily comprised of the bonding material, preferably cobalt (or alloys), of the cemented carbide mass which is infiltrated into layer **19** during the high temperature, high pressure fabrication process. Substrate **15** is preferably comprised of cemented carbide preferably the same as carbide mass **21**.

Filler metal **17** is preferably a high temperature, high strength brazing filler metal having a liquidus greater than the degradation temperature of the particulate mass. The degradation temperature is about 700°C. Acceptable filler metals may be selected from the group of alloys compatible with the materials being brazed. The ability to use such high strength brazing filler metals in accordance with this process enables the formation of a strong bond between the compact and substrate. As defined, the filler metal may be alternatively a metal or alloy used in bonding techniques such as diffusion bonding, hot pressing, resistance welding and the like.

Figure 3.1: Composite Compact Bonded with High Temperature Filler Metal

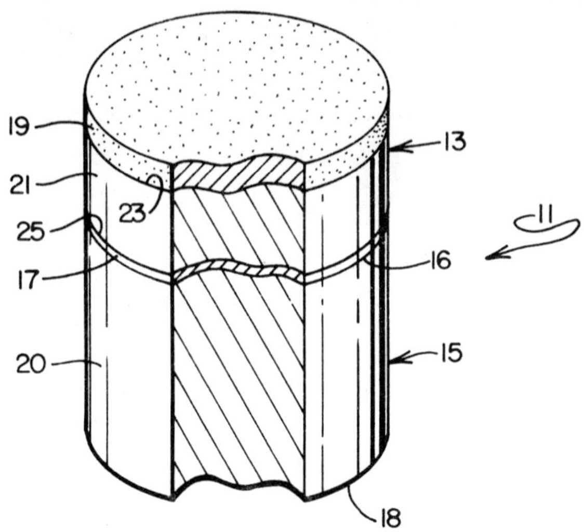

Source: U.S. Patent 4,225,322

Eutectiferous Silicon-Rich Alloy as Binder

The process of *M. Lee, L.E. Szala and R.C. De Vries; U.S. Patent 4,173,614; November 6, 1979; assigned to General Electric Company* relates to the production of a polycrystalline diamond body/silicon nitride substrate composite comprised of a dense mass of diamond crystals bonded to a silicon nitride substrate. The diamond crystals are bonded together and to the silicon nitride substrate by a silicon atom-containing bonding medium.

One of the technical barriers to a high density (high volume of diamond in a body) diamond base compact made below the diamond stable pressure region has been the development of a suitable binder material which will infiltrate the capillaries of a densely packed fine particle size diamond powder. The binder must form a thermally stable strong bond with diamond and should not graphitize or excessively react with the diamond.

The present process utilizes a eutectiferous silicon-rich alloy which infiltrates well through the capillaries of a compressed mass of diamond crystals and which wets the crystals to form a strong cemented diamond body. In addition, the infiltrating alloy forms a strong bond in situ with a silicon nitride substrate. The process also utilizes pressures substantially below those required by the diamond stable region to produce a polycrystalline diamond body/silicon nitride substrate composite in a number of configurations and a wide range of sizes. It is useful as an abrasive, cutting tool, nozzle or other wear-resistant part.

Briefly stated, the process for preparing a polycrystalline diamond body/silicon nitride composite includes a hot-pressing step and comprises placing within a protective container or cup a mass of solid eutectiferous silicon-rich alloy, or solid components for producing eutectiferous silicon-rich alloy, a mass of diamond crystals and a silicon nitride substrate. The mass of diamond crystals is intermediate and in contact with the substrate and the mass of solid eutectiferous silicon-rich alloy, or with at least one of the components for providing eutectiferous silicon-rich alloy.

The container and its contents are disposed within a pressure transmitting powder medium that transmits applied pressure substantially undiminished and remains substantially unsintered during the hot pressing. Sufficient isostatic pressure is applied to the container and its contents via the powder medium to stabilize the dimensions of the container and the contents substantially uniformly producing a shaped substantially isostatic system of powder-enveloped container wherein the density of the resulting compressed mass of diamond crystals is higher than 70% by volume of the volume of the compressed diamond crystals.

The resulting isostatic system is hot pressed producing fluid infiltrating eutectiferous silicon-rich alloy and infiltrating the fluid eutectiferous silicon-rich alloy through the interstices of the compressed mass of diamond crystals and into contact with the contacting surface of the substrate. The hot pressing is carried out at a temperature below 1600°C under a pressure sufficient to infiltrate the fluid silicon-rich alloy through the interstices of the compressed mass of diamond crystals. The solid eutectiferous silicon-rich alloy, or solid components for the alloy are used in an amount sufficient to produce sufficient fluid eutectiferous silicon-rich alloy at the hot-pressing temperature to fill the interstices of the compressed mass of diamond crystals and contact the surface of the substrate.

The hot pressing is carried out at an atmosphere which has no significant deleterious effect on the diamond crystals or on the infiltrating fluid silicon-rich alloy or on the silicon nitride substrate. The hot pressing converts less than 5% by volume of the diamond crystals to nondiamond elemental carbon. The nondiamond carbon or the surfaces of the diamond crystals react with the fluid infiltrating silicon-rich alloy forming carbide. Sufficient pressure is maintained on the resulting hot-pressed, substantially isostatic system during cooling to at least substantially maintain the dimensions of the hot-pressed system.

The resulting polycrystalline diamond body/silicon nitride substrate composite is recovered wherein the diamond crystals are present in an amount of at least 70% by volume of the volume of the bonded polycrystalline diamond body.

The diamond crystals used in the process can be natural or synthetic, i.e., man-made. They range in size in largest dimension from about 1 to 1,000 microns, and the particular size or sizes used depend largely on the particular packing or density of diamond crystals desired and also on the particular use of the resulting body. Preferably, the size-graded crystals range from 1 to 60 microns, and preferably within this size range, about 60 to 80% by volume of the total mass of crystals are of the larger size portion of the range, about 5 to 10% by volume are medium sized, with the balance constituting the small-size crystals or particles.

In the process, the starting solid eutectiferous silicon-rich alloy, i.e., the term alloy herein including intermetallic compound, is comprised of silicon and a metal (alloying metal) which forms a silicide with the silicon. Preferably, the present eutectiferous silicon-rich alloy is comprised of silicon and a metal selected from the group consisting of cobalt, chromium, iron, hafnium, manganese, molybdenum, niobium, nickel, palladium, platinum, rhenium, rhodium, ruthenium, tantalum, thorium, titanium, uranium, vanadium, tungsten, yttrium, zirconium and mixtures thereof.

The starting eutectiferous silicon-rich alloy of the process is a solid at room temperature and contains more than 50 atomic percent, but less than 100 atomic percent of silicon. Usually, it contains a maximum of about 99.5 atomic percent silicon depending largely on the specific effect that the alloying metal has on the resulting silicon-rich alloy. The starting solid silicon-rich alloy is eutectiferous in that it contains some eutectic structure and can be of hypoeutectic, hypereutectic or of eutectic composition.

Using Figure 3.2a as an example, the eutectic **2** is an alloy of specific composition which, under equilibrium conditions, on cooling freezes at constant temperature forming a solid of at least two phases, and which on heating melts completely at the same constant temperature, this constant temperature being referred to as the eutectic temperature, also given at **2**. The eutectic is the composition at which two descending liquidus curves **3** and **4** meet at eutectic point **2**, and therefore it has a lower melting point than its neighboring hypoeutectic or hypereutectic compositions. The liquidus is a curve or line on a phase diagram representing under equilibrium conditions the temperatures at which melting ends during heating of the silicon alloy or freezing begins during cooling thereof. Specifically, the present starting solid eutectiferous silicon-rich alloy is one of the series of alloys on a eutectic horizontal **1**, i.e., the horizontal passing through the eutectic point, and which extends from any alloy whose composition lies to the left of the eutectic on an equilibrium diagram and which contains some eutectic structure, i.e., hypoeutectic, to any alloy whose composition lies to the right of the eutectic on the equilibrium diagram and which contains some eutectic structure, i.e., hypereutectic alloy.

Figure 3.2: Preparing Polycrystalline Diamond Body/Silicon Nitride Substrate Composite

a.

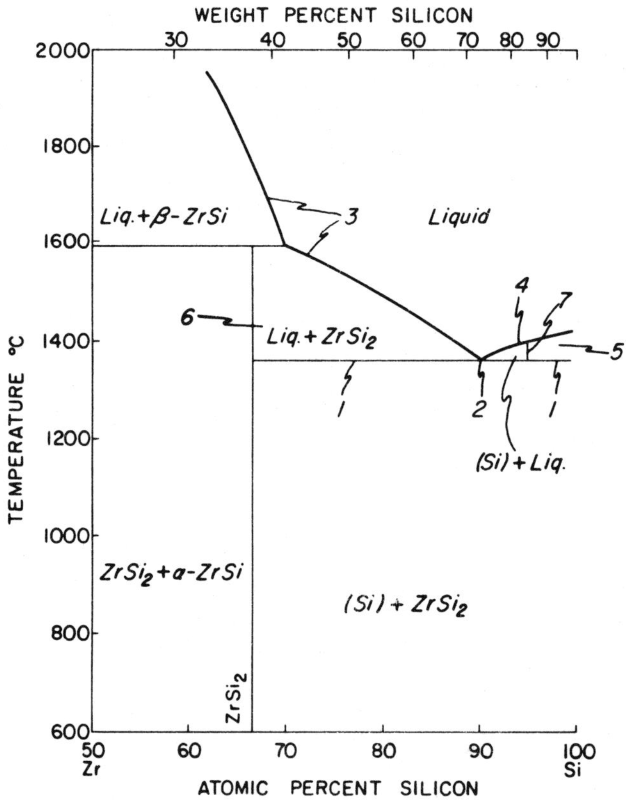

b.

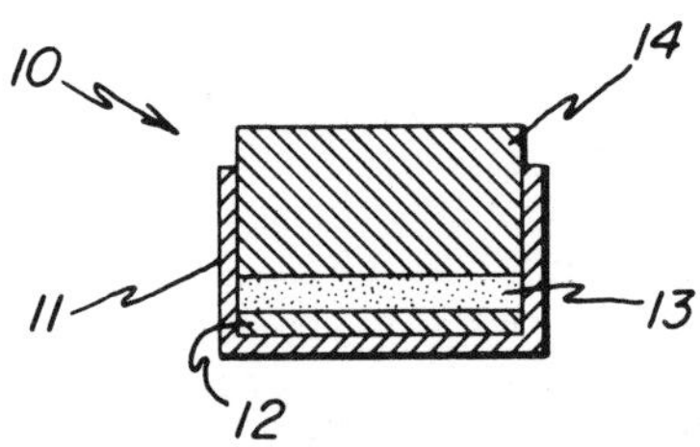

(continued)

Figure 3.2: (continued)

c.

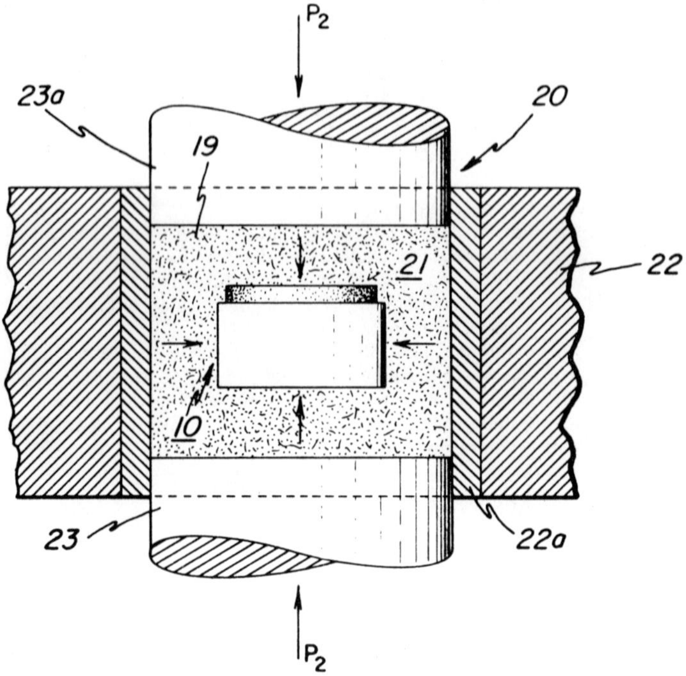

d.

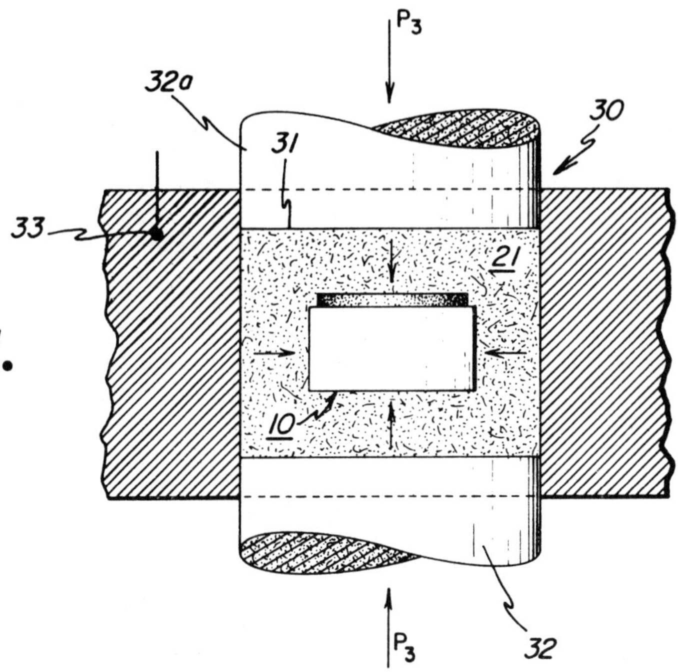

(continued)

Figure 3.2: (continued)

e.

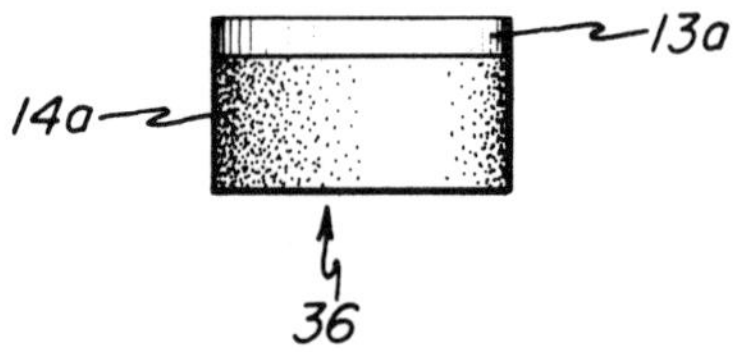

(a) Portion of a silicon zirconium alloy phase diagram showing the equilibrium diagram for eutectiferous silicon-rich zirconium alloy useful in the process.
(b) Cross-sectional view of a cell for carrying out infiltration of silicon-rich alloy.
(c) Sectional view through an apparatus for applying isostatic pressure to the cell.
(d) Sectional view through graphite mold for hot pressing.
(e) Polycrystalline diamond body/silicon nitride substrate composite.

Source: U.S. Patent 4,173,614

The starting solid silicon-rich alloy may or may not be of the same composition as the infiltrating silicon-rich alloy. If all of the starting solid silicon-rich alloy becomes fluid at the hot-pressing temperature then it will have the same composition as the infiltrating silicon-rich alloy. However, if only a portion of the starting silicon-rich alloy, i.e., hypoeutectic or hypereutectic, becomes fluid at the hot-pressing temperature, the starting alloy does not have the same composition as the fluid infiltrating silicon-rich alloy, and in such instance the infiltrating silicon-rich alloy will be more rich in silicon than the starting hypoeutectic alloy but less rich in silicon than the starting hypereutectic silicon-rich alloy.

Using Figure 3.2a as an example, the composition of the present infiltrating eutectiferous silicon-rich alloy and its melting temperature is found on liquidus curves **3** and **4** and includes eutectic point **2**. The area **5** defined by **1, 2** and **4** is comprised of a solid phase (Si) and a liquid phase, i.e., liquid infiltrating alloy phase, with the amount of solid phase increasing and the amount of liquid phase decreasing correspondingly as the distance to the right from eutectic point **2** along horizontal **1** is increased, i.e., as the amount of silicon in the alloy is raised from that contained in the eutectic. Likewise, the area **6** defined by **1, 2** and **3** is comprised of a solid phase $ZrSi_2$ and a liquid phase, i.e., liquid infiltrating alloy phase, with the amount of solid phase increasing and the amount of liquid phase decreasing correspondingly, as the distance to the left from eutectic point **2** along horizontal **1** is increased, i.e., as the amount of silicon in the alloy is lowered from that contained in the eutectic.

In carrying out the process, the desired composition of the infiltrating eutectiferous silicon-rich alloy and its melting temperature are found as a point on the liquidus curves including the eutectic point on the phase diagram for the silicon-rich alloy, and the hot-pressing temperature is the temperature at which such desired infiltrating silicon-rich alloy composition is fluid, i.e., sufficiently flowable to infiltrate through the compressed diamond mass. When a starting solid silicon-rich alloy is used that has the same composition as that of the desired infiltrating alloy, the hot-pressing temperature is the temperature at which the alloy is fluid, which ranges from 10°C to preferably a maximum of about 100°C higher than the melting point of the alloy, but, if desired, hot-pressing temperatures higher than this preferred maximum are useful depending largely upon the particular alloy used. However, hot-pressing temperatures higher than 1600°C are not useful since they tend to graphitize the diamonds excessively.

However, when the starting alloy does not have the same composition as that of the desired infiltrating alloy, but when it is heated to the melting point of the desired infiltrating alloy, it produces such infiltrating alloy as a liquid phase, then the hot-pressing temperature is a temperature at which such infiltrating alloy phase is produced in fluid form, i.e., about 10°C higher than the melting point of the infiltrating alloy phase.

Using Figure 3.2a as an example, for a specific infiltrating alloy of hypereutectic composition, its melting point is found in liquidus line **4**. For example, if the desired infiltrating hypereutectic alloy contains 95 atomic percent Si, its melting point is found on liquidus line **4** to be about 1400°C, as shown by line **7**. When the starting silicon-rich alloy is of the same composition as the desired infiltrating alloy as shown by line **7**, all of the starting alloy would melt at the melting temperature of 1400°C, and the fluid or hot-pressing temperature would range from 1410° to preferably 1510°C, or if desired up to but below 1600°C. However, when the starting silicon-rich alloy is any hypereutectic alloy to the right of line **7**, on horizontal line **1** in the equilibrium diagram, the hot-pressing temperature is the temperature at which the desired infiltrating 95 atomic percent Si-5 atomic percent Zr alloy is produced in fluid form, which would be about 1410°C.

For the preferred group of silicon-rich alloys, the eutectic melting point ranges from 870°C for SiPd eutectic alloy, i.e., about 56 atomic % Si, to 1410°C for SiMo eutectic alloy composition, i.e., about 97 atomic % Si. As shown in the figure, the SiZr eutectic alloy **2** contains 90.4 atomic % Si and has a eutectic melting temperature of 1360°C. The major phase of the solid silicon-rich eutectic alloy of the process is almost pure silicon.

In the arrangement shown in Figure 3.2b, cell **10** consists of cup **11** (right circular cylindrical wall with bottom). Within the cup are disposed a disc **12** of eutectiferous silicon-rich alloy, a mass **13** of diamond crystals in contact with silicon-rich alloy and a thick plug **14**, e.g., a cylinder of polycrystalline silicon nitride substrate fitting closely into the cup and acting as a closure.

The cup is made of a material which is substantially inert during the hot-pressing step, i.e., a material which has no significant deleterious effect on the properties of the diamond body. Such a material can be a nonmetal, such as compressed hexagonal boron nitride, but preferably it is a metal selected from the group consisting of tungsten, yttrium, vanadium, tantalum and molybdenum.

No free space should be left within the plugged cup which would allow an intermixing or free movement of the contents therein so that the contents, at least substantially as initially positioned, are subjected to the substantially isostatic pressure of the cold-pressure step.

Cell **10** is subjected to a cold-pressing step as shown in Figure 3.2c, which is carried out at room or ambient temperature, whereby only sufficient pressure need be applied to produce a dimensionally stabilized substantially isostatic system. Specifically, the cell is placed in the cylindrical core of pressure mold **20** surrounded by mass **19** of very fine particles, preferably -400 mesh, and more preferably ranging in size from about 2 to 20 microns of a pressure transmitting powder medium which remains substantially unsintered under the pressure and temperature conditions of the process, such as hexagonal boron nitride and silicon nitride. This pressure transmitting particulate or powder medium provides for the application of approximately or substantially isostatic pressure to the cell, whereby the cell and its contents are dimensionally stabilized, i.e., densified, substantially uniformly producing a shaped substantially isostatic system of powder enveloped cell wherein the density of the resulting compressed layer of crystals is higher than 70% by volume of the volume of compressed crystals.

Pressure mold **20** (ring **22** and pistons **23, 23a**) may be made of tool steel and, if desired, ring **22** may be supplied with a sintered carbide sleeve **22a** as shown to permit the application of pressures as high as 200,000 psi. Pressures higher than 200,000 psi provide no significant advantage. Within the confines of piston **23**, sleeve **22a** and piston **23a** pressure preferably in the range of 20,000 up to about 100,000 psi, and usually up to about 50,000 psi, is exerted on the pressure transmitting powder medium by the pistons actuated in the conventional manner until the applied pressure becomes stabilized as is done in the conventional powder packing technology.

When the cold-pressing step is completed, either one of the pistons is withdrawn and the resulting consolidated substantially isostatic shaped system **21** is forced out of liner **22a** and into a hole of identical diameter in graphite mold **30** (Figure 3.2d), the transferred system now being contained within the wall of hole **31** between graphite pistons **32, 32a**. The graphite mold is provided with thermocouple **33** to provide an indication of the temperature being applied to the dimensionally stabilized substantially isostatic system. The mold, with the substantially isostatic system so contained, is placed inside a conventional hot-pressing furnace (not shown). The furnace chamber is evacuated, or at least substantially evacuated, causing evacuation of system **21** including cell **10**, providing the system and cell with a substantial vacuum in which the hot-pressing step can be carried out. However, if desired, at this point nitrogen or hydrogen or an inert gas such as argon can be led into the furnace chamber to provide the furnace chamber as well as chamber **21** including the interior of the cell **10** with a suitable hot-pressing atmosphere. While pistons **32, 32a** apply a uniaxial pressure, i.e., the hot-pressing pressure, to system **21**, the temperature thereof is raised to a temperature at which silicon-rich alloy disc **12** produces fluid infiltrating silicon-rich alloy.

In the hot-pressing step, the hot-pressing temperature should be reached quickly and held at such temperature under the hot-pressing pressure usually for at least 1 minute to insure satisfactory infiltration through the interstices of the diamond crystal mass. Generally, a hot-pressing time period ranging from 1 to 5 minutes is satisfactory. Since conversion of diamond to nondiamond elemental carbon depends largely on time and temperature, i.e., the higher the temperature and the longer the time at such temperature the more likely the conversion to nondiamond elemental carbon, the hot-pressing step must be carried out before 5% by volume of the diamond is converted to nondiamond elemental carbon and this is determinable empirically. Conversion of 5% or more by volume of diamond to nondiamond elemental carbon is likely to result in elemental nondiamond carbon phase being left in the final product, which would have a significantly deleterious effect on its mechanical properties.

It is during this hot-pressing step that it is particularly important that substantially isostatic conditions be maintained so that when the silicon-rich alloy is converted to the fluid state this fluid will not be able to pass between mass **13** and cup **11** and escape to any significant extent, but will be forced to move throughout the mass of diamond crystals.

When the hot-pressing step is completed, at least sufficient pressure should be maintained during cooling of the hot-pressing system so that the cell is subjected to substantially isostatic pressure sufficient to preserve its dimensional stability. Preferably, the system is allowed to cool to room temperature. The cell is then removed from the system and the composite **36** (Figure 3.2e) is recovered comprised of polycrystalline diamond body **13a** bonded in situ directly to substrate **14a**. Adherent metal, if any, from the protective container and any squeezed out excess silicon alloy at the outside surfaces of the composite can be removed by conventional techniques such as grinding.

The process of *M. Lee, L.E. Szala and R.C. De Vries; U.S. Patent 4,171,339; October 16, 1979; assigned to General Electric Company* is similar to the process described in the preceding patent except that a diamond body/silicon carbide substrate composite is prepared by substituting a silicon carbide substrate for the silicon nitride substrate.

The present silicon carbide substrate is a polycrystalline body having a density ranging from 85 to 100% of the theoretical density of silicon carbide. Silicon carbide density given herein is the fractional density based on the theoretical density for silicon carbide of 3.21 g/cc. A silicon carbide polycrystalline body having a density less than about 85% is not useful because it would not have the required mechanical strength for most applications, for example, for use as a tool insert. Ordinarily, the higher the density of the silicon carbide body, the higher is its mechanical strength.

In related work *M. Lee, L.E. Szala and R.C. De Vries; U.S. Patents 4,168,957; September 25, 1979 and 4,124,401; November 7, 1978; both assigned to General Electric Company* prepare a silicon-bonded polycrystalline diamond body using only diamond crystals and the eutectiferous silicon-rich alloy or the components thereof as starting materials.

Example: A cast silicon disc weighing 330 mg was placed within a zirconium sleeve in a molybdenum cup. About 500 mg of fine diamond powder, particle size ranging from 1 to about 60 microns with at least 40 wt % of the diamond powder being smaller than 10 microns, was packed on top of the silicon disc. A molybdenum cup slightly larger in diameter than the original cup, i.e., the cup containing silicon and diamonds, was placed over the opening of the original cup as a cover.

The resulting container was then packed in hexagonal boron nitride powder, as shown in Figure 3.2c, and the whole charge was pressed at room temperature, i.e., cold-pressed, in a steel die to about 80,000 psi subjecting the container and its contents to substantially isostatic pressure until the pressure became stabilized, producing a dimensionally stabilized shaped substantially isostatic system of powder-enveloped container. From previous experiments it was known that in the resulting pressed assembly, i.e., in the resulting shaped substantially isostatic system of powder-enveloped container, the density of the diamond crystals was higher than 75% by volume of the compressed diamond mass. Also, the amount of silicon present was about 80% by volume of the compressed diamond mass.

The resulting assembly **21** of powder-enveloped container was then hot-pressed, i.e., it was pushed into a graphite mold of the same diameter size as the steel die, as shown in Figure 3.2d, which was placed within an induction heater. The interior of the container was evacuated and a nitrogen atmosphere introduced therein by evacuating the heater to about 10 torrs before back filling it with nitrogen. A pressure of about 5,000 psi was applied to the pressed assembly and maintained thereon by the graphite die which was then heated to a temperature of 1500°C in about 7 minutes by the induction heater. As the assembly was heated, the pressure also went up to about 10,000 psi due to the expansion of the system. When the temperature reached about 1350°C, the piston **23a** dropped and the pressure dropped to about 5,000 psi, indicating that silicon-rich zirconium alloy had formed, become fluid, and had proceeded to infiltrate through the compressed diamond mass.

The pressure was raised to the maximum hot-pressing pressure of 10,000 psi and when the temperature reached 1500°C, the assembly was maintained at the maximum hot-pressing temperature of 1500°C under 10,000 psi for 1 minute to insure complete infiltration by the alloy of the smaller capillaries of the compressed diamond mass. The power supply was then turned off but no additional pressure was applied. This provided a firm pressure at high temperature but reduced pressure at low temperature providing adequate geometric stability, i.e., this maintained the dimensions of the hot-pressed assembly until it was sufficiently cool for handling.

The resulting polycrystalline diamond body was recovered by grinding and grit blasting away can metal, i.e., molybdenum cup and remainder of zirconium sleeve, and excess silicon at the outside surface and faces of the body. The resulting integral polycrystalline diamond body had the shape of a disc about 0.115 inch thick. It appeared to be well infiltrated and bonded.

X-ray diffraction analysis of the cleaned face through which the alloy entered showed it to be comprised of diamond, silicon carbide and elemental silicon, indicating that the silicon carbide and elemental silicon were present in an amount of at least 2% by volume of the body. However, the x-ray diffraction analysis did not detect nondiamond elemental carbon phase.

Examination of the fractured cross-sectional surfaces of the disc showed that the fracture was transgranular rather than intergranular, i.e., it had fractured through the diamond grains rather than along grain boundaries. This indicates that the bonding medium was highly adherent and was as strong as the diamond grains or crystals themselves.

Examination of the fractured surfaces showed them to be porefree and the bonding medium was uniformly distributed throughout the body. Examination of the polished cross-sectional surface showed no strings of holes formed from diamond fragment pullout, illustrating the strong bonding therein and its usefulness as an abrasive. The density of the diamond crystals was about 81% by volume of the polycrystalline diamond body.

A photomicrograph of the polished surface, magnified 690 times, showed a white phase. X-ray spectral analysis of this phase showed that it consisted of zirconium and silicon indicating that this phase was zirconium silicide.

Tungsten Carbide Binder

A sintered compact for use in a cutting tool and a method of producing the same are disclosed by *A. Hara and S. Yazu; U.S. Patents 4,231,762; November 4, 1980; and 4,171,973; October 23, 1979; both assigned to Sumitomo Electric Industries, Ltd., Japan*. The compact comprises 95 to 20 volume % of diamond finer than 1 micron in size and the balance binder also finer than 1 micron selected from the group of WC, (MoW)C, WC base cemented carbide and (MoW)C base cemented carbide.

The method comprises pulverizing a diamond powder by using cemented carbide balls and a pot having a cemented carbide lining, mixing the diamond powder with a powder abraded from the balls and pot to produce a powder mix finer than 1 micron containing 95 to 20 volume % of diamond, heat-treating the powder mix in vacuum so as to degas, and hot-pressing the powder mix under high pressure at high temperature within the stable range of diamond.

Example: Ultrafine diamond powder having a particle size finer than 1 micron was pulverized in a solvent of acetone by using WC-7% Co cemented carbide balls and a pot having the same cemented carbide lining. The amount of diamond was 5 g, but after 40 hours of ball-milling, the powder weighed 8.3 g. This increase was caused by the fine cemented carbide powder abraded from the balls and pot.

The diamond amount contained in the powder mix was estimated to be 80 volume %. As a result of examination by a scanning electron microscope, it was confirmed that all the particles in the mix were finer than 1 micron.

The powder mix was cold-pressed into a disc having a thickness of 1.5 mm and a diameter of 10 mm. The disc was treated in a vacuum furnace at 1000°C to degas and thereafter was sintered under 55 kbar at 1370°C for 10 minutes by means of a superpressure apparatus for making a diamond.

The sintered compact was lapped by a diamond paste, and the structure thereof was examined to be found that the compact comprises ultrafine diamond and WC particles less than 1 micron in size.

The compact was cut. A piece was brazed to a steel shank, and a cutting edge was ground by a diamond wheel. For comparison, a cutting tool of the same shape was made from a marketed sintered diamond compact. The ground cutting edges were observed. The cutting edges of the marketed compact had a number of chippings with a width of about 10 microns, which might be caused during grinding.

A copper alloy commutator of an electric motor was cut by both of the cutting tools at a cutting speed of 400 m/min, depth of cut 0.5 mm, and a feed of 0.05 mm per revolution. After cutting 2,000 pieces by the cutting tool of the process, the surface roughness of a workpiece was 1.3 microns at maximum, while a workpiece cut by the cutting tool of marketed compact had initial surface roughness of 2.6 microns at maximum, and after cutting 500 pieces the surface roughness reached 3.9 microns.

Monolithic Carbon Material with Diamond Structure

L.F. Vereschagin, E.N. Yakovlev, V.N. Slesarev, V.A. Stepanov, A.Y. Preobrazhensky, T.D. Varfolomeeva and L.E. Shterenberg; U.S. Patent 4,196,181; April 1, 1980 provide a process for transforming the nondiamond carbon-bearing material into a microcrystalline monolithic material crystallized to the diamond structure. The aggregates consist of intergrown grains varying in size from 0.1 to a few microns and characterized in that the monolithic material has a preset geometric shape which corresponds in shape and size to the desired finished article, and which may be manufactured in any desired number of identical items. The material has a density above 3.55 g/cc, an abrasion resistance above 100,000, an oxidation temperature in the air above 700°C, and a hardness such that it can scratch the hardest face of the diamond, i.e., the octahedral surface, a lattice constant a_0 = 3.566 Å at a temperature of 25°C and a pressure of 1 atmosphere and a preset surface finish of from 10 to 60 microns.

The equipment used for applying the high pressure to the nondiamond carbon-bearing material is a modification of the device used for building up high pressure covered by French Patent 7,102,157. This modification is shown in Figure 3.3a.

Figure 3.3: Apparatus for Forming Diamond-Structured Material

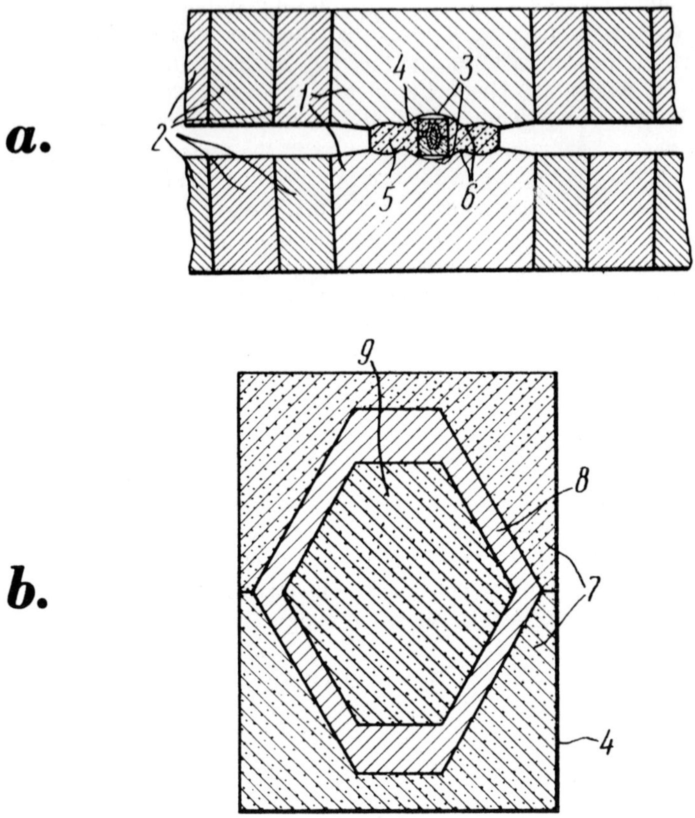

Source: U.S. Patent 4,196,181

The chamber illustrated in the figure consists of two halves and is formed by two centrally-mounted punches **1** opposing each other, pressed into a structure of concentric binding rings **2** ensuring a sufficient axial support for the punches and protecting them against destruction in the process of transforming the nondiamond carbon-bearing material into a microcrystalline monolithic material of carbon having a diamond structure. The best material for the punches was found to be a hard alloy based on tungsten carbide (94% WC, 6% Co). The design modification of the punches used in the process consists in the particular shape of the working surface, and in the dimensions and shape of hollows **3** on this surface. An alloy of steel with a minimum tensile strength of 150 kg/cm^2 hardened to 48 to 50 R_c has proven to be a satisfactory material for the binding rings which support the punches.

Another improvement in the process relates to the method of sealing. The high pressure chamber is sealed with dense limestone referred to as a lithographic stone which is simultaneously used as the medium for conveying the pressure to a reaction cell **4**. The lithographic stone is used for making a pellet **5** whose face surfaces repeat the configuration of the mold made from the surface of the hollows of the punches. A channel located in the center of the pellet, and perpendicular to its surface, accommodates the reaction cell. The height of the pellet has been selected experimentally so that when the pressure inside the reaction cell reaches the value required for transforming the nondiamond carbon-bearing material into a microcrystalline monolithic material of carbon with a diamond structure, the distance between the projections **6** on the working surfaces of the punches is equal to **1** mm, approximately. When the punches are brought together and compressed, the material of the pellet forms a shaped seal between them which makes it possible to reliably lock the high pressure chamber at all changes in the pressure and temperature and, in particular, during the pulse heating of the precompressed reaction cell.

The reaction cell is placed into the pellet between the faces of the punches. In the capacity of the compressible element the reaction cell comprises a heater **7**, a catalyst **8** and a carbon-bearing nondiamond material, mostly graphite, which serves as a pattern **9** for the article of the microcrystalline monolithic material of carbon with a diamond structure whose shape and surface finish depend on the shape and surface finish of the pattern. The location and interaction of the individual parts of the reaction cell are shown in Figure 3.3b.

The heater consists of two parts which, taken together, form a round hollow cylinder closed at the ends, the height of which is equal to, or larger than, the height of the pellet while its diameter ensures tight fitting of the heater into the pellet. It is practicable that the heater should be made of graphite.

The catalyst used in the process is in the form of a fine-grained powder. The catalyst is placed inside the heater filling the entire free space around the pattern. The catalyst should be selected from the known metals, alloys and compounds which are used in transforming carbon-bearing materials under the effect of high pressure and temperature into diamond, e.g., the metals of Group VIII of the periodic system, their alloys, carbides and metal-carbide systems.

The pattern of the nondiamond carbon-bearing material is made mostly from graphite which is shaped like the finished article and the surface of the pattern is machined to a surface finish which is one or two classes higher than the desired surface finish of the finished article of the microcrystalline monolithic material of carbon having the diamond structure. The pattern made in this way is dipped into the powdered catalyst inside the hollow heater and placed centrally with relation to the heater and the powdered catalyst inside.

The chamber ensures the building up of the pressure required for transforming the material into a microcrystalline monolithic material of carbon with a diamond structure. The chamber is placed between the anvil plates of a suitable press (not shown in the drawing) then the halves of the chamber are brought together in the direction perpendicular to the chamber plane and, as a result, the pellet and the reaction cell are compressed to a point in which a high pressure is reached inside the reaction cell.

The material in the reaction cell is subjected to high pressure, that of at least about 80 kbar. It is essential that the pressure should be applied uniformly to the entire surface of the pattern so as to preserve its shape and condition during the course of compression.

The rise of the temperature in the high pressure chamber inside the reaction cell is achieved by passing an electric current through the conducting section of the reaction cell. Heating is ensured by the Joule heat produced simultaneously in the graphite pattern, the powdered catalyst and the heater when an electric pulse passes through them. The electric current is supplied to the reaction cell through the chamber punches, the press anvils and the copper busbars (not shown in the drawing) connected to a source of power supply. The electric contact between the punches and the reaction cell is ensured in the course of compression of the reaction cell in the chamber, with the contact surfaces being formed by the faces of the reaction cell and the corresponding parts of the working surfaces and hollows of the punches directly resting on the faces of the reaction cell. The contact resistance has become constant close to a pressure of approximately 10 kbar. The temperature inside the reaction cell can be raised quickly and for a short duration by means of an electric circuit into which a battery of capacitors is discharged so that the discharge current passes through the reaction cell and raises its temperature.

After compression, the nondiamond carbon-bearing pattern immersed into the powder catalyst is heated by an electric current pulse to at least about 1500°C in the course of from 0.1 to 10 seconds. This period of time is sufficient for forming the microcrystalline monolithic material consisting of carbon having a diamond structure.

In related work, *L.F. Vereschagin, E.N. Yakovlev, T.D. Varfolomeeva, A.Y. Preobrazhensky, V.N. Slesarev, V.A. Stepanov and L.E. Shterenberg; U.S. Patent 4,089,933; May 16, 1978; assigned to Institut Fiziki Vysokikh Daleny Akademii Nauk SSSR, U.S.S.R.* provide a method of producing polycrystalline diamond aggregates in which a carbon-containing material is given a required shape, then it is surrounded by a catalyst in the form of a powder and subjected to the action of pressure of at least about 80 kbar and simultaneously heated to a temperature of at least 1500°C by using an electric pulse of 0.1 to 10.0 seconds duration.

In earlier work *L.F. Vereschagin, A.Y. Preobrazhensky, V.A. Stepanov and V.S. Voblikov; U.S. Patent 3,949,062; April 6, 1976* provide a method for producing polycrystalline diamond aggregates of a predetermined shape. The process comprises subjecting a workpiece of a nondiamond carbonaceous material, in the presence of a powdered catalyst, to a pressure of at least 80 kbar and heating, by a pulse of electric current, to a temperature of at least 1500°C for a period of 0.1 to 10 seconds, sufficient for the formation of a polycrystalline diamond aggregate of a predetermined shape. The workpiece has the same shape as the polycrystalline diamond aggregate of predetermined shape and dimensions exceeding those of the polycrystalline diamond aggregate of a predetermined shape by a contraction coefficient value of the nondiamond carbonaceous material during its transformation into the polycrystalline diamond material, and the powdered catalyst particles have dimensions less than the minimum dimensions of the workpiece.

Hence, the method for producing polycrystalline diamond aggregates of a predetermined shape according to the process makes it possible to produce both massive polycrystalline aggregates of a simple geometric shape, e.g., octahedrons, cylinders, spheres, and the like, and shaped articles such as screw-cutting dies, draw plates, drills, bearings, nuts, screws, and the like, and to use polycrystalline diamond as a structural material.

Work Hardening

The work hardening of diamond crystals as described by *R.C. De Vries, F.P. Bundy and R.H. Wentorf, Jr.; U.S. Patent 4,181,505; January 1, 1980; assigned to General Electric Company* is accomplished by filling a volume solely with diamond crystals (i.e., no bonding medium is employed) or solely with diamond crystals embedded in a powdered material selected from the group consisting of diamond and cubic boron nitride and then subjecting the filled volume to the simultaneous application of high pressure and temperature in a region defined on the carbon phase diagram. Pressure and temperature conditions employed in this process are incapable of bringing about crystal-to-crystal bonding of diamond (or of cubic boron nitride powder, if employed) of any consequence. Thus, after the temperature and pressure have been reduced, the work-hardened diamonds are readily separated for use in diamond tools, e.g., saws or single-point tools.

The pressures and temperatures employed may range, for example, from 10 kbar at 1200°C to 60 kbar at 900°C. When embedding a larger diamond crystal to be work hardened in powdered material as noted above, brittle fracture thereof is inhibited. The region of plastic deformation in which the work hardening of diamonds may be conducted includes temperature/pressure conditions both in and out of the diamond stable region. Preferred operation is in that portion of the region of plastic deformation below a pressure of 55 kbar and at a temperature of less than 1500°C. The size designation for diamond crystals as expressed herein is for the largest linear crystal dimension.

Figure 3.4 represents the phase diagram of carbon having defined thereon the region of plastic deformation in which the process may be practiced. Operational techniques for simultaneously applying both high pressures and high temperatures in such apparatus are well known to those skilled in the superpressure art.

Figure 3.4: Phase Diagram of Carbon

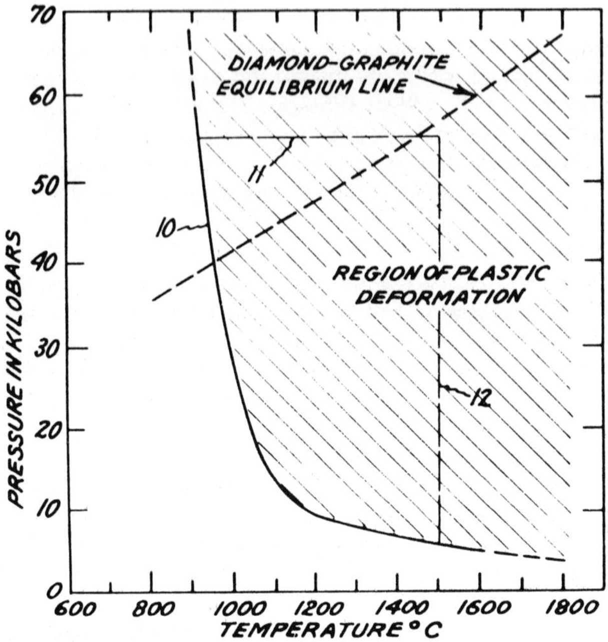

Source: U.S. Patent 4,181,505

Inhibition of brittle fracture of a given diamond in order to maximize recoverable yield during work hardening thereof may be accomplished as follows:

(a) Diamonds in the size range from about 5 to 250 μm should constitute the entire charge filling the reaction vessel; each crystal provides the requisite support for adjacent crystals; use of a gradation of sizes in the upper end of this range in a given charge is preferable, but not critical.

(b) Diamonds in the size range from about 250 μm to 5 mm should be embedded in diamond or CBN powder, the combination of larger crystal and powder constituting the entire charge filling the reaction volume.

(c) Diamonds in a size range (from about 250 to 500 μm) may be work hardened using either (a) or (b) above.

Silicon as Binder

In the process of *M. Lee, L.E. Szala and R.C. De Vries; U.S. Patent 4,167,399; September 11, 1979; assigned to General Electric Company* a cavity of predetermined size is formed in a compressed pressure transmitting powder medium. A mass of diamond crystals in contact with a mass of elemental silicon are disposed within the cavity and additional pressure transmitting powder is placed over the cavity and its contents producing a powder-enveloped cavity. Pressure is applied to the powder medium, resulting in substantially isostatic pressure being applied to the cavity therein and its contents sufficient to dimensionally stabilize the cavity and its contents. The resulting shaped substantially isostatic system of powder-enveloped diamond and silicon is hot pressed to liquefy and infiltrate the silicon through the interstices between the diamond crystals producing, upon cooling, an adherently bonded integral body.

Example: The apparatus used in this example was substantially the same as that shown in Figure 3.5. Hexagonal boron nitride powder ranging in size from about 2 to 20 microns was packed into a die and a cylinder used as a mold was pressed into the powder as shown in Figure 3.5a by **19a** and **9**. The cylinder was made of cemented metal carbide and was about 0.35 inch in diameter and 0.25 inch in thickness. The axis of the cylinder was approximately lined up with the central axis of the die.

In this embodiment, after the cylinder was inserted in the powder, additional hexagonal boron nitride powder was placed in the die covering the cylinder completely, and the resulting powder-enveloped cylinder was pressed at room temperature under a pressure of 50,000 psi. Piston **23a** was then withdrawn and piston **23** was used to push the resulting pressed powder-enveloped cylinder partially out of the die. The exposed portion of the pressed powder was removed leaving the cylinder partially exposed. The cylinder was then withdrawn leaving the cavity it had impressed therein.

A 140 mg disc of silicon having a diameter the same as the inner diameter of the cavity was placed in the bottom of the cavity. About 250 mg of size-graded diamond powder, ranging from 1 to 60 microns with about 40 wt % being smaller than 10 microns, was packed on top of the silicon disc.

A disc of hot-pressed hexagonal boron nitride powder of the same diameter as the inner diameter of the cavity was placed within the cavity on top of the diamond powder to insure that the surface of the resulting polycrystalline diamond body would be flat.

The entire mass was then pushed into the center of the die by piston **23a** which was then withdrawn. An additional amount of hexagonal boron nitride powder was added to the die to cover the hot-pressed disc of hexagonal boron nitride resulting in the cavity and contents being enveloped by hexagonal boron nitride as shown by **19** in Figure 3.5b. The resulting charge was then pressed at room temperature, i.e., cold-pressed, in the steel die under a pressure of 80,000 psi, subjecting the cavity and its contents to substantially isostatic pressure until the pressure became stabilized producing a dimensionally stabilized shaped substantially isostatic system of powder-enveloped cavity and contents.

From previous experiments it was known that in the resulting pressed assembly, i.e., in the resulting shaped substantially isostatic system of powder-enveloped cavity and contents, the density of the diamond crystals was higher than 75% by volume of the compressed diamond mass. Also, the amount of silicon present was about 40% by volume of the compressed diamond mass.

The resulting pressed assembly **21** of powder-enveloped cavity and contents was then hot-pressed, i.e., it was pushed into a graphite mold of the same diameter size as the steel die, as shown in Figure 3.5c, and placed within an induction heater. The interior of the cavity was evacuated and a nitrogen atmosphere introduced therein by evacuating the heater to about 10 torrs before back filling it with flowing dry nitrogen. A pressure of about 5,000 psi was applied to the pressed assembly and maintained thereon by the graphite die, which was then heated to a temperature of 1500°C in 7 minutes by the induction heater. With this heating the pressure reached about 10,000 psi due to the expansion of the entire system.

At 1500°C the piston and the pressure dropped to about 5,000 psi indicating that the silicon had melted and become fluid and had infiltrated through the diamond mass. The pressure was raised back to 10,000 psi where it was maintained for 1 minute at 1500°C to insure that the silicon had completely infiltrated through the diamond mass. The power supply was then turned off but no additional pressure was applied. This provided a firm pressure at high temperature but reduced pressure at lower temperature providing adequate geometric stability. At room temperature the resulting polycrystalline diamond body was recovered.

After removing surface scales of hexagonal boron nitride powder, the resulting integral polycrystalline diamond body had the shape of a disc with a diameter of 350 mils and a thickness of 50 mils.

The polycrystalline disc had substantially smooth flat surfaces and appeared to be well infiltrated by the bonding medium. Optical examination of the disc, magnified 100 times under a microscope, showed it to be porefree.

Using a hammer and wedge, the disc was fractured substantially in half. Examination of the fractured cross-sectional surfaces of the disc showed that the fracture was transgranular rather than intergranular, i.e., it had fractured through the diamond grains rather than along the grain boundaries. This indicates that the bonding medium was highly adherent and was as strong as the diamond grains or crystals themselves. Also, the fractured surfaces were porefree and the bonding medium was uniformly distributed throughout the body.

A cross-sectional fractured surface of the disc was polished on a cast iron scaife. Examination of the polished surface showed no strings of holes formed from diamond fragment pullout illustrating strong bonding therein and its usefulness as an abrasive.

Figure 3.5: Apparatus for Preparing Polycrystalline Diamond Body

a.

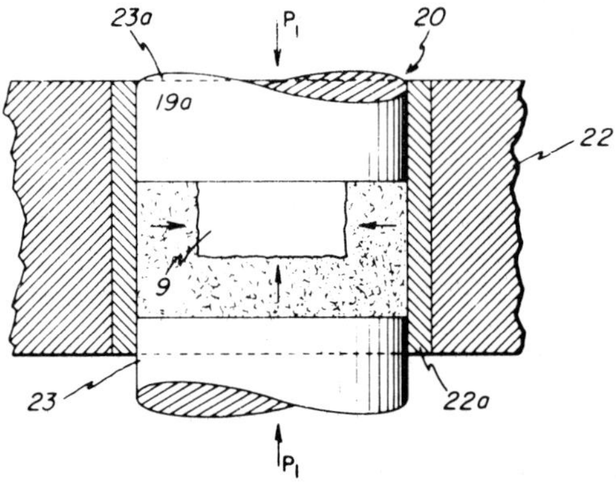

b.

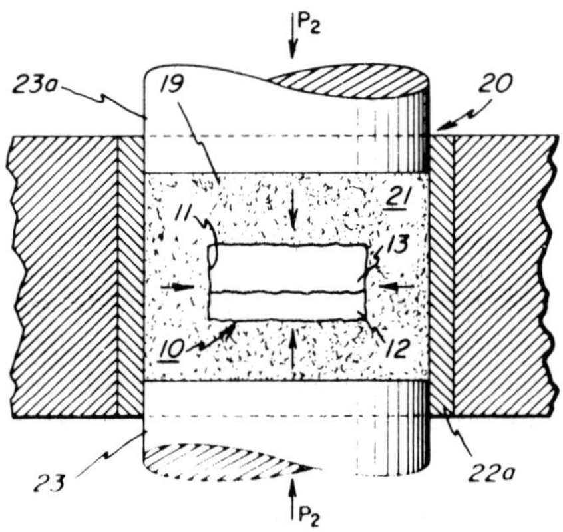

c.

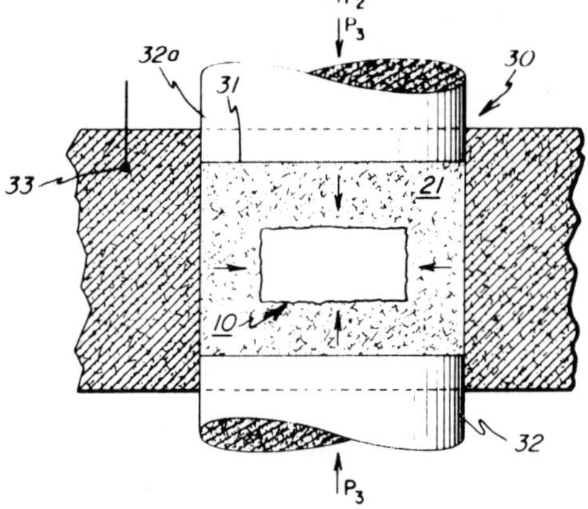

Source: U.S. Patent 4,167,399

In related work by *M. Lee, R.C. De Vries, L.E. Szala and R.E. Tuft; U.S. Patent 4,151,686; May 1, 1979; assigned to General Electric Company* a mass of diamond crystals contacting a mass of elemental silicon are confined within a pressure-transmitting medium. The resulting charge assembly is subjected to a pressure of at least 25 kbar causing application of isostatic pressure to the contacting masses which dimensionally stabilizes them and increases the density of the mass of diamond crystals.

The resulting pressure-maintained charge assembly is heated to a temperature sufficient to melt the silicon and at which no significant graphitization of the diamond occurs whereby the silicon is infiltrated through the interstices between the diamond crystals producing, upon cooling, an adherently bonded integral body.

Example: The apparatus used in this example was substantially the same as that shown in Figures 3.5a and 3.5b. Hexagonal boron nitride powder ranging in size from about 3 to 20 microns was packed into a cylindrical steel die with a 0.46 inch internal diameter, and a cylinder used as a mold was pressed into the powder.

The cylinder was made of cemented metal carbide and was about 0.35 inch in diameter and 0.25 inch in thickness. The axis of the cylinder was approximately lined up with the central axis of the die.

In this embodiment, after the cylinder was inserted in the powder, additional hexagonal boron nitride powder was placed in the die covering the cylinder completely, and the resulting powder-enveloped cylinder was pressed at room temperature under a pressure of about 50,000 psi. The top piston was then withdrawn and the bottom piston was used to push the resulting pressed powder-enveloped cylinder partially out of the die. The exposed portion of the pressed powder was removed leaving the cylinder partially exposed. The cylinder was then withdrawn leaving the cavity it had impressed therein.

A cast commercially pure silicon disc of about 90 mg having a diameter almost the same as the inner diameter of the cavity was placed in the bottom of the cavity. About 240 mg of size-graded diamond powder, ranging in particle size from about 1 to 50 microns with at least 25 wt % of the particles having a size smaller than 8 microns, was packed on top of the silicon disc.

A disc of hot-pressed hexagonal boron nitride powder of almost the same diameter as the inner diameter of the cavity was placed within the cavity on top of the diamond powder to insure that the surface of the resulting polycrystalline diamond body would be flat.

The entire mass was then pushed into the center of the die by the top piston which was then withdrawn. An additional amount of hexagonal boron nitride powder was added to the die to cover the hot-pressed disc of hexagonal boron nitride powder resulting in the cavity and contents being enveloped by hexagonal boron nitride. The resulting charge was then pressed at room temperature in the steel die under a pressure of about 50,000 psi, as shown in Figure 3.5b, subjecting the cavity and its contents to substantially isostatic pressure confining the contacting masses therein, until the pressure became stabilized producing a dimensionally stabilized shaped substantially isostatic system.

This pressed mass, i.e., charge assembly, when it was pushed out of the steel die, had a diameter of about 0.46 inch and a thickness of 0.450 inch. The charge assembly was then inserted and centrally positioned in a reaction chamber. The chamber had a diameter of 0.46 inch and pressed salt cylinders were placed below and on top of the charge assembly to fill in the remaining space of the chamber. The reaction vessel was placed in a high pressure-high temperature apparatus and at ambient temperature was pressed to about 55 kbar. When the pressure became substantially stabilized, electric current was applied through the graphite heating element of the reaction vessel. The temperature of the reaction chamber reached about 1500°C within approximately 2 to 3 minutes. The reaction chamber was kept at 1500°C for 30 minutes, then the power was turned off. After the reaction vessel cooled to about room temperature, the pressure was released and the reaction vessel was removed from the apparatus.

The recovered polycrystalline diamond body was a well-bonded disc of about 0.045 inch in thickness but the disc was cracked slightly, probably due to inadequate support by the hot-pressed boron nitride disc.

Using a hammer and wedge, the disc was fractured substantially in half. Examination of the fractured cross-sectional surfaces of the disc showed that the fracture was transgranular rather than intergranular, i.e., it had fractured through the diamond grains rather than along the grain boundaries. This indicated that the bonding medium was highly adherent and was as strong as the diamond grains or crystals themselves.

A cross-sectional fractured surface of the disc was polished on a cast iron scaife. Optical examination of the polished cross section of the disc showed a well-bonded microstructure which was substantially porefree and which also was free of nondiamond elemental carbon phase. The examination of the polished surface also showed no strings of holes formed from diamond fragment pullout illustrating strong bonding therein and its usefulness as an abrasive. The diamond density was determined to be about 90% by volume of the disc.

A portion of the disc was crushed and the powder was analyzed for its crystalline structure by x-ray diffraction technique which showed it to be comprised of diamond in substantial amount with silicon carbide and trace of silicon, indicating that the silicon carbide and elemental silicon were present in an amount of at least 2% by volume of the body. However, the x-ray diffraction analysis of the crushed body did not detect elemental nondiamond carbon phase.

Another portion of the disc was subjected to a jet of silicon carbide particles to evaluate its abrasion resistance. The particle size of the silicon carbide was about 50 microns and the pressure of particle-carrying gas was about 80 psi. The rate of feed of silicon carbide particles was substantially high. The estimated eroded volume of the disc after 2 minutes of jet impingement was 0.03×10^{-3} cc. Under identical test, for a comparison, a hot-pressed silicon carbide (96% dense) showed 0.16×10^{-3} cc of eroded crater and a cemented tungsten carbide with 6 wt % cobalt binder showed 0.15×10^{-3} cc of eroded crater. These tests show that the present polycrystalline diamond body has a significantly high wear resistance which makes it useful as an abrasive.

Sintering Without an External Load

V.N. Bakul, I.F. Vovchanovsky and N.V. Tsypin; U.S. Patent 4,164,527; August 14, 1979 prepare superhard articles by a method which consists of placing a mixture of powdered hard-alloy matrix material and diamond grains into a hard-alloy envelope formed from a hard-alloy material, preferably of exactly the same composition as the hard-alloy matrix material of the mixture, with the thickness of the hard-alloy envelope over the entire area thereof exceeding by at least 3 to 4 times the maximum space between any two adjacent diamond grains within the mixture. The mixture in the envelope is then subjected to sintering in a free condition thereof, without any external mechanical loads being applied thereto.

Such a technical solution eliminates any danger of graphitization of the diamonds in the mixture by increasing the rate of the process and also avoids nonuniform shrinkage of the composition, which in the present case is compressed by the outer envelope in the course of sintering.

Example: The realization of the method of making superhard articles may be explained by the following method of manufacturing an insert with a diameter of 10 mm and a height of 10 mm for reinforcing a dressing tool. The starting material is a hard-alloy matrix material of WC-Co (e.g., an alloy containing 94% of tungsten carbide and 6% of cobalt). The insert is shaped in a metallic press-mold which can be used many times.

The press-mold is composed of a die **1** (Figure 3.6a), a lower plunger **2** and a yokelike support **3**. Initially a hard-alloy matrix material **4** is placed onto the bottom of the chamber of the press-mold in an amount adequate for shaping a hard-alloy envelope **5** (Figure 3.6b).

The envelope is then pressed by an upper plunger **6**. Separately, a briquette **7** of superhard material is prepared. The briquette consists of powdered hard-alloy matrix material of WC-Co and diamonds, taken in a ratio of 4:1 respectively. After the plunger is extracted from the press-mold, the briquette of the superhard material is placed into the envelope (Figure 3.6c). After that, the envelope is covered with a top layer **8** (Figure 3.6d) of the matrix material to form a closed shell, the plunger is set into position and the article-insert is pressed.

Then, the aforementioned free sintering of the composition is carried out. To this end, the shaped briquette under the hard-alloy envelope is freely placed into alumina powder within a special graphite container currently used for sintering hard-alloy articles. Heating is carried out in a hydrogen furnace by pushing the container through the heating zone of the furnace. At a temperature of 1400° to 1460°C the hard-alloy mixture is sintered.

The hard-alloy envelope surrounding the briquette has a higher contraction coefficient than that of the diamond-containing mixture within the briquette, thus, in the course of sintering, the hard-alloy envelope compresses the briquette providing thereby the forces for preventing the cracking of the sintered article.

Figure 3.6: Sintering Without an External Load

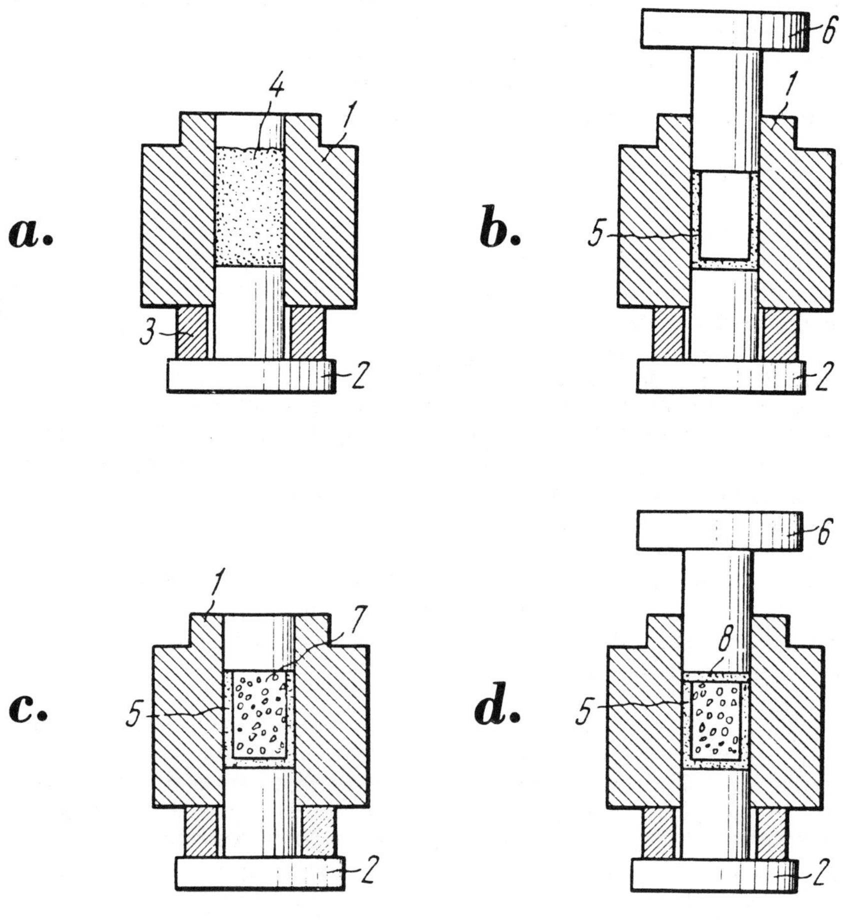

Source: U.S. Patent 4,164,527

Copper-Titanium or Copper-Zirconium Intermetallides as Binder

L.F. Vereschagin, A.A. Semerchan, T.T. Gankevich, M.E. Dmitriev and V.P. Modenov; U.S. Patent 4,142,869; March 6, 1979 provide a compact-grained diamond material comprising powdered diamond particles bound by a metal binder filling the spacing between the diamond particles. The function of the metal binder is performed by copper-titanium or copper-zirconium intermetallides or their mixtures, whereas the diamond particles, which form the base of the diamond material are coated with a titanium or zirconium carbide film which forms an uninterrupted cellular structure over the entire volume of the material.

In the proposed material the diamond powder content, i.e., the diamond particle content, is equal to or greater than 65% by volume. The size of the diamond particles is between 1 and 400 microns. The term "intermetallide" stands for chemical compounds, for example, Ti_2Cu, TiCu, Ti_2Cu_3, $TiCu_3$, Zr_2Cu, ZrCu, Zr_2Cu_3, $ZrCu_3$, whose composition is, or is close to, stoichiometric.

An important feature of the proposed compact-grained diamond material comprising copper-titanium and copper-zirconium binders is the presence therein of thin films of titanium or zirconium carbide, while films coat individual diamond particles constituting the material. Such films are formed at temperatures of 1000° to 1200°C, and under a pressure of 10 to 30 kilobars as a result of diamond carbon saturating a layer of a molten copper-titanium or copper-zirconium binder contacting with diamond crystals. After crystallization this binder layer turns into a carbide film, which fact has been corroborated by x-ray and electron microscopic examination.

The process of manufacturing the proposed compact-grained diamond material is as follows. Diamond powder is placed in a graphite container. Intermetallide binder is put in the same container in the form of a compact slug or cake compressed from chips of the intermetallide. The intermetallide binder may be a mixture of two intermetallides, for example, Ti_2Cu and TiCu. The binder slug or cake is so placed with respect to the diamond powder that both are continuously in contact with each other.

The optimum volume ratio between the diamond powder and binder is between 80:20 and 70:30, although other ratios are possible, with a proviso that the diamond content is equal to or higher than 65% of the volume of the material.

The graphite container and its contents are placed in a high-pressure chamber where a pressure upward of 1 kilobar is developed. Under this pressure the container and its contents are heated to a temperature sufficient to melt the binder. The molten binder impregnates the compressed diamond powder and is at the same time saturated with diamond carbon. The process is continued for 0.5 to 1 minute, after which the heating is stopped and the pressure is gradually brought down to normal (this takes 2 to 3 minutes to avoid cracking of the baked material).

It should be noted that due to a high cubic concentration of diamond particles the proposed material does not easily lend itself to machining. It is therefore expedient that the material be given the desired shape and size during production.

Zirconium Layer Bonding Diamond Compact to Cemented Carbide Backing

R.D. Mitchell; U.S. Patent 4,063,909; December 20, 1977 describes a compact of cubic boron nitride or diamond abrasive particles or a mixture thereof present in an amount of at least 50% by volume, bonded into a hard conglomerate, at least one surface of the compact having bonded thereto a layer of metal which is a high temperature braze metal capable of wetting the abrasive compact. The compact is substantially free of deteriorated abrasive particles, i.e., substantially free of graphite in the case of diamond compacts and substantially free of hexagonal boron nitride in the case of cubic boron nitride compacts.

The high temperature braze metal is preferably a transition metal such as titanium, chromium, manganese, vanadium, molybdenum, platinum, iron, cobalt, or nickel or an alloy containing one or more of these metals. The preferred metals are stated to be titanium, a copper/titanium alloy, or a copper/tin/titanium alloy.

The layer of high temperature braze metal is utilized to bond the compact to a suitable support which may be a cemented tungsten carbide support. Such bodies may be used as tool inserts for the grinding of hard materials.

R.D. Mitchell; U.S. Patent 4,108,614; August 22, 1978 has found that excellent bonding between the diamond compact and a cemented metal carbide backing can be achieved if the high temperature braze metal is zirconium. Not only is a very strong bond achieved between compact and backing, but also the reproducibility of this strong bond is readily achieved.

Thus, the process provides an abrasive body comprising a diamond compact bonded to a cemented carbide backing by means of a continuous zirconium layer, the thickness of which is less than 0.5 mm, the diamond compact being substantially free of graphite and having a bonding matrix selected from a metal of Group VIII of the Periodic Table or an alloy containing one or more such metals and at least 70% by volume of diamond particles, and the cemented carbide backing being selected from cemented tantalum carbide, cemented tungsten carbide and cemented titanium carbide and mixtures thereof.

The thickness of the zirconium layer will in general be at least 10 microns. The thickness of the layer is preferably in the range of 50 to 500 microns. There will be some alloying between the metal bonding matrix of the compact and the zirconium layer. Similarly, there will be some alloying of the bonding matrix of the cemented carbide backing and the zirconium. In referring to the thickness of the zirconium layer, any alloy portions are included in this thickness.

Preferably the diamond content of the compact will be in the range of 80 to 90% by volume. The compact has a metal bonding matrix which is a solvent for diamond growth. Particularly suitable metals are cobalt, iron or nickel or alloys containing one or more of these metals. With such a bonding matrix there will be a certain amount of diamond intergrowth or direct diamond-to-diamond bonding as compact manufacture, as is known in the art, is carried out under conditions of temperature and pressure in the diamond stable region of the carbon phase diagram.

The abrasive body is made in the following manner. A mass of starting material suitable for making the abrasive body is placed in the reaction capsule of a conventional high temperature/high pressure apparatus suitable for diamond growth. The mass comprises a zirconium foil or layer of zirconium powder on top of a mass of carbide molding powder or preformed cemented carbide body. On top of the zirconium are placed the ingredients necessary for compact manufacture, i.e., diamond plus metal for the bonding matrix. The metal for the bonding matrix can either be in powdered form and admixed with the diamond particles or can be in the form of a layer of metal powder or metal foil on the zirconium. In the latter case, a mass of diamond particles is placed on the layer of bonding metal. The contents of the reaction capsule are then subjected to elevated temperature and pressure conditions in the diamond stable region of the carbon phase diagram.

The preferred elevated temperature conditions are 1400° to 1600°C and the preferred elevated pressure conditions are 50 to 75 kilobars. These elevated conditions of temperature and pressure are maintained for at least 3 minutes and generally about 10 to 30 minutes. The temperature of the contents of the reaction capsule is then allowed to return to ambient and the pressure released. The body which is formed in the capsule can be recovered from the capsule using conventional recovery techniques.

Epitaxial Coating Method

D.V. Fedoseev, B.V. Deryagin, V.N. Bakul, A.I. Prikhna, V.P. Varnin, S.P. Vnukov, V.K. Gerasimenko, J.I. Nikitin, N.V. Tsypin and A.V. Hochko; U.S. Patent 4,104,441; August 1, 1978; assigned to Institut Sverkhtverdykh Materialov SSR and Institut Fizicheskoi Khimii Akademii Nauk, U.S.S.R. describe a method of preparing a polycrystalline diamond member, wherein diamond particles having a coating are placed in a high pressure and high temperature chamber to be sintered therein for a time sufficient to provide a strong bond between diamond particles thus forming a single compact body.

This method is characterized in using diamond particles of practically any size including those having a minimum size of less than 1 micron; epitaxially depositing both diamond and diamond-graphite coating on the particle; and sintering the resulting particles having diamond-graphite coating in contact with one another in such a manner that, as a result a combined action of mechanical pressure and heating, phase conversions occur in the coating material in the zone of contact between the particles with the formation of diamond zones binding the particles into a single compact body which is substantially free from graphite inclusions.

The resulting compact member is characterized in the absence of marked graphite inclusions, and diamond cleavage between the particles results in a considerable improvement of the strength of the compact member.

According to one embodiment of the process, the method of preparing a polycrystalline diamond member is characterized in that the diamond particles are selected to have substantially identical size, and the process of epitaxial deposition of coating in a flow of a carbonaceous medium is interrupted by a predetermined moment prior to the formation of a pure graphite layer on the particles.

Example: A diamond-graphite layer was deposited in the following manner. Seed diamond powder with grading 1/0 (grain size less than 1 μm) was placed in a reactor which was evacuated and heated at 950°C. Methane under a pressure of 3 mm Hg was passed over the powder to deposit thereon diamond and diamond-graphite layers. Two graphite electrodes were placed within the diamond powder so as to measure variations in electric resistance of the powder. The value of electric resistance at the beginning of the test was 10 Mohm. During the growing of diamond layer and formation of diamond-graphite layer the value of electric resistance of the powder decreased. The deposition was interrupted when the resistance was 0.06 Mohm. Then the powder was poured from the reactor and weighed. On the basis of the weight increment thus measured and specific surface of the starting powder, which was determined beforehand, the thickness of diamond and diamond-graphite layer was calculated to be about 10 Å in this example.

The diamond powder having a grading 1/0 with epitaxially deposited diamond and diamond-graphite coating of a thickness of 10 Å was placed into a tubular graphite heater having the following dimensions: 7 mm outside diameter, 4 mm inside diameter and 4 mm length, and sealed at both sides with graphite discs. The reaction vessel thus formed was then subjected to the action of pressure of 77 kbar at 1600°C and allowed to stay under these conditions for 60 seconds with subsequent lowering of temperature and pressure.

As a result, a polycrystalline diamond member was obtained which had the shape of a cylinder of 3.5 mm diameter and about 4 mm in length. The polycrystalline member weighed 0.5 carat.

For further use, the resulting polycrystalline diamond member was mechanically fastened in a mandrel and machined on a diamond-electrolytic machine to obtain a blade-type cutting tool.

The tests of tools of polycrystalline diamond members prepared as described above showed that in machining brass of LS69 grade (speed, 400 m/min; longitudinal feed, 0.02 mm per revolution; cutting depth, 0.10 mm) wear resistance at the trailing edge was 0.044 mm after 1,000 min of operation which corresponded to the distance of 500 km. Roughness of machined surfaces was of 8a-8b classes. The resulting polycrystalline diamond member left traces on the surfaces of a diamond single crystal.

Coating with Polycrystalline Metal Whiskers

According to the process of *H.J. Schladitz; U.S. Patent 4,046,302; September 6, 1977* there is provided a method of manufacturing a body of conglomerated particles, comprising coating the particles with a network of polycrystalline metal whiskers with a diameter in the range of 0.05 to 1 μm, heating the coated particles to a temperature at which the whiskers sinter together at their areas of contact, mixing the coated particles with filaments, and metallically bonding the filaments and the coated particles at their areas of contact.

Preferably, the whiskers have a diameter in the range of 0.05 to 0.5 μm, with an optimum average diameter of 0.3 μm. The filaments and the coated particles can be joined together at their areas of contact by sintering, by the deposition of metal by means of thermal decomposition of a metal compound, or by electroless metal deposition.

The polycrystalline metal whiskers consist preferably of Fe, Ni, Mo, W or Cr, or alloys of these metals. For particularly stringent demands it is possible to use dispersion hardened Fe whiskers.

The matrix in which the coated particles are incorporated may be formed from inorganic filaments of coal, glass, quartz, or Al_2O_3, or mineral wool (of asbestos for example). Alternatively, organic filaments may be used, for example, polyacrylonitrile, polyamide or polyimide. Preferably all these filaments have a diameter between 1 and 20 μm.

The method initially produces a porous conglomerate body which can have a pore volume of up to 95%. It is usable in this form for various applications for which porosity is required, inter alia for hydrogen storage, for collecting electrodes and fuel cells, for catalysts and for filters. However, when the need for porosity is less than the requirement for high mechanical strength, in a further step of the process the originally porous body can be compressed to the desired density by hot or cold pressing. Alternatively, it may be impregnated with metals or synthetic resins, for example, epoxy resin or polyimide resin. This technique is suited for the production of grinding discs incorporating diamond or carbide particles, or of self-lubricating bearing materials incorporating solid lubricant particles.

In another particularly preferred technique for the manufacture of grinding tools, the diamond or hard material particles coated with whiskers are sintered or metallized with filaments on a substrate.

The temperature at which the metal whiskers coating the particles are sintered together at their areas of contact is preferably in the range of 400° to 600°C. If the following stage of the process involves the metallic joining of the coated particles with thick whiskers by further sintering, that is carried out in the preferred range of 600° to 1100°C.

Low plus High Melting Metal Powders as Binder

W.R. Pratt and J.H. Hawkes; U.S. Patent 3,984,214; October 5, 1976; assigned to Federal-Mogul Corporation provide a diamond abrasive article preferably in the form of a grinding wheel, the work-contacting portion of which has a resinoid body containing metal coated diamond particles and also containing powdered metal of two kinds, one of the powdered metals having a relatively low melting point and the other having a substantially higher melting point such as tin and copper, respectively.

The low melting point metal powder in the matrix mixture should be one having a melting point below 660°C, and preferably below 250°C. The metal of higher melting point is preferably one having a melting point higher than 950°C. Alloys having melting points within these ranges may also be used. The choice of which metal (or alloy) to use will depend partly on whether the metal, or alloy, will be readily wet by the lower melting point metal or alloy. The metallic coating on the diamond particles should also be wetted by the lower melting point metal or alloy. Consequently, the choice of the low melting point metal powder or alloy depends partly on what metal is used to coat the diamond particles.

A specific example for making an abrasive article according to the process is to prepare a bond mix made up of copper powder, tin powder and phenolic resinoid binding material in the proportions of 8.41 g of copper powder (43.5% by wt), 8.41 g of tin powder (43.5% by wt), and 2.63 g of phenolic resinoid bonding material (13.5% by wt). To this bond mix, having a weight of 19.45 g, there was added 8.12 g of copper coated diamond powder. The diamonds were dibursed approximately uniformly through the bond mix. The mix with the diamonds therein was then molded around the rim of a previously partially cured core or disc of an aluminum-bearing phenolic molding compound. The molding was accomplished substantially as described in U.S. Patent 2,073,590, using a molding temperature of approximately 177°C, for 30 minutes at a pressure of about 2 tons per square inch. The work was then stripped from the mold, and further cured by heating it to about 177°C in an oven for 4 hours to complete the cure of the resin binder. The finished wheel was a 6-inch diameter wheel, one-sixteenth of an inch thick.

It was observed that the cutting operation was exceptionally good and the work being ground remained cooler than when grinding with competitive wheels which contained only high melting point metallic fillers.

Each Grain in Metallic Setting

O.B. Lindstrom; U.S. Patent 3,955,324; May 11, 1976 provides a grinding tool having a plurality of distinct abrasive particles embedded in a synthetic resin. Each of the abrasive particles has a plurality of abrading grains which are embedded in a metal phase with a metallic layer separating and binding adjacent surfaces of the grains. Each individual grain is thus kept in a metallic setting which is structurally reinforced by the other grains present in the particle with the metal phase between adjacent grain surfaces producing a certain elasticity.

The ratio of the volume of the metal to the volume of the abrading grains in the abrasive particles falls within the range of 1:4 to 20:1. The ratio of the volume of the abrasive particles to the volume of the resin falls within the range of 1:3 to 20:1. The abrading grains are of a size as measured in the longest tip-to-tip direction that falls within the range of 0.37 to 0.350 mm. The abrasive particles are of a size as measured in the longest tip-to-tip direction that falls within the range of 0.2 to 1 mm.

The individual abrasive particles can be produced by mechanically breaking (e.g., crushing and/or grinding) larger bodies that have been produced by any of several methods used in the preparation of conventional metal bonded discs.

One example of a method for the preparation of such bodies is the following die-casting process. The mold is advantageously made of an aluminum alloy containing copper and magnesium as the primary alloying elements together with smaller amounts of iron, nickel and titanium.

The abrading grains are initially mixed with a metallic filler. The metallic filler may comprise, for example, a mixture of chips of the same composition as the mold. These filler chips are then mixed with powdered antimony with the same volume ratio between the chips and the antimony being preferably 2:1. The metallic filler is screened to a size corresponding to the grain size of the diamonds so that the diamonds can be distributed uniformly throughout the entire mixture.

The die-casting alloy (or "carrier") preferably comprises about 65% of a conventional die-casting zinc alloy which contains copper and aluminum as the principal alloying elements and includes about 32% cadmium and 3% commercial silver. The primary requirement for the die-casting alloy is a melting point which is lower than the melting point of the metallic filler. It will, therefore, be apparent that other die-casting alloys which meet the required melting point conditions can be used.

The mixture of diamonds (and optionally other abrading grains) and the metallic filler material is then introduced into the mold. This mixture, which has been prepared by tumbling, can be uniformly dispersed into the mold by rotating the mold. After this rather loosely packed mixture has been evenly distributed within the mold, the mold is introduced into a die-casting machine wherein the molten die-casting metal is added so as to fill in the pores between the diamonds and the filler material. The aggregate is then cooled and the mold removed.

The bodies formed in this manner are then crushed so as to produce abrasive particles of the desired shape and size for casting into a grinding tool.

Bonding at Temperatures and Pressures Outside Diamond-Stable Region

H.T. Hall; U.S. Patent 3,913,280; October 21, 1975; assigned to Megadiamond Corporation provides a polycrystalline diamond composite having diamond-to-diamond bonds formed between adjacent diamond particles. The composite is produced by mixing diamond particles with a sintering aid and creating direct diamond-to-diamond bonds under temperature and pressure conditions at which diamond is normally unstable. Historically, it has been essential to form diamond compacts within a diamond-stable region, as shown in the figure below.

Figure 3.7: Bonding at Temperatures and Pressures Outside Diamond-Stable Region

a.

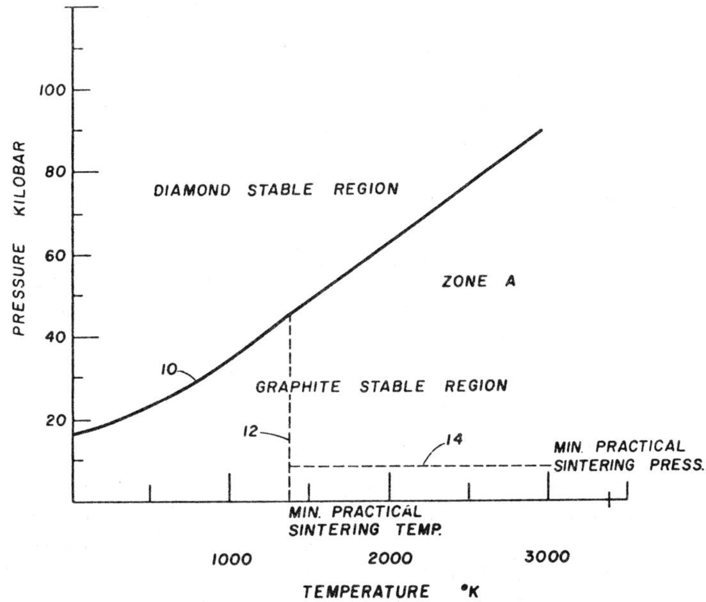

b.

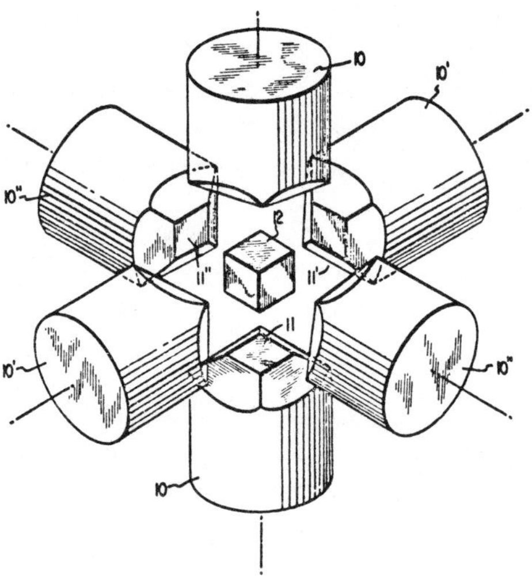

(continued)

Figure 3.7: (continued)

c.

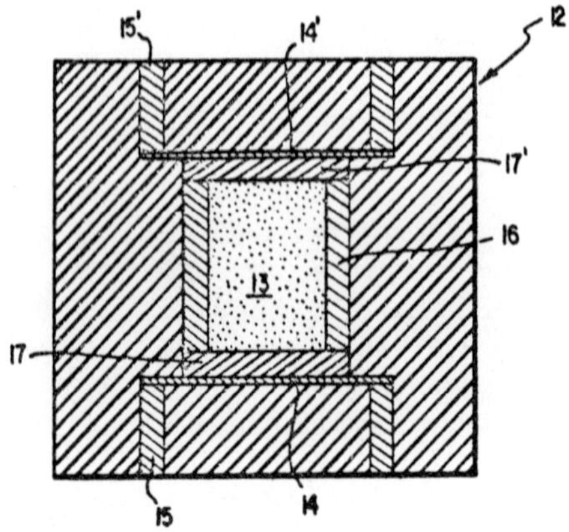

Source: U.S. Patent 3,913,280

Figure 3.7a is a graphical representation of the pressure-temperature zones in which diamond is stable and in which diamond normally reverts to graphite. Line **10** is the thermodynamic equilibrium line, calculated and experimental, for the diamond-graphite system. Above this line carbon is stable as diamond and below the line graphite is the stable form. The minimum stabilization pressure at any given temperature may vary somewhat from that given by the graph since it depends to a certain extent on the size, purity, and surface characteristics of the diamond particles present in the powder and also on the nature of the sintering aid.

Line **12** represents the approximate minimum temperature at which sintering takes place. It should be pointed out that the position of this line represents the temperature at which sintering takes place at a sufficiently rapid rate to make it economical. While line **12** may represent a lower temperature, sintering may take place so slowly as to be impractical. Time for satisfactory sintering is reduced as temperature increases. Accordingly, line **12** could be situated at a higher temperature which would have the effect of decreasing the sintering time.

Line **14** represents the approximate minimum pressure region in which diamond sintering will take place. The actual minimum pressure required for sintering varies somewhat with the size of the article. The area identified as zone **A**, defined by lines **10, 12** and **14** represents suitable sintering conditions for forming a diamond compact by sintering diamond powder according to this process.

The diamond particles are combined with sintering aids selected from the list of metals consisting of aluminum, beryllium, hafnium, magnesium, molybdenum, niobium, rhenium, titanium, tungsten, and zirconium. Other suitable sintering aids have been found to be boron, nitrogen and silicon. Included in this list of sintering aids are the refractory borides, carbides, nitrides, oxides, and silicides.

The proportion of sintering aid relative to the amount of diamond is not particularly critical. However, better bonding is effected when the amount of sintering aid present in the final compact does not exceed 50% by volume.

Reference is now made to Figures 3.7b and 3.7c directed toward the preferred embodiment of the process. Sintering takes place within a sample cube **12**. The sample cube is preferably made of pyrophyllite and is illustrated in longitudinal cross section in Figure 3.7c. The sample cube is designed for use in a cubic press, which consists of six tungsten carbide anvils, in matching pairs **10, 10'** and **10''**, with equal area square faces **11** and 45° sloping shoulders.

The anvils, electrically insulated from each other, are aligned for rectilinear movement along three mutually perpendicular coordinate axes and synchronized in their motion by an anvil guide mechanism. Each anvil is attached to and powered by support apparatus (not shown) including a double-acting hydraulic ram affixed to a base; the bases are fastened together by an arrangement of twelve tierods forming the outline of a regular hexahedron. The thrust of the six rams simultaneously moves the synchronized anvils **10** toward the symmetry center of the press about a cube-shaped pyrophyllite cell **12** having square faces approximately 60% greater in area than the anvil faces and parallel to the corresponding anvil faces. Further advance of the anvils extrudes and compresses pyrophyllite between the 45° sloping shoulders of the anvils generating pressure within the cell. This pressure is transmitted to the sample **13** undergoing sintering in a sample container **16** with end caps **17** usually also serving as an electrical resistance heating element and in electrical contact with the anvil.

Metallic Binder as Separate Entity

L.F. Vereschagin, A.A. Semerchian, V.P. Modenov, T.T. Bocharova and M.E. Dmitriev; U.S. Patent 3,912,500; October 14, 1975 provide a process for producing diamond-metallic materials by impregnating diamond powder with a metallic binder under the action of pressure and heat, wherein use is made of a system composed of diamond powder-metallic binder and the latter being disposed as a separate entity in any portion of the diamond powder but in contact therewith. This is subjected to a pressure of at least 10 kbar for closely packing the diamond powder, whereafter the system with the packed diamond powder is subjected to the simultaneous action of a pressure of at least 10 kbar and heating at a temperature sufficient for melting the metallic binder. The metallic binder melts and fills, under pressure, the diamond powder.

The metallic binder may be disposed in the system by various methods, e.g., under a layer of diamond powder, above a layer of diamond powder, or along the side face of a diamond briquette, but by all means contacting the binder by any portion of the diamond briquette.

Under the term "metallic binder" as used herein is intended any metal or alloy having a boundary angle of wetting of less than 90° in respect to diamond. Such metals may be nickel, cobalt, iron, manganese, chromium, and the like. As the metallic binder use may be made of alloys of metals wetting diamond, as well as alloys of metals wetting diamond with metals nonwetting diamond or with nonmetals.

The binder volume should constitute 10 to 30% of the volume occupied by compressed diamond powder. If a binder is employed in an amount exceeding these limits, it is permitted that a layer of another powder with solid particles be disposed on the opposite side of a diamond-powder briquette such as boron carbide. This layer will take up excess binder after the melting thereof thus impregnating the diamond-powder briquette.

The process makes it possible, depending on the grain size of the diamond powder and binder employed, to produce diamond metallic materials possessing extremely high mechanical properties: hardness of 95 to 96 units HRA (for diamond grains of 3 to 5 μ size; HRA-Rockwell hardness, scale A), compression strength of 450 to 550 kg/mm^2, and elastic modulus of 55,000 kg/mm^2 (for diamond grains of 10 to 15 μ size). When a cone made of a diamond-metallic material (with diamond grain size 3 to 5 μ) is pressed in a tungsten-carbide alloy of the composition 94% WC and 6% Co, the diamond-metallic material withstands contact pressure of up to 10,000 kg/mm^2. In addition to high mechanical characteristics, the diamond-metallic materials produced in accordance with the process possess satisfactory viscosity.

An important feature of the process for producing diamond-metallic materials resides in a possibility of manufacturing big articles up to 10 cc in volume or even more. An article volume is defined by the dimensions of a reaction volume of a pressure chamber and, since pressures required for performing the process are relatively small, the dimensions of a reaction volume may be easily increased over 10 cc.

Graphite-Free Diamond Particles

According to the process of *R.J. Caveney; U.S. Patent 3,879,901; April 29, 1975; assigned to De Beers Industrial Diamond Division Limited, South Africa*, there is provided a compact comprising substantially graphite-free diamond particles having a continuous coating of titanium or molybdenum held in a matrix compatible with the titanium or molybdenum coating. The term "compatible" means that the matrix must be such as to allow diffusion of the titanium or molybdenum into it.

The matrix may be an alloy selected from the group of Fe/Ni, Ni/Co/Cr/Fe, Fe/Si, Ti/Si, Ti/Ni and Ti/Fe alloys, but is preferably one of the following alloys: Ni/Co/Cr/Fe 34/18/14/5, Fe/Ni 12/5 or Fe/Si 63/2. All of the abovementioned ratios are weight for weight.

The matrix may also advantageously be tungsten carbide bonded with a transition metal of Group VIII, preferably cobalt. The transition metal is preferably present in the amount of about 10% by wt of the tungsten carbide.

The thickness of the titanium or molybdenum coat would normally be of the order of 1,000 to 2,000 Å, but coats of greater thickness can also be used.

Diamond compacts having a variety of matrices were prepared in the following manner. In all cases, diamond grit having a continuous titanium or molybdenum coating and the metal powders necessary to make the desired alloy were weighed into a plastic bottle. The metal powders and grit were then mixed by placing the bottle in a milling machine for about 20 minutes. The mixture was placed in a graphite mold and loaded into a standard synthesis capsule. If the synthesis volume was not completely filled by the mold, compacted graphite was added to the capsule to fill the balance of the volume. The mixture was then subjected to compaction for about 5 minutes at a pressure of about 60 kbar and a temperature between 1200° and 1400°C. These temperatures and pressures are in the diamond-stable zone. Using the abovementioned method a number of diamond compacts were manufactured. The diamond content of the compacts was 65% volume per volume.

Tests indicated that the compacts of the process have superior abrasion resistance than a standard compact. Of the results, the most significant were those obtained for the compacts having Fe/Ni, WC bonded with 10% Co, Ni/Co/Cr/Fe, and Fe/Si matrices.

Electroless Plating

T.P. Christini, A.L. Eustice and A.H. Graham; U.S. Patent Reissue 29,285; June 28, 1977; assigned to E.I. Du Pont de Nemours and Company have discovered a method of concurrently depositing, by the electroless-plating technique, as a disperse phase, particulate diamond in composite with Ni(B), Ni(P), Co(B), Co(P) and other metals and metal alloys singly, or as mixtures of any two or more of these substances together, as well as metallic copper as the continuous phase or matrix. The coatings produced are highly adherent, relatively nonporous and possessed of a truly remarkable abrasive wear resistance. In addition, the diamond particle pull-out characteristics are very good, so that objectionable detritus is not carried over into the surrounding environment which could possibly act as an abrasive agent to gall or otherwise damage the fine finish of bearings or other metal-to-metal contact surfaces.

The combination of properties displayed by the diamond composites of the process are such that they have great potential value as abrasion-resistant surfaces for both dry and wet service, writing instrument nibs, cutting tool surfaces, piston ring and other sliding contacting surfaces, textile wear surfaces and other extremely demanding uses.

The electroless alloy-diamond composite coatings of the process can be deposited by a wide variety of plating techniques ranging from simple rack plating, wherein articles are supported by a rack, to barrel plating, wherein articles are introduced as free bodies into a rotating bath container, which can have its axis horizontally disposed or somewhat inclined. In addition, articles can be tumble plated.

The diamond powder to be composited (0.5 to 100 g as desired) is first blended with about 200 to 400 ml of the plating bath in a high-speed mixer to break up agglomerates, wet all of the particles and form a concentrated slurry containing a uniform dispersion of particles. The slurry is then slowly added to the plating vessel, where the powder particles are kept in suspension by mechanical agitation and/or bath circulation. The quantity of diamonds maintained in the suspension, while most commonly in the range of 1 to 10 g/ℓ of bath, can range up to as much as 40 g/ℓ, the upper limit being only that the bath must remain sufficiently fluid to be capable of ready agitation and circulation.

CUBIC BORON NITRIDE

Direct Conversion of PBN to CBN

F.R. Corrigan; U.S. Patent 4,188,194; February 12, 1980; assigned to General Electric Company provides a process for making a sintered polycrystalline cubic boron nitride (CBN) compact by direct conversion of a pyrolytic, preferably substrate nucleated, boron nitride (PBN). Direct conversion takes place by subjecting the PBN to temperatures and pressures within the CBN stable region. The crystallites of the compact produced by this process are preferentially oriented. Also, such compacts have a high thermal conductivity between about 2 and 9 W/cm-°K and thus are useful as a heat sink for electronic devices. The size of the CBN crystallites of the compact may be between about 1×10^3 to 1×10^5 Å.

The direct transformation of PBN is carried out at pressures in excess of 60 kbar and preferably about 65 kbar, and at temperatures of 1800° to 2000°C and higher within the CBN stable region. High pressure/high temperature conditions are maintained for times sufficiently long to allow for transition of the PBN to the cubic phase and development of strong intercrystalline bonding. Substrate nucleated PBN is the preferred starting material with which superior results are achieved.

One preferred embodiment of a conventional belt apparatus is shown in Figure 3.8a, with the reaction cell **11** disposed therein and is fully described in U.S. Patent 2,941,248.

In Figure 3.8a, apparatus **17**, suitable for use in this process, includes a pair of cemented tungsten carbide punches **19** and **21** and an intermediate belt or die member **22** of the same material. Die member **22** includes an aperture **25** in which the reaction cell **11** is positioned. Between punch **19** and die **22** and between punch **21** and die **22** there are gasket/insulating assemblies **27, 29**, each comprising a pair of thermally insulating and electrically nonconducting pyrophyllite members **30** and **31** and an intermediate metallic gasket **32**.

Figure 3.8: Direct Conversion of PBN to CBN

a.

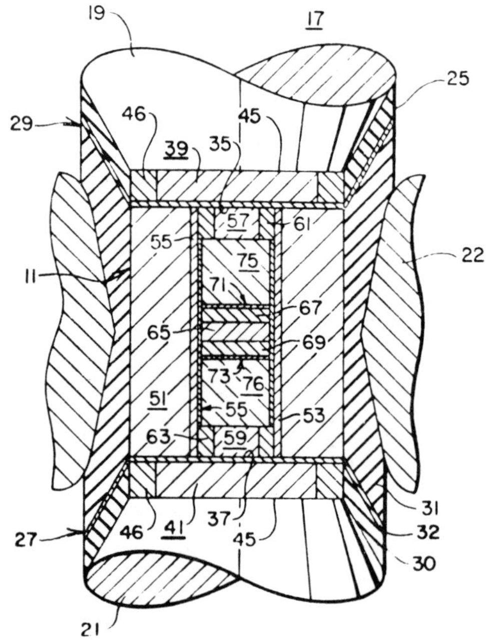

b.

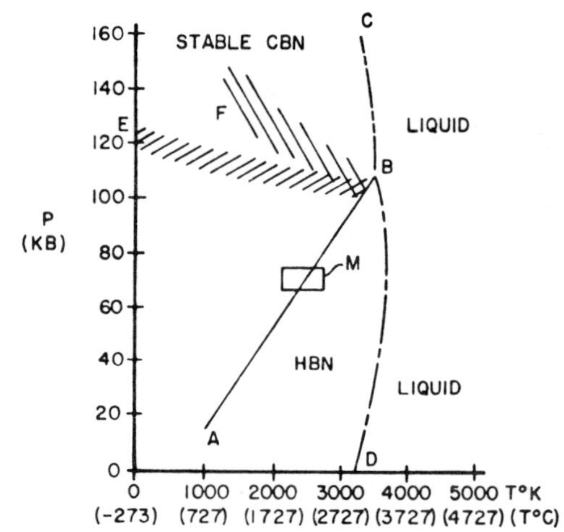

Bundy-Wentorf BN Phase Diagram

(continued)

Figure 3.8: (continued)

c.

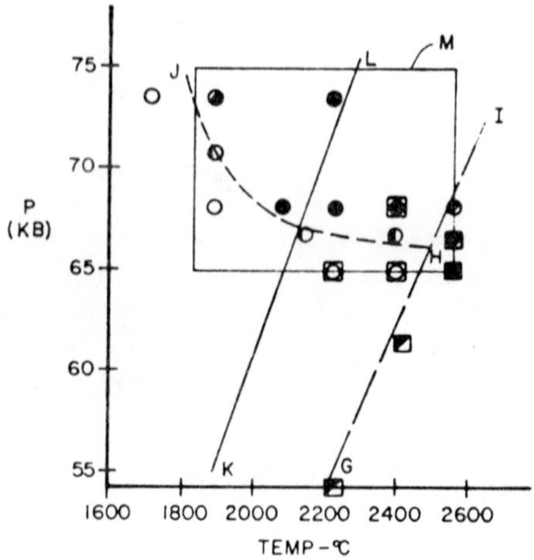

○ Conversion runs - shading = degree of conversion
□ Reconversion runs - shading = degree of reconversion

Source: U.S. Patent 4,188,194

The reaction cell includes a lava cylindrical bushing **51**. The bushing may alternatively be made of glass, soft ceramic, talc, steatite or forms of pyrophyllite or soapstones. Positioned concentrically within and adjacent to the bushing is a graphite resistance heater tube **53**. Within the tube there is in turn concentrically positioned a contamination shield tube **55** of tantalum. Disposed outwardly of the shield tube and closing each end of the heater tube are heater tube plugs **57** and **59** of hot pressed boron nitride or other high temperature insulating material. The plugs are surrounded by electrically conductive carbon sleeves **61, 63**, respectively.

Electrically conductive metal end discs **35, 37** are utilized at each end of the graphite heater tube to provide electrical connection thereto. Adjacent each disc is an end-cap assembly **39, 41** each of which comprises a pyrophyllite plug or disc **45** surrounded by an electrically conducting steel ring **46**.

A disc **65** of PBN starting material is centrally disposed within the shield tube. A pair of spacer discs **67, 69** of carbon are disposed one on each side of disc **65**. A pair of contamination shield discs **71, 73** of tantalum are disposed outwardly of the spacer discs. Outwardly of shield discs **71, 73** and closing the respective ends of shield tube **55** are a pair of electrically insulating discs **75, 76** of hot pressed BN.

The shield tube and shield discs **71, 73** serve to prevent diffusion of or act as a getter for contaminants, which have been found to interfere with the conversion and sintering process, into the sample during conversion under high pressure/high temperature conditions.

Rectangle **M** shown in both Figure 3.8b and Figure 3.8c shows the general relationship of the preferred operating region shown in Figure 3.8c relative to the complete Bundy-Wentorf phase diagram (Figure 3.8b).

Figure 3.8c shows the results of a series of direct conversion and reconversion experiments in the lower pressure region. The direct conversion experiments were run on PBN samples in Figure 3.8a type cells for 10-minute heating times. The reconversion experiments were run on CBN cluster compacts previously prepared by direct conversion of PBN to CBN, also in a Figure 3.8a type cell for 10-minute heating times. In Figure 3.8c conversion of PBN to CBN was obtained within the region **JHI** and reconversion of CBN to HBN was obtained at temperatures above the line **GHI**. Also shown is a section **KL** of the CBN/HBN equilibrium line **AB** from the Bundy-Wentorf diagram. The results indicate that the CBN stability region extends beyond the Bundy-Wentorf equilibrium line **KL**. Although partial conversion of PBN to CBN was obtained at temperatures of 1850° to 1900°C, in practice temperatures above 2000°C were found necessary to obtain CBN compacts suitable for machining applications.

In selecting a PBN starting material for practice of this process, it has been discovered that substrate nucleated PBN should be used in order for the transformation to proceed and for large, strong, well-sintered masses to be consistently produced. If continuously renucleated PBN is used as the starting material, conversion is inhibited.

Binder of Carbide of a Group IV or V Metal

S.S. Ordanian, V.S. Lysanov, V.V. Digonsky, V.A. Blinkov, L.A. Pekelis, V.I. Denisova, V.I. Unrod, S.V. Sinelnikov and M.P. Chestnokova; U.S. Patent 4,186,022; January 29, 1980; assigned to Vsesojuzny Nauchno-Issledovatelsky Institut Abrazivov i Schlifovania and Leningradsky Politekhnichesky Institut Imeni Lensoveta, U.S.S.R. provide a superhard composite material comprising cubic boron nitride and a binder consisting of carbide and diboride of a metal selected from Group IV or V of the Periodic Table, taken in eutectic ratio, the components being contained therein in the following amounts, in percent by weight: cubic boron nitride 28 to 80, preferably 50 to 70; carbide and diboride of a metal selected from Group IV or V taken in eutectic ratio, 72 to 20, preferably 50 to 30.

The superhard composite material, according to the process, should preferably contain carbide and diboride of titanium or vanadium. Such binder permits the pressure and temperature at which the caking process is conducted to be lowered while preserving high mechanical properties and electrical conductivity of the material.

The superhard composite material according to the process is produced by mixing the components. The resultant mass is caked under a pressure of 35 to 55 kbar and at a temperature of 1400° to 1900°C. This temperature is achieved by passing electrical current through a carbon-graphite heater, or directly through the mixture.

From Hexagonal Boron Nitride Using Zinc Catalyst

V.I. Farafontov, M.I. Sokhor, N.G. Kushkova, V.V. Digonsky and V.S. Lysanov; U.S. Patent 4,148,863; April 10, 1979; assigned to Vsesojuzny Nauchni-Issledovatelsky Institut Abrazivov I Shlifovania, U.S.S.R. describe a method of preparing polycrystalline cubic boron nitride. In the method hexagonal boron nitride is subjected to the action of pressure 40 to 90 kbar and temperature of 1200° to 2400°C in the presence of a catalyst which is a zinc compound.

The proposed catalyst diffuses easily into the bulk of hexagonal boron nitride at high temperatures and pressures, ensuring a complete transformation of the boron nitride into a cubic modification. The catalyst, remaining in the bulk of polycrystalline cubic boron nitride, affects to a lesser degree the weakening of polycrystals than the catalysts used heretofore. This ensures high wear resistance of polycrystalline cubic boron nitride employed in cutting tools. It is expedient to use a zinc compound zinc oxide, hydroxide, nitride or amide, or mixtures thereof in amounts of 0.1 to 12 wt %.

This method makes it possible to increase considerably the wear resistance of polycrystalline cubic boron nitride used in tools for working hardened steels and alloys.

Example 1: Powders of hexagonal boron nitride (98 wt %) and zinc oxide (2 wt %) are mixed. The mixture obtained is pressed into briquettes and placed into a high-pressure chamber where it is subjected to the action of 65 kbar and 1600°C. The final product is polycrystalline cubic boron nitride with wear resistance of 0.12 mm.

Example 2: A powder of hexagonal boron nitride is introduced into a catalyst which is an aqueous solution of zinc chloride. The obtained mixture is treated with an aqueous solution of ammonia, then evaporated and roasted at 420°C. The prepared mixture containing 98 wt % of hexagonal boron nitride and 2 wt % of zinc oxide is treated by following the procedure described in Example 1. The final product is polycrystalline cubic boron nitride with wear resistance of 0.08 mm.

Sintering with Boride of Ti, Zr or Hf

A sintered compact of boron nitride with high density form having high oxidation resistance, water resistance, compressive strength and hardness can be obtained by sintering wurtzite-structured boron nitride together with a boride of titanium, zirconium or hafnium, or with mixtures thereof, under a condition of high temperature and high pressure according to *S. Saito, A. Sawaoka and*

M. Araki; U.S. Patent 4,132,554; January 2, 1979; assigned to Nippon Oil and Fats Co., Ltd., Japan.

In the first aspect of the process, a mixture of wurtzite-structured boron nitride and at least one of titanium boride, zirconium boride and hafnium boride is sintered for 1 to 60 minutes under a temperature-pressure condition within a parallelogram formed by connecting points **A** (1400°C, 80 kbar), **B** (1300°C, 55 kbar), **C** (1650°C, 55 kbar) and **D** (1750°C, 80 kbar) in the temperature-pressure diagram shown in Figure 3.9a.

Figure 3.9: Temperature-Pressure Diagrams Showing Regions Used in Sintering

a.

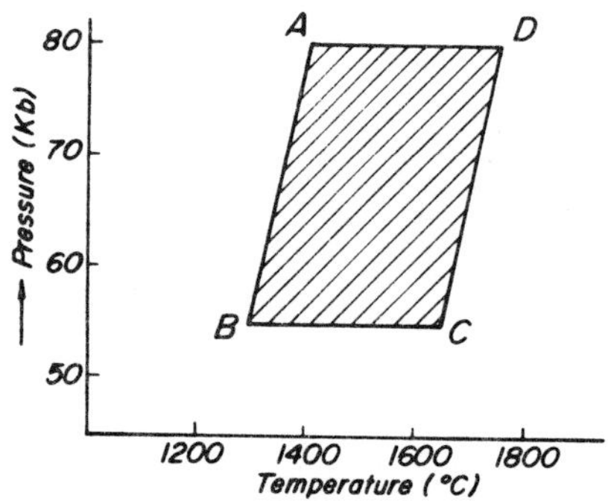

b.

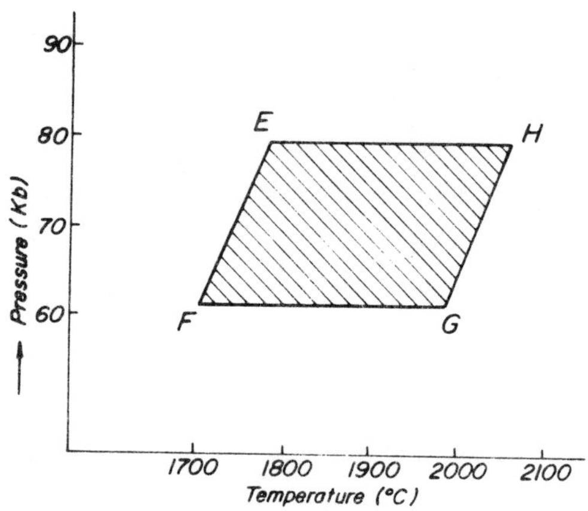

Source: U.S. Patent 4,132,554

Under the abovementioned sintering condition (shown in Figure 3.9a), wurtzite-structured boron nitride is sintered without losing wurtzite structure, and a sintered compact having high hardness, oxidation resistance and compressive strength can be obtained. The sintered compact, due to its high compressive strength, is particularly suitable for cutting metals having a rough surface. However, when metals having a higher hardness are cut, it is preferable to use a sintered compact of boron nitride having a higher hardness.

When a mixture of wurtzite-structured boron nitride and the abovedescribed boride is sintered at a temperature higher than the temperature of the above-described temperature-pressure condition, wurtzite-structured boron nitride is partly converted into zinc blende-structured boron nitride, and a sintered compact having a remarkably high hardness, which consists mainly of a mixture of wurtzite-structured boron nitride and zinc blende-structured boron nitride in a wide mixing ratio can be obtained. This is the second aspect of the process. That is, in the second aspect, a mixture of wurtzite-structured boron nitride and at least one of titanium boride, zirconium boride and hafnium boride is sintered for 1 to 60 minutes under a temperature-pressure condition within a parallelogram formed by connecting points **E** (1770°C, 80 kbar), **F** (1700°C, 62 kbar), **G** (1980°C, 62 kbar) and **H** (2050°C, 80 kbar) in the temperature-pressure diagram shown in Figure 3.9b.

The wurtzite-structured boron nitride to be used as a starting material in the process may be any wurtzite-structured boron nitrides obtained by converting low density boron nitride or amorphous boron nitride into the wurtzite-structured boron nitride by the shock compression method or under a static super-high pressure.

Silicon Carbide Substrate

The process of *M. Lee and L.E. Szala; U.S. Patent 4,110,084; August 29, 1978; assigned to General Electric Company* comprises placing in a metal container a layer of aluminum or aluminum alloy or metal for providing the aluminum alloy, a layer of a mass of cubic boron nitride (CBN) crystals, the amount of the aluminum or aluminum alloy ranging from 25 to 60% by volume of the mass of CBN crystals, and a silicon carbide substrate with one face or side of the substrate facing all of the layers and preferably in direct contact with the layer of CBN crystals. Preferably the container is vibrated to densify the CBN mass to contain CBN crystals in excess of 70% by volume of the CBN mass.

The container and its contents are placed within a pressure transmitting powder medium. Pressure is applied at room temperature to the powder medium ranging from 20,000 to 200,000 psi, thereby subjecting the container and its contents to substantially isostatic pressure which stabilizes the dimensions of the container and its contents producing an isostatic shaped system formed of powder-enveloped container. The isostatic system is evacuated, a nitrogen gas atmosphere introduced into the evacuated system, then simultaneously pressure and heat are applied to the resulting isostatic system, the pressure ranging from 500 to

100,000 psi and the heat ranging in temperature from the critical wetting temperature of the aluminum or aluminum alloy up to about 1450°C. As a consequence, the aluminum or aluminum alloy in liquid form infiltrates through the voids in the mass of cubic boron nitride crystals and penetrates into the contacting face of the silicon carbide substrate diffusing around the grains and through the pores therein to a depth of at least about twice the thickness of the layer of cubic boron nitride crystals. The resulting system is cooled under sufficient substantially isostatic pressure to maintain the dimensions of the container and contents producing an adherently bonded integral abrasive composite.

Transition Metal-Aluminum Alloy Matrix

The method of *R.C. De Vries and J.F. Fleischer; U.S. Patent 3,918,931; November 11, 1975; assigned to General Electric Company* comprises the following steps.

> Mixing HBN powder with a powdered metallic phase to produce a homogeneous mixture, the atomic content of the metallic phase consisting essentially of aluminum atoms, atoms of a metal selected from the group consisting of chromium and manganese, and atoms of at least one metal selected from the group consisting of iron, cobalt and nickel, the weight percent of HBN in the mixture being in the range of 10 to 50
>
> Pressing the mixture into some predetermined shape
>
> Simultaneously subjecting the mixture to an operating temperature and operating pressure within the stability region of CBN defined by the use of the selected metallic phase, the operating temperature being high enough to render molten all of the metallic phase, the period of time of simultaneous temperature and pressure application being sufficient to permit dissolution of all of the HBN in the molten metallic phase and the precipitation of CBN crystals therefrom
>
> Quenching the resulting CBN/alloy system to about room temperature while maintaining the operating pressure
>
> Reducing the pressure to atmospheric pressure
>
> Recovering the preshaped abrasive body consisting essentially of CBN crystals distributed in a transition metal-aluminum alloy matrix

The minimum pressure for the conversion of HBN to CBN has been found to be about 45 to 47 kbar regardless of the specific metallic phase (unalloyed mix or preformed alloy) employed, but the minimum temperature varies. Thus, once a metallic phase formulation has been selected it is preferable to determine the pressure-temperature stability region for CBN for that given formulation. Such a phase diagram is routinely determinable for a given metallic phase formulation by one skilled in the high temperature-high pressure art.

Semi-Isostatic System

In the process of *M. Lee; U.S. Patent 3,982,911; September 28, 1976; assigned to General Electric Company* cubic boron nitride crystals of graded particle size are chemically cleaned and disposed in a protective metal cup together with a mass of substrate source material and a concentration of metal for providing an alloy for infusing and wetting the walls of capillary-size voids. The substrate source material is selected from the group consisting of carbide powder containing metal bonding agent, metal bonded carbide, and substrate metal (the latter having a linear thermal expansion coefficient of less than 10×10^{-6} inch per inch degree centigrade and having a melting point at the operating pressure greater than the operating temperature and being selected from the group consisting of molybdenum, niobium, platinum, chromium, vanadium, tungsten, tantalum, titanium, silicon, rhodium, ruthenium, zirconium and alloys thereof). The alloy-providing metal may be either a preformed alloy or a plurality of alloy-forming metal components. The CBN content is increased in density (as by vibration of the metal cup and its contents under light pressure application) to greater than 70% by volume, after which the metal cup is placed within a finely-divided particulate semi-isostatic pressure transmitting system, e.g., very fine hexagonal boron nitride powder.

Pressure in the preferred range of from 20,000 to 100,000 psi is applied semi-isostatically to the system to dimensionally stabilize the metal cup and its contents. The semi-isostatic system containing the metal cup and its contents is then transferred from the pressure mold to a heating mold. Heat, up to 1500°C, and pressure, e.g., about 10,000 psi, are simultaneously applied to the semi-isostatic system in order to exceed the critical wetting temperature of the infusing and wetting alloy (performed or made in situ). The molten alloy enters capillary-size interstices between the CBN crystals to bond them together. The CBN crystals also become directly bonded to the substrate, e.g., a sintered carbide mass, some of which material is forced in between CBN crystals at the substrate/CBN interface. The semi-isostatic system is cooled and the strong composite CBN/substrate body is recovered therefrom.

High Melting Additives plus Calcium and/or Magnesium Borides

J.M. Kovalchuk, V.S. Lysanov, L.I. Feldgun, M.A. Varzanov and G.O. Gomon; U.S. Patent 3,894,850; July 15, 1975 describe a superhard composite based on borazon (CBN) with high melting metallic additives comprising borazon to the extent of from 85 to 97% by wt in the form of a dendritic structure of interpenetrating crystalline grains measuring 10^{-3} to 10^{-4} cm, as well as magnesium and/or calcium borides and oxides to the extent of from 2 to 5% by wt and high melting additives to the extent of from 1 to 10% by wt.

Silicon carbide, diamond, corundum, boron carbide and borosilicocarbide, taken separately or in combinations, may be used as the additives. The choice of the additives is determined by the particular application of the superhard material. Thus, for example, for working hard-to-machine nonferrous metals and their alloys, a preferred additive would be diamond and/or corundum as well as silicon carbide and/or borosilicocarbide. For hard-to-machine nickel-base alloys, diamond and/or boron carbide may be recommended as the additive.

The superhard composite may be produced by subjecting a mixture composed of hexagonal boron nitride, high melting hard compounds and calcium and/or magnesium borides, to a pressure of 85 to 95 kbar and a temperature in the range of 2000° to 2200°C.

Example: A homogeneous mixture is prepared to comprise 85 wt % hexagonal boron nitride, 10 wt % diamond powder of grain size 5 to 10 μm, and 5 wt % magnesium diboride. The mixture is briquetted, charged to the reaction space of a lithographic stone container disposed interiorly of a high-pressure chamber, and exposed for 2 minutes to a pressure of 85 kbar and temperature of 2000°C.

The resultant superhard composite has a dendritic structure of interpenetrating crystalline grains of borazon measuring 10^{-3} to 10^{-4} cm and amounting to 84% by wt of the composite. The diamond particles amounting to 10 wt % and the particles of magnesium borides and oxides amounting in total to 5 wt % serve to reinforce the structure of the superhard composite produced, hence ensuring high strength. With the foregoing proportions of the components, the superhard composite has a compressive strength of about 300 kg/mm^2 and heat conductivity of about 1 W/cm-°K.

Use of Shock-Wave Treated Materials

V.P. Alexeevsky, A.V. Bochko, S.S. Dzhamarov, D.M. Karpinos, G.G. Karjuk, I.P. Kolomiets, A.V. Kurdjumov, M.S. Pivovarov, I.N. Frantsevich and V.V. Yarosh; U.S. Patent 3,876,751; April 8, 1975 provide a method of producing polycrystalline boron nitride by compressing a mixture having as one of the components highly defective wurtzite boron nitride obtained by a shock-wave method, wherein the mixture contains, in combination with the wurtzitelike boron nitride, less than 50 wt % of graphitelike boron nitride subjected to a shock-wave treatment. The mixture is compressed and heated at temperature and pressure values corresponding to the region of the phase diagram for boron nitride which is defined by the graphitelike boron nitride compact modifications of boron nitride equilibrium line and the cubic boron nitride wurtzitelike boron nitride equilibrium line.

The mixture of the starting components is preferably subjected to a pressure of about 50 to 300 kbar and temperatures of about 20° to 3000°C. The crystals of boron nitride obtained by the method according to the process are compact (void ratio of 2 to 5%), they have a pycnometric density of 3.25 to 3.38 g/cc, while their microscopic hardness is within the range of 4,000 to 6,000 kg/mm^2 as compared to 7,000 to 8,000 kg/mm^2 for polycrystalline cubic boron nitride. The resulting material has greater plasticity as compared to that of polycrystalline cubic boron nitride. Tools made of this compact polycrystalline material have a longer service life in machining hardened steel and chilled iron under impact loads.

Furthermore, the mixture is preferably protected against direct contact with a pressure transmitting medium. This allows for obtainment of crystals of boron nitride having the wurtzite structure due to the fact that eventual impurities which could nucleate the seeding crystals of cubic boron nitride are prevented from penetrating into the mixture.

Example: A starting mixture containing 10 wt % of graphitelike boron nitride and 90 wt % of wurtzitelike boron nitride obtained by a shock-wave treatment of the starting graphitelike boron nitride with a shock-wave front pressure of about 120 kbar was poured into a graphite heater, compressed under a pressure of 80 kbar and heated at 1600° to 1800°C for 1 to 2 minutes to obtain strong compact black cakes with a void ratio of 2 to 3%, exhibiting both wurtzite and cubic boron nitride structures with a microscopic hardness of 5,000 to 6,000 kg/mm^2.

Cutting tools made of such polycrystalline boron nitride had a life of 60 to 80 minutes until the wear of the trailing edge Δh of 0.4 mm when turning hardened steel (HRc 60-65) at a cutting speed of 80 to 100 m/min, with a longitudinal feed s of 0.07 mm/rev and a cutting depth t of 0.2 mm; the surface finish was of the 8-9 class.

Carbide-tipped tools could not work under such cutting conditions (a life of 0.5 to 1 minute).

Alloy Matrix of Limited Reactivity with CBN

E.L. Fontanella; U.S. Patent 3,868,234; February 25, 1975; assigned to General Electric Company has found that CBN crystals can be bonded together using an alloy as the metallic bonding medium. In the preparation the selected alloy may be generated in situ or preformed alloys may be used. The alloy may be a two, three, four (or greater number) component system.

Pressures during preparation may range from subatmospheric to pressures greater than atmospheric but less than about 1 kbar. The cubic boron nitride content may extend to less than 70% by volume of the total amount of metal and CBN crystals.

It has been found that by employing slow heating and cooling at low pressures, tenacious alloy coatings may be applied to CBN crystals serving as a matrix to hold these crystals together as an abrasive body. Nickel aluminum alloys and cobalt aluminum alloys have been found to be particularly effective.

In addition to alloys of these metals, alloys may be used wherein a first metal is selected from the group consisting of aluminum, silicon, chromium, tungsten, vanadium, niobium, tantalum, titanium, zirconium, lanthanum and other rare earth metals and together therewith any metal is used which, when molten, will wet the first metal and form a two, three, four (or greater) component alloy system that is homogeneous on solidification, the alloy having a melting point below about 1500°C.

Functionally, the alloy system used as the bonding medium should be capable of reducing any thin B_2O_3 glass film that might be coating the CBN. Utilization of at least one of the specifically mentioned metals as a component of the alloy system assures the presence of an element that forms a boride or nitride but the ideal alloy system is one that exhibits a finite but limited reactivity with CBN. Alloy systems in which too much of an active metal (one that forms a highly stable nitride and/or boride) is present may convert too much of the CBN to the active metal boride or nitride.

Thus, when very small (e.g., 1 to 5 μm) crystals of CBN are to be bonded together it is important either to select an alloy system that does not react to too great an extent with CBN (or else many crystals may be destroyed) or to select an alloy system that forms a diffusion zone with the CBN of material that introduces highly desirable properties to the system. Also, this process enables the covering of individual CBN crystals with metal that is bonded to the CBN crystal through a diffusion zone. Such alloy-covered CBN crystal may be placed in a metal-bonded, resin-bonded or ceramic-bonded abrasive system for the preparation of various cutting and polishing tools (e.g., flexible belts).

Except, perhaps, in the case of tool steels, the component of lesser (or least) concentration should be present in a quantity ranging from at least 1 to 5% by wt of the alloy system, although larger quantities may be used.

The ability to utilize various tool steels as a matrix for CBN compacts is a particularly important aspect of this process. Such CBN compacts may be age hardened for maximum performance in machining, drilling, grinding, etc. Thus, tool steels may readily be selected that have various useful specific properties to be made available in the finished article, e.g., a CBN/tool steel tool insert. Tool steels may be selected such that the tool steel in a CBN/tool steel composite may be oil hardened or air hardened and may be made shock resisting, highly wear resistant, highly resistant to heat softening, etc. by treatments at temperatures significantly below 1500°C.

Example 1: A small number of CBN crystals together with nickel powder were loaded into a Ta boat. The assembly was heated up slowly while maintained in a vacuum of about 4×10^{-4} torr. When the Ni-Ta eutectic was reached, the temperature was held for about 5 minutes and then the system was slowly cooled by reducing the temperature in increments of about 50°C every 3 to 4 minutes. The vacuum was not broken until the boat had cooled. Total time for heating and cooling was about 45 minutes. The system was allowed to cool at atmospheric pressure. The Ni-Ta alloy that had been generated had wetted and adhered to the CBN crystals. The coated CBN crystals were bonded together in the alloy matrix.

Example 2: Example 1 was repeated using cobalt powder, CBN crystals and a Ta boat. About 1 to 5 wt % of the Ta reacted with the Co and the Co-Ta alloy bonded well to the crystals forming a CBN alloy-bonded compact.

DIAMOND AND/OR CBN

Bonded to Silicon Carbide Substrate

The composite of *C.R. Morelock; U.S. Patent 4,242,106; December 30, 1980; assigned to General Electric Company* is an integral, self-supporting shaped body comprised of a diamond and/or cubic boron nitride crystal-containing body component and a supporting silicon carbide substrate component free of the diamond or cubic boron nitride crystals of the crystal-containing body component. The polycrystalline body component of the composite is the work or load-bearing component.

Specifically, in this composite, the polycrystalline body component is comprised of crystals selected from the group consisting of diamond, cubic boron nitride, and mixtures or combinations thereof, adherently bonded together by a bonding medium comprised of silicon carbide and elemental silicon. The crystals range in size from submicron up to about 2,000 microns. The density or volume of the crystals ranges from about 1% by volume to about but less than 80% by volume of the polycrystalline body component.

The bonding medium is present in an amount ranging up to about 99% by volume of the polycrystalline body component. The bonding medium is distributed at least substantially uniformly through the polycrystalline body component, the portion of the bonding medium in contact with the surfaces of crystals consisting essentially of silicon carbide. The polycrystalline body component is at least substantially pore-free.

Example: An equilateral triangular piece of the silicon carbide substrate, 0.10 inch thick and 0.625 inch in length on each side, was used. The substrate had flat triangular surfaces.

At room temperature, DTA curing agent was mixed with 0.01 gram of Epon 828 resin in an amount of 10% by weight of the Epon 828, and this mixture was then admixed with 0.17 gram of 210/230 mesh (equivalent to 74/62 microns) carbon-coated diamond crystals and 0.02 gram crushed carbon cloth to form a uniform mixture.

The resulting crystal-containing mixture was spread on one of the flat triangular surfaces of the silicon carbide substrate forming a continuous coating thereon. The resulting preform was placed in a 100°C oven for 1 hour to cure, i.e., harden, the Epon 828. The resulting cured crystal-containing layer of the preform had a thickness of about 0.025 inch.

The preform, which was held together by the cured resin, was fitted onto a graphite mold, wherein all of the surfaces had been sprayed with hexagonal boron nitride.

The associated structure or apparatus for carrying out the silicon infiltration is described below. Specifically, a solid graphite cylinder was used with a cavity drilled into its upper end portion for holding silicon and a cavity drilled in its lower end portion for use as a mold cavity. The mold cavity was drilled to approximately correspond to the shape and dimensions of the preform. Two infiltration holes, about 50 mils in diameter, were drilled substantially perpendicular between the two cavities connecting them. The preform was placed within the mold cavity with the exposed flat triangular surface of the substrate portion of the preform facing the upper cavity. Any free space remaining in the mold cavity was filled with hexagonal boron nitride powder. A wick of elemental nondiamond carbon fiber, about 500 mils long and about 40 mils in diameter, was positioned within each infiltration hole to extend into the upper cavity, but to be just in contact with the surface of the substrate portion of the preform in the mold cavity. The mold cavity was then closed with a graphite plate forming a mold which was not airtight and therefore could be ventilated without a vent hole. The graphite plate was held in place by carbon filament or graphite bars. 0.4 gram of granular elemental silicon was placed in the upper cavity.

The resulting associated structure was placed in a graphite vacuum furnace which was evacuated to about 0.1 torr and maintained at about 0.1 torr during silicon infiltration and subsequent furnace-cooling to room temperature. The residual gases in the furnace were nonoxidizing.

The furnace was heated to about 1500°C and maintained at such temperature for 10 minutes. The power was then cut off and the structure was furnace-cooled to room temperature.

The amount of infiltrating silicon was sufficient to completely infiltrate through the resulting mixture of crystals and elemental nondiamond carbon, and its interface with the substrate, and completion of the silicon infiltration in the given time was based on experience and results with similar runs.

The resulting composite was recovered from the mold without sticking. The remains of the infiltration wicks were machined off the substrate component of the composite. The resulting composite was an integral equilateral triangular piece which had a form, i.e., shape and dimensions, not significantly different from that of the preform. Specifically, the total thickness of the composite was about 0.125 inch, and the thickness of the polycrystalline diamond coating, i.e., polycrystalline diamond body component of the composite was about 0.025 inch.

Laminated Compacts

W.I. Wilson; U.S. Patent 4,229,186; October 21, 1980 provides a laminated abrasive body which is in effect a thick compact comprising a plurality of diamond or cubic boron nitride compacts laminated together, joining of adjacent compacts taking place by means of a layer of a metal or metal alloy braze and the thickness of the laminate exceeding 5 mm and the longest dimension of the body transverse to the thickness exceeding 2 mm. The process also covers methods of making such laminated bodies by forming alternate layers of braze and compact or compact-forming material and then forming the laminate under elevated temperature and pressure.

Example 1: A laminated abrasive body consisting of a plurality of diamond compacts laminated together was produced in the following manner. Alternating layers of a diamond/cobalt powder mixture and zirconium foils were placed in the reaction capsule of a conventional high-temperature/high-pressure apparatus. The diamond/cobalt mixture contained 80% by volume diamond particles and 20% by volume cobalt powder and had an average particle size of about 40 microns. The mixture layers had a depth or thickness of about 2 to 3 mm and the zirconium foils were each of the order of 100 microns in thickness. The reaction capsule was placed in the reaction zone of the high temperature/high-pressure apparatus and its contents were exposed to a temperature of the order of 1600°C and a pressure of the order of 55 to 60 kilobars. These conditions were maintained for about 10 minutes. The temperature was allowed to return to substantially ambient conditions and the pressure was released. Recovered from the capsule was a laminated abrasive body consisting of a plurality of disc-shaped diamond compacts laminated to each other through zirconium layers. The body in the direction of lamination had a thickness of about 15 mm and a diameter of 5 mm.

Example 2: A laminated abrasive body consisting of a plurality of cubic boron nitride compacts laminated together was produced in the following manner.

Six disc-shaped cubic boron nitride compacts, each having a thickness of about 2 mm and a diameter of 4 mm were produced using known techniques. Each compact had a bonding matrix containing silicon nitride and aluminum and consisted of 85% by volume cubic boron nitride, the balance being the matrix.

A composite was formed of alternating layers of cubic boron nitride compacts and powdered copper/tin/titanium alloys. Each layer of powdered alloy had a thickness of about 100 microns. The composite was placed in a vacuum of 10^{-4} torr and exposed to a temperature of 750°C for a period of 15 minutes. Recovered from the vacuum was a laminated abrasive body consisting of a plurality of cubic boron nitride compacts laminated together and having a dimension in the direction of lamination, i.e., thickness of 12 mm.

Increased Covering by Support Means

W.I. Wilson; U.S. Patent 4,219,339; August 26, 1980 describes an abrasive body for machining a workpiece including an abrasive compact comprising a mass of abrasive particles such as diamond and cubic boron nitride, present in an amount of at least 70% by volume of the compact, bonded into a hard conglomerate; support means covering more than 50% of the surface of the compact and bonded to the compact. The compact presents a formation, salient in relation to the support means, capable of machining the workpiece. Preferred configuraations for the body are those in which the compact is sandwiched between two supports and a cylinder of compact inside a supporting sleeve.

In the abrasive body the compact is supported more comprehensively than is the case with similar bodies of the prior art. It is preferred that at least 75% of the surface of the compact be covered by the support means.

In the abrasive bodies of the process, the compact may be attached to the support means directly without the presence of an interposed metal or metal alloy braze layer in the manner illustrated and described in U.S. Patents 3,743,489 and 3,745,623. Alternatively, the compact may be attached to the support means through a metal or metal alloy braze layer. The braze layer may be a low temperature braze or a high temperature braze of the type described and illustrated in British Patent 1,489,130. In the case of cubic boron nitride compacts, when a braze layer is provided it is preferably a copper/tin/titanium braze. In the case of diamond compacts, when a braze layer is provided it is preferably zirconium.

The braze layer, when provided, is generally less than 0.5 mm, and typically in the range 50 to 500 microns, in thickness.

The support means is made of a high rigidity material such as high modulus steel of the M or T series or a cemented carbide. The support means is preferably made of a cemented carbide.

Example: A cubic boron nitride compact was made by subjecting a mixture of 80% by weight cubic boron nitride particles and 20% by weight powdered aluminum/cobalt (50/50 weight percent) alloy to conditions of temperature and pressure at which cubic boron nitride is crystallographically stable, i.e., a temperature of the order of 1600°C and a pressure of the order of 55 to 60 kilobars. These elevated conditions were maintained for a period of 3 to 5 minutes. The temperature was allowed to return to ambient conditions and the pressure was released. The compact produced was in the form of a circular disc.

This disc was then cut into three 60 degree segements. A cemented tungsten carbide support (cobalt binder), also in the form of a 60 degree segment, was then attached to each of the two major surfaces of each segment. This was achieved by means of a copper/tin/titanium brazing alloy. The brazing alloy in powdered form was deposited on a major surface of a segment, a cemented tungsten carbide support then placed on the powdered alloy, and the whole heated to a temperature of about 900°C under vacuum for a period of about 10 minutes. One of the cemented carbide supports of each segment was polished away, starting from the pointed end of the segment and working backwards.

Core of High Rigidity Material

According to *C. Phaal; U.S. Patent 4,215,999; August 5, 1980* there is provided an abrasive core comprising an abrasive compact and a core of high rigidity material embedded in the compact, the compact comprising a mass of abrasive particles, present in an amount of at least 70% by volume of the compact, bonded into a hard conglomerate, the core of high rigidity material being located inside of and extending through the compact. The core of high rigidity material is preferably located substantially in the center of the compact.

In one preferred form of the abrasive body, the compact is cylindrical and the core of high rigidity material is located axially in the compact. An abrasive body having such a configuration may be used, for example, as a bearing surface or as a cutting or grinding wheel. The core of high rigidity material provides the axis for the body and the compact surround provides a hard and durable bearing, cutting or grinding surface. The high rigidity material may be a hard steel such as a steel of the M or T series.

The preferred high rigidity material is a cemented carbide. Suitable cemented carbides are, for example, cemented tungsten carbide, cemented titanium carbide, cemented tantalum carbide and mixtures thereof. Such carbides, as is known in the art, have a metal bonding matrix usually consisting of cobalt, nickel, iron or a mixture thereof. The metal bonding matrix is usually provided in an amount of 3 to 25% by weight of carbide.

The compact is preferably a diamond or cubic boron nitride compact. The core of high rigidity material may be fastened or bonded to the compact either directly or through a bonding layer.

Example: A cylindrical core of cemented tungsten carbide (15% by weight cobalt binder phase) was placed in the reaction capsule of a conventional high-temperature/pressure apparatus of "belt" type and surrounded with a mixture of diamond and cobalt powder. The powdered mixture had an average particle size of 40 microns. The diamond was provided in an amount of about 80% by weight of the mixture, with the cobalt comprising the remainder of the mixture. The contents of the reaction capsule were then exposed to a temperature of the order of 1600°C and a pressure of about 55 to 60 kilobars and these elevated conditions maintained for about ten minutes. The temperature and then pressure were allowed to return to ambient. Recovered from the reaction capsule using conventional techniques was a diamond compact having a centrally located slug of cemented tungsten carbide. The compact was of a cylindrical shape, having an outer diameter of 10 mm and a thickness or length of 4.0 mm. The cemented tungsten carbide was machined away and this left a cylindrical compact, having a centrally located hole, which was useful as a bearing surface.

Metal Bonding Matrix of Borided Cobalt

There is provided by *V.R. Conradi; U.S. Patent 4,142,872; March 6, 1979* a metal-bonded abrasive body comprising diamond or cubic boron nitride abrasive particles held in a metal bonding matrix. The metal-bonding matrix consists of cobalt, present in an amount of at least 50% by weight of the matrix, and boron in the form of cobalt borides substantially uniformly distributed through the matrix.

The cobalt may be in the form of substantially pure cobalt in which event the metal-bonding matrix will consist substantially only of cobalt and boron in the form of cobalt borides. Alternatively, the cobalt may be in the form of a cobalt-containing alloy wherein the cobalt is the major constituent of the alloy. The metal-bonding matrix consists, in effect, of cobalt which has been borided.

The borided-cobalt matrix of the process is tough and hard and can be produced at temperatures below 1000°C. The boron will generally be present in an amount of 0.5 to 3% by weight of the bonding matrix. The abrasive particle content will generally constitute 5 to 15% by volume of the body.

In a typical body, the cobalt constitutes 70 to 90% by volume of the body, the abrasive particles 5 to 15% by volume of the body, with other additives and fillers such as tungsten carbide being optionally provided. The metal-bonded abrasive body may be a saw segment, the rim of a metal-bonded grinding wheel, a stope coring bit, an impregnated bit, or like body.

It is preferred that the cobalt or alloy powder is borided before it is mixed with the abrasive particles. The reason for this is that gases are generally produced during the boriding process and these gases tend to leave pores in the abrasive body if the boriding takes place simultaneously with the body-forming sintering step.

The boriding powder may be any well known in the art. Boriding powders usually consist of boron carbide (B_4C) alone or in combination with other boron-containing substances such as potassium borofluoride. A typical boriding powder is 50% potassium borofluoride and 50% boron carbide.

Example: A segment was made for a stope coring bit in the following manner. Powdered cobalt was mixed with about 4% by weight of a commercially available boriding powder Degussa "G27." The mixture was heated to a temperature of about 900°C and this temperature was maintained for about 60 minutes. The borided powder was then mixed with diamond particles. The diamond particles constituted about 10% by volume of the mixture. The diamond-containing mixture was placed in a suitable mold and sintered at a temperature of about 950°C. A segment was recovered from the mold which was found to have a Rockwell C hardness of 60 to 95. This was very much harder and tougher than a similar segment made in the conventional manner without the boriding step where the Rockwell B hardness was found to be about 90 to 100 (which is about 8 to 10 on the Rockwell C scale). Furthermore, the use of a borided cobalt enabled the sintering to take place at a low temperature of 950°C. To achieve the same bond hardness without the use of a borided cobalt, it is necessary to use other metals which can be sintered only above 1030°C at which temperatures synthetic diamond tends to graphitize.

Matrix of Refractory Substance and Solvent Substance

According to the process of *F.R. Bell; U.S. Patent 3,944,398; March 16, 1976* there is provided an abrasive compact comprising an effective quantity of abrasive particle selected from cubic boron nitride, diamond and mixtures thereof bonded together by a matrix comprising a refractory substance and a solvent substance which is capable of dissolving the abrasive particle to at least a limited extent. The matrix may also contain a hard material resulting from the interaction of the refractory substance with the solvent substance during compact manufacture.

The refractory substance may be a boride, nitride or silicide refractory substance. A particularly suitable refractory substance is silicon nitride, particularly silicon nitride in the α form. The refractory substance may also be magnesium oxide, spinel or alumina either alone or in combination with a nitride, silicide or boride refractory substance.

The refractory substance may also be: silicon nitride powder mixed with silicon, boron and boron carbide, all in powdered form; silicon nitride powder mixed with magnesium oxide powder; silicon nitride powder mixed with alumina and/or magnesium oxide, both of which are in powdered form; silicon nitride powder mixed with powdered spinel; a substance (in crushed form) resulting from the heat treatment of silicon nitride with alumina and magnesium oxide; or a substance (in powdered form) resulting from the heat treatment of silicon nitride with spinel.

The solvent substance must be such as to be capable of dissolving the abrasive particle to at least a limited extent. When a mixture of diamond and cubic boron nitride is used, a mixture of solvent substances will also be used. Such substances are well known in the art.

In the case of cubic boron nitride, particularly preferred solvent substances are aluminum, lead, tin, magnesium and lithium and alloys containing one or more of these metals.

The various ingredients for the mixture are preferably in finely powdered form. Preferably, the average particle size of the ingredients is less than 40 microns. In order to produce the compact, the mixture is subjected to conditions of temperature and pressure at which the abrasive particle is crystallographically stable and the solvent substance acts as a solvent.

Example: A powdered mixture of 8 micron powdered silicon nitride (α form), commercially available aluminum powder and 2-6 micron cubic boron nitride powder was cold compacted and placed in a tantalum casing. The mixture consisted of 80.5 volume percent cubic boron nitride, the balance of the mixture being made up of 68 weight percent silicon nitride and 32 weight percent aluminum. The casing was placed in the reaction zone of a conventional high-temperature/pressure apparatus. The pressure of the reaction zone was raised to 55 kilobars and the temperature was raised to 1300°C. Compact formation resulted and after release of the temperature and pressure the compact was recovered in the conventional manner.

The abrasive properties of the recovered compact were compared with a commercially available P 10 tungsten carbide compact and with a compact consisting of about 90 volume percent cubic boron nitride in a 10 volume percent cobalt/aluminum matrix made according to techniques of the prior art. The compacts were incorporated in single point tools and their abrasive properties tested in the turning of mild steel. It was found that the compact of the process showed 100 times less wear than the tungsten carbide compact and 4 times less wear than the cubic boron nitride/cobalt-aluminum compact.

Moreover, no smearing of the aluminum on the workpiece was observed with the compact of the process which is attributable to the fact that the silicon nitride reacted with the aluminum to form a hard material. It is believed that the strength of the compact of the process is also attributable, at least in part, to the fact that there was a certain amount of intergrowth between the cubic boron nitride particles.

ALUMINUM OXIDE

Reactive Metal Oxide with Specified ΔG of Formation

N.P. Suh, P.D. Fillion and S.H. Lee; U.S. Patent 4,217,113; August 12, 1980; assigned to Massachusetts Institute of Technology describe a composition formed from aluminum oxide and/or a solid solution and/or multiphase composition containing aluminum oxide and a metallic alloy containing a metal reactive with aluminum oxide. The reactive metal is characterized by a ΔG of formation per mol of oxygen of its oxide which is equal to or greater than the ΔG of formation per mol of oxygen of aluminum oxide. The composition has a microstructure comprising the oxide of the relative metal and/or solid solutions containing the oxide and aluminum oxide and/or solid solutions containing aluminum oxide, each having a grain size less than about 10 microns, and a metallic phase. When used as a cutting tool the composition contains less than about 25 weight percent of the metallic alloy. The composition is prepared by hot pressing or sintering the aluminum oxide particles and the metallic alloy in a nonoxidizing environment like a vacuum or nonreactive gas under conditions to effect reaction of the reactive metal with aluminum oxide.

Example: This example illustrates the process and sets forth the improved results obtained with the cutting tools formed. Mixtures of aluminum oxide and nickel-30% zirconium alloy were ball milled for three days. The compositions formed contained 10 weight percent of the alloy based upon the weight of the aluminum oxide-alloy mixture.

The compositions were hot pressed in a mold having a cylindrical configuration for about 5 minutes at a temperature of 1400° to 1600°C and a pressure of 5,000 to 10,000 psi. Hot pressing was conducted in a vacuum of less than about 10^{-3} torr. Hardnesses of over 2,000 kg/mm^2 are obtained, with higher resistance to fracture during cutting tests than pure ceramic tools.

The cutting results obtained with the composition formed herein are compared with the results obtained with the conventional GE 210 tool (TiC in Ni-Mo formed by sintering) and the conventional GE-030 tool (Al_2O_3, 10-15 weight percent TiC formed by hot pressing). The results in the table were obtained in cutting 4340 steel with a 0.050 inch depth of cut, 0.005 inch per revolution feed rate and a speed of 1,000 surface feet per minute in a lathe.

Tool	Crater Depth (inch)	. . . . . Time. . . . . . Minutes	Seconds
GE 210	0.00056	4	-
	0.00085	7	45
	0.0012	11	20
	0.0014	14	30
Al_2O_3-(Ni, 30%-Zr alloy)	0.00023	4	5
	0.0003	4	25
	0.00034	5	-
	0.00045	8	45
	0.00048	9	15
	0.00055	10	15
	0.0007	12	40
	0.00074	14	5
GE-030	0.00034	5	40
	0.00038	6	25
	0.00043	6	35
	0.00104	12	50
	0.0011	13	5
	0.00132	27	20

Improved Thermal Shock Resistance

A sintered body comprising alumina as a main ingredient is the only sintered ceramic body now available for producing cutting tools which can be used at a cutting speed of more than 200 m/min. However, it is susceptible to breakage in milling. Various attempts have therefore been made to increase its strength by adding a carbide such as TiC, WC or Mo_2C, a nitride or nitridecarbide such as TiN, ZrN or TiCN, or a boride such as TiB_2 to the Al_2O_3. However, such ceramics have low resistance to thermal shock, and are suspectible to breakage in milling. These defects are attributed to this low thermal conductivity. Among the aforesaid additives for Al_2O_3, WC has a thermal conductivity of 0.45 cal/cm-sec-°C, the highest among them. Addition of WC, therefore, ought to increase the thermal conductivity of the sintered ceramic body and, therefore, its thermal shock resistance. Actually, however, the result is much the same as in the case of adding TiC which has a thermal conductivity of 0.04 cal/cm-sec-°C.

Extensive investigations of Al_2O_3-WC ceramic bodies have led to the discovery that because (i) fine pores with a size of less than 0.5μ are liable to occur in such a sintered ceramic body, (ii) a part of the carbon in WC reacts with Al_2O_3 to form brittle W_2C in the crystals, and (iii) the grain growth of Al_2O_3 is vigorous, a decrease in strength occurs, thermal shock resistance is not enhanced, and the result is much the same as in the case of using other additives having a lower thermal conductivity.

Y. Yamamoto, K. Sakurai and H. Tanaka; U.S. Patent 4,204,873; May 27, 1980; assigned to NGK Spark Plug Co., Ltd., Japan provide a sintered ceramic body containing Al_2O_3 as a main ingredient, which is suitable for producing cutting tools which can be used at relatively high cutting speeds and which are less susceptible to breakage and have high thermal shock resistance.

The sintered ceramic body for use in cutting tools has an average crystal grain size of not more than 2μ and comprises 60 to 95% by volume of (a) Al_2O_3 and the remainder (5 to 40%) being (b) WC, provided that it may contain W_2C in which case the intensity ratio $I_{W2C}(101)/I_{WC}(111)$, of x-ray diffraction using CuK α rays is not more than about 0.5, preferably less than about 0.2, and (c) TiN, the volume ratio of component (b) to (c) being from about 5:95 to 95:5.

In producing the sintered body, the components (a), (b) and (c) are first selected in predetermined proportions and mixed. The starting powdery mixture is then sintered at a high temperature and a high pressure. Sintering is usually performed by a uniaxial hot press in a graphite mold or by a hot hydrostatic press. The sintering temperature is generally from about 1600° to 1900°C, and the sintering time is generally from about 5 to 30 minutes. Generally, better sintered bodies are obtained at higher pressures. Usually, the pressure is about 100 to 300 kg/cm^2 in a hot hydrostatic press. The hot hydrostatic press is suitable for obtaining uniform compact sintered bodies on a mass-production basis.

The method in accordance with this process is characterized by adding carbon powder in the preparation of the starting powder mixture. It has been found that the addition of carbon powder serves to prevent the conversion of a large amount of WC to W_2C during sintering.

Sintering Cold Pressed Powder

E. Rudy; U.S. Patent 4,022,584; May 10, 1977 provides a composition of material comprising aluminum oxide and refractory transition metal diborides, with additions of magnesium oxide and, optionally, transition metal carbides, nitrides, carbonitrides and iron group metals for sintering and grain morphology control which is particularly useful as a cutting tool for steels, cast iron, and hard-to-machine materials, such as superalloys. In contrast to the TiC-Al_2O_3 cermets of the prior art which have to be hot pressed for optimum properties, a large portion of the cermets of the process can also be fabricated, without detriment to performance, by sintering the cold pressed powder compacts in an inert furnace atmosphere at temperatures between 1650° and 1820°C.

The aluminum oxide comprises at least 48 weight percent of the composition, the refractory transition metal diborides comprise from 5 to 50 weight percent of the composition, the refractory transition metal carbides and nitrides comprise from 0 to 35 weight percent of the composition, the magnesium oxide comprises between 0.3 and 3 weight percent of the composition and the alloying metals comprise from 0 to 6 weight percent of the composition.

The ingredients used in the preparation of the cermets of the process are powders of the oxides and the prealloyed borides, carbides, nitrides, and carbonitrides of the transition metals; the iron group metals are added as elemental powders. Whatever the manner employed in preparing the composition of the cermet, it is important to select conditions which insure a uniform distribution of the constituents in the alloy blend. This applies in particular to the magnesium oxide.

Aside from the routine fabrication variables, choice of the ingredients and alloying additions and their grain size distributions, as well as milling (mixing) and sintering conditions, strongly influence microstructure and phase constituents and, as a result, the properties of the sintered cermet.

Although no generally valid guidelines can be given because each individual application has its own special requirements and the fabrication method and parameters to be chosen are dependent on the gross composition as well as the compositions and grain size distributions of the individual ingredients, the following observations, which are based on measured property and performance data of the cermets of the process, may serve as guidelines.

The best performing cermets are those in which the oxide and the metallic phase form two continuous, interpenetrating lattices, the oxide has a very small grain size, and in which the metallic phase has a dual grain structure, consisting of smaller grains ($<1.5\mu$) and larger grains up to 8 microns in diameter, and constitutes preferably between about 28 and 35 volume percent of the cermet.

Among the cermets described in this process the combinations containing, besides aluminum oxide and magnesium oxide, only the Group IV transition metal diborides, in particular TiB_2, have the best sintering properties, are stable under vacuum sintering conditions, and also were found to have the highest bending strength values of any of the cermets studied. However, the oxide-boride composites are also prone to rapid grain growth when heated to more than 50° to 70°C above the normal sintering temperatures of 1725°C. At boride levels less than 30 volume percent, excessive grain coarsing can cause disruption of the three-dimensional continuity of the boride lattice, indicated by the loss of electrical conductivity of the cermet and deterioration of the strength levels to less than 60,000 pounds per square inch. Discontinuity of the boride skeleton also results in a relief of the compressive stresses on the aluminum oxide phase, and the characteristic fringe pattern, observed along the aluminum oxide grain boundaries in structures with two interpenetrating skeletons under polarized light, is absent in cermets with isolated diboride grains.

Grain growth stability of the alloy phases is significantly improved by additions of carbides and nitrides, but such additions also necessitate higher sintering temperatures and, generally, an inert gas furnace atmosphere during sintering as such additions can cause undesirable interactions with the aluminum oxide under vacuum at temperatures above 1450° to 1500°C.

At high concentrations of stoichiometric, or near-stoichiometric, carbides, the pressure-sintering (hot pressing) technique is to be preferred because the lower temperatures required for complete densification aid in diminishing such undesirable interactions between the carbide and the oxide phase.

Lowering of the carbon activity in monocarbides with large homogeneity ranges, such as TiC, by decreasing the carbon content decreases the interaction tendency and also improves the sintering behavior; the properties and performance of such cermets fabricated by cold pressing and sintering under inert gas equaled those fabricated by hot pressing, and generally were superior to tools fabricated from the stoichiometric carbides by either of the two methods. For reasons not yet understood, strength, and performance as tools, of cermets prepared from nitrides with large nitrogen deficiencies were inferior to those fabricated from the stoichiometric or near-stoichiometric compounds.

Hot-Press Forming at High Temperature and Pressure

H. Tanaka, Y. Yamamoto; U.S. Patent 3,886,254; May 27, 1975; assigned to NGK Spark Plug Co., Ltd., Japan provide a method of making a ceramic sintered body, comprising preparing a powder mixture consisting of 40 to 80% by weight of alumina with or without replacing not more than 2% by weight by a grain growth inhibitor, 16 to 40% by weight of titanium carbide and 4 to 20% by weight of metallic titanium, the total being 100% by weight, and hot-pressing the mixture. The mixture is not subjected to a calcination in nonoxidizing atmosphere at a temperature of 600° to 1000°C as previously proposed but it is directly subjected to hot-press forming at a high temperature under a high pressure.

Omitting the calcination in the nonoxidizing atmosphere applied in the above mentioned prior art method, the process provides a ceramic sintered body which is particularly available for cutting tool tips, bearings or other articles which must have a high hardness, a strong toughness, and an excellent wear-resistant property.

High Proportion of Al_2O_3

The process of *K.-H. Schuller and H. Nink; U.S. Patent 3,871,891; March 18, 1975; assigned to Rosenthal Stemag Technische Keramik AG, Germany* uses the well-known hot-pressing method, in which a mixture of at least 98% Al_2O_3 powder of commercial grade and of a carbon-containing plasticizer, solvent or diluent, whose premature evaporation is prevented, is introduced into a normal hot press. This method is characterized by the succession of the following operating steps:

(a) Mixing 100 parts by weight of at least 98% Al_2O_3 powder of commercial fineness with 10 to 50 parts by weight of a carbonaceous plasticizer, solvent and the like:

(b) optionally, pressing a blank;

(c) heating the mixture in a hot press to a temperature range between 1280° and 1500°C and pressing with a holding time of at least 5 minutes at the highest temperature; and

(d) cooling.

The carbonaceous agent is preferably a polyvinyl alcohol (PVA) dissolved in an inert solvent. After step (a), the slightly dried mixture can be passed through a screen to separate lumps or to reduce them in the screen.

Instead of applying surface pressure as is generally customary, i.e., in the cold state, it is possible to apply the pressure only when the appropriate maximum temperature has been attained.

The heating rate should not be below about 50°C/min to prevent premature evaporation of the organic agent mentioned in step (a). The Al_2O_3 powder can also be 100% pure.

This method produces well-defined materials with a high Al_2O_3 content, and a high degree of purity which are characterized by a greatly reduced material loss per hour under mechanical surface stress. With this "wear-resistance" are combined or automatically linked other valuable properties of the materials, e.g., fine and uniform grain, great hardness, practically theoretical density and high surface quality.

CEMENTED CARBIDES

Dewaxing in Nitrogen Atmosphere

A method of making cemented titanium carbide base cutting tools with increased tool life is disclosed by *D. Moskowitz; U.S. Patent 4,230,462; Oct. 28, 1980; assigned to Ford Motor Company.*

Substantially unoxidized carbide and binder metal powders are ground to a submicron particle size in the presence of a low-boiling oxidation-inhibiting ingredient covering the particles of the powders. A low-boiling pressing aid is added.

The ground powders are then subjected to air drying to evaporate the oxidation inhibiting liquid ingredient and the pressing aid consisting preferably of polyethylene glycol. After compacting the waxed ground particles to a predetermined configuration, the compact is first heated to volatize the wax at an appropriate temperature and critically in the presence of a nitrogen-based atmosphere which may contain 0 to 10% H_2. The dewaxed compact is then heated to a higher second temperature level for effecting sintering in the presence of a vacuum.

Although the process is premised upon utilizing starting materials which are substantially unoxidized, it is almost impossible to obtain an absolute unoxidized condition of such fine-sized particles; therefore, it has been found convenient to add a small amount of a reducing gas to the nitrogen-based atmosphere, which preferably includes about 5 to 10% hydrogen. The small amount of hydrogen serves to reduce any small amount of oxides that may have been formed during the grinding operation (which may include any of the oxides of molybdenum or nickel).

Example: This example illustrates the tool life improvement that can be obtained for a roughing grade titanium carbide tool which is comprised of 22.5% nickel, and 10% molybdenum for serving as the binder for the titanium carbide.

One tool was prepared according to the preferred method of this process which comprises mixing 22.5% nickel, 10% molybdenum and the remainder titanium carbide, along with sufficient acetone to cover the powders in a ball mill. The mixture was subjected to ball milling for a period of 96 hours rendering a particle size that was submicron. Four percent Carbowax binder was added to the ground mixture, which was then air dried. The waxed mixture was then compacted at 10 tsi and subjected to a dewaxing heat treatment under an atmosphere consisting of 95% nitrogen and 5% hydrogen. The other tool was similarly prepared except that a pure hydrogen atmosphere was employed during dewaxing. Each of the resulting tools was used in a machining operation; the 95 $N_2/5H_2$ dewaxed tool showed a tool life improvement of 40%.

Molybdenum-Tungsten-Carbonitride with Tungsten Carbide Structure

P. Ettmayer and R. Kieffer; U.S. Patent 4,212,671; July 15, 1980; assigned to Sandvik Aktiebolag, Sweden have found that a tungsten-molybdenum-carbonitride having the tungsten carbide structure can be obtained under certain conditions. Thus, a way has been disclosed concerning how to substitute for the expensive tungsten carbide in cemented carbides by means of an economical and easily available material without any loss of the desired material properties.

The process relates to sintered hard metal or cemented carbide consisting of one or more hard materials or principles in an amount of 70 to 97% by weight and a binder metal or a binder alloy of the iron group metals of the periodic system in an amount of 3 to 30% by weight. It is characterized in that the hard material portion consists of at least 20% of molybdenum-tungsten-carbonitride having the structure of the tungsten carbide. Usually, the hard material portion is at least 30% of the carbonitride, but often the amount is more than 50%.

The molybdenum-tungsten-carbonitride with the structure of tungsten carbide has normally a nitrogen content of 0.05 to 0.5% N by weight.

The carbonitride has a composition within the limits $(Mo_{0.9}W_{0.1})$ (C,N) and $(Mo_{0.05}W_{0.95})$ (C,N), preferably between $(Mo_{0.7}W_{0.3})$ (C,N) and $(Mo_{0.4}W_{0.6})$ (C,N). It also is possible to substitute a part of the molybdenum and tungsten atoms in the molybdenum-tungsten carbonitride by other atoms, e.g., Cr, Nb, Ta.

The other hard principles include carbides of the transition metals as well as their mixed crystals, nitrides of the transition metals as well as their mixed crystals, or mixed crystals of the carbides and nitrides of the transition metals. Usually, the amounts of these materials are less than 70% and preferably below 50% by weight. A desirable amount has been 5 to 50% by weight.

According to the process the molybdenum-tungsten carbonitride with the structure of tungsten carbide can be prepared by heating molybdenum and tungsten in a nitrogen-containing atmosphere, preferably at 1000° to 1300°C possibly in bound or combined form, together with carbon in a ratio being sufficient for formation of the monocarbide.

As starting material, mixtures of molybdenum or tungsten metal powders with carbon can be provided in such a ratio that the sum of the gram atom numbers of molybdenum and tungsten is the same as the gram atom number of the carbon.

The starting material can also be mixtures of the carbides of molybdenum (Mo_2C) and of tungsten (W_2C and/or WC) added together with carbon in such a ratio that the total carbon content in gram atom amounts is the same as the sum of the gram atom numbers of the metals molybdenum and tungsten.

According to another embodiment of the method there are used as starting materials, mixtures of the ammonium salts of the molybdic and tungstic acids with carbon, the amount of carbon being so adjusted, that after the removal of oxygen there remains sufficient carbon for forming, with the metals and nitrogen, a carbonitride having the composition (Mo, W)(C,N).

Furthermore, it is possible to use mixtures of the oxides of molybdenum and tungsten with carbon as starting materials. The amount of carbon is so adjusted that after the reduction of the oxide there still remain sufficient amounts of carbon for forming, with the metals and nitrogen, a carbonitride, having the composition (Mo,W)(C,N).

The starting materials can also be mixtures with carbon of jointly precipitated hydrate-acids of molybdenum and tungsten. The amount of carbon is so adjusted that after the reduction of the oxides there remain sufficient amounts of carbon so as to be formed a carbonitride with the composition (Mo,W)(C,N).

In each of the mentioned method embodiments, the mixing ratio of molybdenum and tungsten in the mixed carbonitride, the ratio being expressed in gram atoms, is preferably between 9:1 to 1:9, particularly 4:1 to 1:3.

To prepare the cement carbide the powder mixtures are sintered in a conventional manner to a dense, pore-free material. As binder metals, cobalt or nickel in amounts of 3 to 30% are preferable. As sintering atmosphere the normal protective gases, as for example hydrogen, noble gas or nitrogen or their mixtures, are suitable. Also vacuum-sintering is possible. Preferably, the sintering of the cemented carbide is performed in a nitrogen-containing atmosphere, the pressure being up to 200 atmospheres, particularly up to 50 atmospheres. The sintering temperature should be maintained between 1300° and 1500°C.

Tungsten Carbide plus Titanium Carbide

The cemented carbide material described by *Y. Soga; U.S. Patent 4,046,517; September 6, 1977; assigned to Ltd. Dijet Industrial Co., Japan* comprises 10 to 60% by weight of tungsten carbide, 5 to 40% by weight of titanium carbide, 5 to 30% by weight of tantalum carbide, 3 to 20% by weight of titanium nitride and 5 to 20% by weight of an iron family metal such as cobalt, nickel or iron. The cemented carbide material may further contain 5 to 20% by weight of molybdenum and/or molybdenum carbide.

The cemented carbide material having the foregoing composition is more resistant to heat than conventional titanium carbide grades, has increased hardness while substantially retaining the desired flexural strength and is adapted for a wide variety of cutting conditions.

As a tool material for cutting steel or high-grade cast iron, titanium carbide is most useful in reducing the flank wear and crater wear to be encountered. As far as wear is concerned, therefore, it is advantageous to increase the proportion of titanium carbide to the greatest possible extent, whereas the very low thermal conductivity of titanium carbide may give rise to various problems.

To assure effectiveness of titanium carbide material, tungsten carbide, tantalum carbide, niobium carbide, etc. are usable in the form of a solid solution. For example, a preferably solid solution consists of tungsten carbide, titanium carbide and tantalum carbide in the ratio of 5:3:2 or 5:2:3. Usually, such solid solution is admixed with tungsten carbide, tantalum carbide, niobium carbide, cobalt, nickel, iron, etc. to prepare the desired composition, which is then sintered.

However, when the volume proportion of titanium carbide in the composition is in excess of a certain level, portions of the titanium carbide-containing solid solution in contact with each other tend to fuse together to produce large particles during sintering, however thoroughly the composition may be mixed. The size of the enlarged particles is a critical factor which influences tool wear, so that it is desired that the titanium-containing solid solution have a small particle size.

When added in a suitable amount to the composition, titanium nitride suppresses the growth of the particles. More specifically, titanium nitride permits formation of the peculiar structure of titanium carbide-base cemented carbide material in which titanium carbide serves as nuclei, inhibiting the growth of solid solution particles which is predominant with titanium carbide and thereby ensuring formation of fine crystalline particles. As compared with titanium carbide, moreover, titanium nitride has higher resistance to thermal impact and entails reduced heat generation because of its lower coefficient of friction relative to steel. Consequently, the cemented carbides incorporating titanium nitride have higher resistance to thermal impact than usual titanium carbide grades. Use of titanium nitride which assures formation of fine particles gives increased hardness and greatly improved wear resistance to the material obtained. Thus, the material exhibits high cutting performance with a relatively low titanium content and is less susceptible to cracking or chipping when used in a milling operation whether the operation is by a wet or the usual dry method.

With high titanium carbide contents, molybdenum or molybdenum carbide (Mo_2C) is effective in suppressing the growth of particles as is well known. Although titanium nitride is singly useful if it is desired only to suppress the growth of particles, use of 5 to 20% of molybdenum or molybdenum carbide is found to give a material which is very advantageous as a tool material for milling which is an intermittent cutting operation. When molybdenum is not used, the resulting material is useful in a turning operation that is a continuous cutting operation.

Coated Carbide Core

N. Reiter; U.S. Patent 4,019,873; April 26, 1977; assigned to Fried. Krupp Gesellschaft mit beschrankter Haftung, Germany provides a wear-resistant shaped body comprising (1) a core of hard metal body containing at least one carbide as a hard material and at least one binder metal and (2) a surface coating on the core, the surface coating being composed of two superimposed layers including an outer, extremely wear-resistant layer consisting essentially of at least one material selected from the group consisting of aluminum oxide and zirconium oxide and an inner layer of at least one boride, selected from the group consisting of the borides of the elements titanium, zirconium, hafnium, vanadium, niobium, tantalum, chromium, molybdenum and tungsten.

The boride used in the inner layer preferably is at least one diboride selected from the group consisting of diborides of the elements titanium, zirconium, hafnium, vanadium, niobium, tantalum, chromium, molybdenum and tungsten.

The shaped bodies according to the process have particularly advantageous properties if their inner layer has a thickness of 0.5 to 20 μ, preferably 1 to 3 μ, and if their outer layer has a thickness of 0.5 to 20 μ, preferably 2 to 10 μ.

The carbide is a carbide of the type normally used in preparing high-strength cutting materials and includes one or more of the metal carbides of tungsten, titanium, tantalum, niobium, vanadium, zirconium, hafnium and molybdenum. The binder metal used in the core generally is at least one metal selected from the group iron, cobalt and nickel. The relative proportions of carbide and binder metal depend on the end use of the product, and generally, the binder metal can comprise from about 5 to 20% of the core for hard metal bodies intended for use as cutting tools.

The shaped bodies are produced according to known coating processes where the deposition of the boride and oxide layers from the gaseous phase on the hard metal body is particularly advantageous. The inner boride layer generally contains at least one or more borides as the sole component of the inner layer and includes borides such as titanium boride (TiB_2), zirconium boride (ZrB_2), hafnium boride (HfB_2), vanadium boride (VB_2), niobium boride (NbB_6, NbB, NbB_2), tantalum boride (TaB, TaB_2), chromium boride (CrB, CrB_2, Cr_5B_3), molybdenum boride (MoB, MoB_2) and tungsten boride (WB, W_2B_5 and W_2B). Because of the carbon content of the hard metal core there can be a small amount of carbide of the deposited refractory element in the inner boride layer.

The shaped bodies structured according to the process have the following advantages compared to known layered hard metal bodies. First, the outer layer has very high adhesion because the inner boride layer almost completely prevents diffusion of carbon from the hard metal body core even at high temperatures. Second, the shaped hard metal body does not exhibit formation of brittling eta layers between the boride layer and the core of the hard metal body. In addition, the shaped bodies exhibit substantially improved wear behavior with a smooth cut as well as with an interrupted cut.

Example: An already formed and shaped hard metal body, consisting of 80% tungsten monocarbide, 13% titanium carbide and tantalum carbide and 7% cobalt is used as a core and is treated in a first reaction stage at 1000°C, under a pressure of 50 torrs, for 1 hour, with a gas mixture composed of boron trichloride, titanium tetrachloride and hydrogen. The hard metal body is brought to the reaction temperature of 1000°C by heating it in a protective gas atmosphere. During the subsequent 1 hour treatment with the gas mixture, a TiB_2 layer of 3 μ thickness is produced. In order to apply the TiB_2 layer, a gas mixture of the following composition is used: 1,900 ℓ hydrogen per hour, 20 ml $TiCl_4$ per hour, and 4 g BCl_3 per hour.

In a second reaction stage, the hard metal body, which has been coated with TiB_2, is treated for 3 hours, at 1100°C and atmospheric pressure, with a gas mixture consisting of aluminum trichloride, carbon dioxide and/or carbon monoxide, and hydrogen. In the second reaction stage, an α-aluminum oxide layer of about 5 μ thickness is deposited on the hard metal body. The gas mixture used in the second reaction stage is produced by conducting hydrogen over aluminum trichloride which has been heated to 130°C, and then CO_2 or CO, respectively, is added to this mixture. In the second reaction stage, a gas mixture with the following composition is used, for example: 500 ℓ hydrogen per hour, 5 ℓ carbon dioxide per hour, and 20 g aluminum trichloride per hour. The two reaction stages may be conducted in one furnace or in two separate furnaces.

OTHER COMPOSITES

Alumina-Coated Aluminum Nitride

In a process by *H. Tanaka and Y. Yamamoto; U.S. Patent 4,203,733; May 20, 1980; assigned to NGK Spark Plug Co., Ltd., Japan* a ceramic sintered body composed mainly of AlN, e.g., at least about 97 volume % of AlN, is used as a base material and a thin layer or coating of alumina is formed on the surface of the ceramic sintered body. The formation of the thin alumina layer can be performed by depositing alumina on the surface of the AlN body by chemical vacuum deposition (CVD). Conventional chemical vacuum deposition techniques can be used.

It has been found that by adjusting the average particle diameter of the alumina crystals used to about 2.5 microns or less, the wear of a cutting tool ultimately obtained can be reduced. A better alumina coating can be obtained when the average particle diameter of the alumina crystals is 1 micron or less. The thickness of the layer or coating of the alumina crystals can be about 2 to 30 microns.

Example: 100 parts by volume of commercially available AlN having an average particle diameter of 2 microns was mixed with 1.0 part by volume of WC having an average particle diameter of 0.7 micron. After thorough mixing, the mixture was placed in a graphite mold and hot pressed in an atmosphere of nitrogen at 1750°C under a pressure of 250 kg/cm^2 for 1 hour to form a plate having a thickness of 5 mm and an area of 100 mm x 100 mm.

A rectangular tip, 12.7 x 12.7 x 4.8 mm in size, (nose R: 0.8 mm; amount of chamfering: 0.05 mm) was cut out using a diamond wheel, and machined.

Using an apparatus as described in Figure 3.10, the AlN molded product produced as described above was placed in a stainless steel reaction vessel **5**, and heated to about 1100°C in a heating furnace **6**. Then, H_2, CO and CO_2 gases were respectively fed from gas bombs **3a, 3b, 3c**, and via an $AlCl_3$ evaporating device **4**, were converted to a gaseous mixture of 8% by volume of $AlCl_3$, 30% by volume of CO, 3% by volume of CO_2 and 59% by volume of H_2. The gaseous mixture was passed through the reaction for 4 hours. The reactor was maintained at 20 to 30 torr by a vacuum pump **7**. As a result, an Al_2O_3 crystalline layer having a thickness of 3.0 microns precipitated, and the average particle diameter of the alumina crystals was 1.1 microns.

Figure 3.10: Coating Using Chemical Vapor Deposition

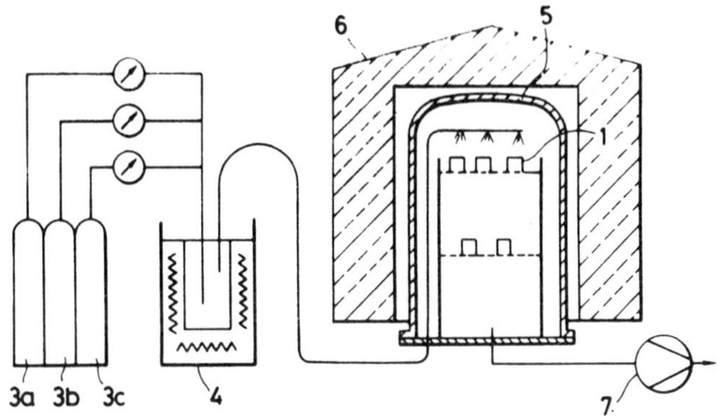

Source: U.S. Patent 4,203,733

Synthetic Slate

M.C. Kuehn; U.S. Patent 4,073,848; February 14, 1978; assigned to Ormite, Inc. describes a solid, slate-like product, a composition and method of producing the same comprising a substantially uniform mixture of at least five parts by weight of a finely divided taconite iron ore material containing at least 1% magnetite therein and an amine-formaldehyde resin such as urea-formaldehyde or melamine-formaldehyde. The mixture, heated under pressure, forms a dense slate-like material.

Example: 5 parts of taconite tailings, analyzed as 2% magnetite, 3% hematite and 0.5% iron silicate as the iron-containing components were passed through a series of standard (Tyler) mesh screens. Material having a mesh size of from 100 to 400 was collected and introduced into a pony mixer. 1 part of an aqueous solution of urea-formaldehyde resin having 60% resin (Borden's CR-585) with a 1.7:1 ratio of formaldehyde to urea and 0.25 part of walnut shell flour (Borden's FM-142) were introduced into the mixture. The mixture was a substantially uniform viscous mass at the end of 30 minutes.

The mixture was placed in a molding apparatus comprising a rectangular mold which was placed in a hydraulic press equipped with upper and lower heating plates. The mold composition was subjected to 300 psi at 250°F for 15 min. The pressure was released and the gaseous products were removed. The mold composition was then subjected to 300 psi at 325°F for 25 min. The mold composition was uniformly cooled by just maintaining contact between the press's heating plates while they were cooling. When the press plates and the mold composition reached 100°F, the molded product was removed. The product obtained was a hard dense slate-like material with a density of 0.25 lb/in^3 (slate = 0.28 lb/in^3) and a Rockwell C hardness of 45 to 60 (slate hardness average 54).

For Cutting Edge of Ground-Engaging Tool

In the process of *L.J. Moen; U.S. Patent 4,066,422; January 3, 1978; assigned to Caterpillar Tractor Co.* a wear-resistant composite material is provided which is particularly adaptable to a ground-engaging tool such as the cutting edge of a motor grader. Such composite material includes a plurality of highly abrasive-resistant particles which are embedded in an iron-boron matrix. The abrasive-resistant particle portion is preferably a relatively low-carbon, chromium-iron-based alloy having a predetermined amount of boron therein. The chemical composition of this alloy in percent by weight is chromium, 25-70%; boron, 6-12%; silicon, <2%; carbon, <2%; and remainder iron.

This combination of elements gives a complex mixture of iron and chromium borides having extremely high hardness values, typically from 1,200 to 1,600 kg/mm Knoop (or above 70 on the Rockwell C hardness scale). Preferably this mixture is formed into semiround or spheroidal particles (extremely hard shot) having diameters within the range of from 0.5 to 2 mm and a melting temperature in the range of from 1232° to 1899°C.

Referring now to the particle carrying matrix portion such matrix has been found to exhibit a relatively significant degree of toughness and ductility, while also retaining a substantially high hardness for an extended wear life. This matrix contains in percent by weight: boron 3-5%, carbon <0.2% and remainder iron.

Preferably the iron-boron matrix is of eutectic composition, wherein the boron is controlled to a level of about 3.8%. The eutectic composition provides an alloy having a relatively fine ferritic microstructure and a high average hardness of from 35-45 on the Rockwell C scale due to boride needles therein. Also, because of this range of boron content, the melting temperature is accurately established within a relatively small range of from 1161° to 1200°C.

An article may be made of the wear-resistant composite material of the process by initially placing a quantity of the chromium-iron-boron shot into a ceramic mold having the desired shape, and then depositing a quantity of the iron-boron alloy material on top thereof for subsequent melting. Preferably, the mix consists essentially of 45 to 70% by volume of the extremely hard shot. Furthermore, it has been found to be particularly desirable to deposit the iron-boron alloy material in the mold in the form of spheroidal shot having substantially the same range of diameters as the extremely abrasive-resistant shot, and with the latter remaining substantially physically unchanged during further processing of the composite material.

The ceramic mold and both forms of the shot are then deposited in the chamber of a furnace, and the chamber is subsequently substantially evacuated and/or filled with a high-purity inert gas such as argon to provide a generally inert type of atmosphere. At this time a relatively limited amount of nitrogen gas may be introduced into the chamber at a very low pressure to protect the furnace and elements of the composite material from vaporization problems. This nitrogen environment particularly inhibits the evaporation of the chromium and boron.

The furnace chamber and materials are subsequently initially preheated at a temperature of approximately 1093°C for a period of approximately 1 hour in order to obtain a uniform temperature thereof. This minimizes the time required to hold an immediately following final heating temperature of approximately 1204°C, which is maintained for approximately 15 to 30 minutes. During the final heating stage the iron-boron alloy shot is melted, with the melt seeping downwardly through gravity to fully infiltrate and encapsulate the chromium-iron-boron shot. This reduced time at final temperature minimizes the erosion of the extremely abrasive-resistant shot by the fully embracing matrix material, and otherwise protects the original physical characteristics thereof.

Lastly, of course, it is to be understood that the furnace is subsequently cooled and the completed composite article removed from the ceramic mold. The composite article of the process may then be joined to a substrate such as the steel cutting edge of a ground engaging tool by brazing.

BINDERS, MATRICES AND FILLERS

Soft Acting Phenol-Formaldehyde Resin Binder

D.I. Harris and L. Staples; U.S. Patent 4,239,503; December 16, 1980; assigned to Norton Company provide a phenol-aldehyde resin bonded abrasive product with the mild grinding properties normally associated only with shellac or alkyd resin bonded abrasive products. This end result is accomplished by utilizing, as the polymeric binder, a powdered novolac resin which contains admixed therewith, a methylene group donor in a quantity so as to provide only 0.3 to 1.5% by weight of methylene groups based on the combined weight of the phenol-aldehyde resin and the methylene group donor. The source of methylene groups is preferably hexamethylenetetramine, an aldehyde like formaldehyde or furfuraldehyde, trimethylol phenol, a one-stage phenol-aldehyde resin, or mixtures of these materials. Prior art phenol-aldehyde resins, by contrast, contain from 3.6 to 7.8% by weight of methylene groups, based on the combined weight of the methylene group donor and the novolac.

When the grinding wheel specification involved contains abrasive grit coarser than 320 mesh (U.S. Standard Sieve Series), then a liquid pick-up agent is generally applied to the abrasive prior to addition of the powdered bond, for the purpose of insuring uniform distribution of the powdered bond. In this case, there can be two sources of methylene groups, viz, the pick-up agent if it is an aldehyde like furfuraldehyde, or, a liquid one-stage resin; and a powdered methylene group donor in the powdered novolac such as those described above. If the grinding wheel being fabricated is 320 mesh or finer, uniform mixings can be made without the aid of a liquid pick-up agent, in which case the methylene group donor is entirely contained in the powdered two-stage resin.

Within the scope of the process are those abrasive products made up of from 44 to 60% by volume of abrasive, 2 to 56% by volume of bond, and 0 to 38% by volume of pores. The bond may incorporate therein from 0 to 30% by weight of filler and from 5 to 30% by weight of a liquid bond pick-up agent, with 40 to 90% by weight of the two-stage resin.

Use of Blowing Agent

U.S. Patent 4,115,077 describes a method of continuously producing resinoid abrasive wheels containing pores uniformly distributed therein. The method comprises the steps of preparing pore forming granules having predetermined mechanical strength and containing a substance thermally decomposable at a baking temperature for the production of resinoid abrasive wheels, kneading the pore forming granules with abrasive grains, a binder, a glass fiber reinforcement and a filler to prepare an abrasive composition, molding the composition into a block, heating the block by a high frequency heater, rolling the block into a sheet by rolls, blanking out circular pieces from the sheet, and baking the circular pieces of the uncured abrasive composition containing the pore forming granules to cure the binder and form pores by the decomposition of at least part of the thermally decomposable substance in the granules. However, since the pore forming granules used in the proposed method contain only a thermally decomposable substance such as starch, the pores produced on baking are relatively small, so that the abrasive wheel obtained still remains to be improved in cutting performance. *H. Fukuda; U.S. Patent 4,226,602; October 7, 1980* describes an improvement on this process.

The improvement comprises: (a) adding a blowing agent selected from the group consisting of sodium hydrogen carbonate, amyl acetate, butyl acetate and diaminobenzene, along with the thermally decomposable substance in the step of preparing pore forming granules; and (b) forming pores by the decomposition of at least part of the thermally decomposable substance contained in the granules containing the blowing agent and by the decomposition of the blowing agent contained in the granules while curing the binder contained in the circular sheet in the baking step, the pores being larger than the pore forming granules and located where the pore forming granules were present.

Polycarbodiimide and Polyurethane Resin Binder

Y. T. Torimae and T. Ishiai; U.S. Patent 4,221,572; September 9, 1980; assigned to Kao Soap Company, Japan have found that an abrasive material which consists essentially of abrasive grains and a polycarbodiimide and polyurethane resin as the binder for the abrasive grains, has good properties, that is, good abrasive grain binding, good thermal resistance and good friction resistance. The hardness of such resin can be controlled easily and the abrasive material can be formed with ease and is suitable for a large-scale production.

The abrasive material consists essentially of abrasive grains and a polycarbodiimide and polyurethane resin as the binder for the abrasive grains, the resin having from 0.5 to 10 gram equivalents of carbodiimide bonding and at least 0.5 gram equivalent of urethane bonding in 1,000 grams of the resin.

The polycarbodiimide and polyurethane resin itself is prepared from a polyisocyanate and a polyol according to the method known to one skilled in the art. The abrasive material of this process is prepared by heating the mixture of a polyisocyanate, a polyol and abrasive grains in the presence of a carbodiimidation catalyst.

Example: A mixture of 200 grams of triol prepared by reacting propylene oxide with glycerin (OH equivalent = 200) and 350 grams of toluene diisocyanate (TDI, NCO equivalent = 87) (the NCO/OH of the mixture equals 4.02) was gradually heated up to 70°C for 30 minutes under stirring and left standing for 1 hour at that temperature. The resulting urethane prepolymer had a viscosity of 900 cp (at 30°C) and a NCO weight percent of 23.1. Thus, substantially all hydroxyl groups were converted to urethane bondings. To 100 grams of the urethane prepolymer, 3 grams of 30 weight percent of a diethylene glycol (OH equivalent = 53) solution of an equimolar mixture of dipotassium mono-2-ethylhexyl phosphate and monopotassium di-2-ethylhexyl phosphate was added. (The NCO/OH ratio became 3.30.)

Fifty grams of the mixture was blended with 300 grams of alumina (abrasive grains) and the blend was stirred for 1 minute at room temperature, was filled in a disk-shaped mold made of metal having a diameter of 13 cm and a thickness of 1 cm, and pressed at room temperature. Then the mold was placed in an oven at 120°C for two hours, was taken out to stand until it had cooled to room temperature; then, the abrasive material was taken out of the mold.

This abrasive material had a bulk density of 2.0, was porous and abrasive grains were dispersed uniformly. The content of carbodiimide bonding in 1,000 grams of the binder resin was 2.80 gram equivalents and that of urethane bonding was 2.44 gram equivalents, respectively.

After being dressed into a disk having a diameter of 12.5 cm, the performance of the polishing material was tested by polishing black rust off steel plate. The test machine was a perfect pressure-type plane grinding machine. The pressure was 5 kg/cm^2. The rotation frequency was 1,500 and 2,500 rpm. The feed speed of the steel plate was 0.1 m/sec.

The polishing material maintained steady polishing ability for 10 minutes at 1,500 rpm and, after dressing, for 10 minutes at 2,500 rpm.

Crosslinked Aromatic Polyimide Matrix

E.R. Lovejoy; U.S. Patent 4,142,870; March 6, 1979; assigned to Abrasives International NV, Netherlands Antilles and U.S. Patent 4,107,125; August 15, 1978; assigned to E.I. Du Pont de Nemours and Company provides an abrasive article of a matrix and filler particles including abrasive particles substantially homogeneously distributed throughout the matrix. The matrix is a crosslinked aromatic polyimide polymeric material comprising a crosslinked chemical combination of (a) a higher molecular weight base resin of linear aromatic polyimide polymer having at least one ketonic carbonyl per repeat unit and a molecular weight of from 3,000 to 300,000 and (b) a lower molecular weight crosslinker resin of linear aromatic polyimide polymer having amine end-groups and a molecular weight of from 500 to 15,000. The mol ratio of crosslinker resin to base resin is in the range of greater than 2 to 40 and the molecular weight ratio of higher molecular weight resin to lower molecular weight resin is from 2 to 50.

Preparation of Coreactant Resins: To prepare base resins and crosslinker resins, 4,4'-oxydianiline (ODA) and 3,3',4,4'-benzophenone tetracarboxylic dianhydride (BTDA) or pyromellitic dianhydride (PMDA) are reacted together in appropriate amounts to yield linear aromatic polyamide acid which is converted to polyimide powder having the following characteristics: Resin (I) has a monomer mol ratio of 2:1, ODA:BTDA and a calculated molecular weight of 929; Resin (II) has a monomer mol ratio of 100:103, ODA:BTDA and a calculated molecular weight of 16,272.

Grinding Wheel Testing: The grinding wheels are tested in dry grinding operation on commercial grinding machines using tungsten carbide workpiece material designated as "C-2" tungsten carbide and containing 94 weight percent tungsten carbide and 6 weight percent cobalt. Test conditions are as follow: wheel speed 3,900 rpm, infeed 0.025 mm/pass, table speed hand controlled. A Cincinnati #1 grinding machine was used.

Example: In this example, coreactant resins (I) and (II) are mixed in weight proportions of 4.08 and 5.92, respectively, to yield a mol ratio of 12.1. As a preliminary experiment, a chip of the crosslinkable polyimide composition is compacted in a mold at a temperature of about 410°C and a pressure of 35 megapascals (5,000 lb/in^2) for 20 minutes. The chip has a glass transition temperature of about 285°C and is not crosslinked. The chip is then heated at about 400°C for about 3 hours to complete the molding and to accomplish the crosslinking reaction. As a result of that heating step the glass transition temperature is increased to about 322°C, indicating that crosslinking has occurred. Glass transition temperatures are determined by thermomechanical analysis.

A second chip is made and compacted at about 390°C under the same pressure and for the same time as above. The glass transition temperature is only about 212°C due to the decreased compacting temperature. A heat treatment at 400°C for about 3 hours results in a crosslinked polyimide material with a glass transition temperature of about 329°C.

To prepare a crosslinkable polyimide composition for making a grinding wheel, the same mixture of base and crosslinker resins is combined with 150 grit (88 to 105 micrometers particle size) copper-coated diamonds and 220 grit (62 to 74 micrometers particle size) silicon carbide in proportion to yield a molded abrasive grinding wheel rim containing 17% diamonds, 8% copper coating, 30% silicon carbide, and 45% coreactant resin mixture. The rim is compacted in a mold at about 410°C under pressure of 35 megapascals for about 20 minutes. The compacted rim is then removed from the mold, allowed to cool, and heated to about 400°C at a rate of about 1.8 degrees per minute. The temperature of 400°C is maintained for about 3 hours. The rim is affixed to an aluminum core and tested in accordance with the conditions of the above test, yielding a grinding ratio of 1.8.

The wheel, rim and core, is exposed to an additional heat treatment of 435°C for about 1 hour to increase crosslinking. The wheel, retested under the same conditions, is then found to have a grinding ratio of 17.9.

Nonfoamed, Friable Polyurethane Substrate

M.A.F. Hiles; U.S. Patent 4,049,396; September 20, 1977; assigned to National Research Development Corporation, England provides a friction or abrasive composition comprising a dispersion of friction or abrasive particles in a polymeric substrate that is a polyurethane which becomes friable at a temperature below its softening point.

The polyurethane substrate is one that becomes friable, that is to say, the polyurethane suffers a degradation such that its resistance to abrasion is reduced when heated to a temperature below its softening point. Simple tests on any individual polyurethane can be made to determine whether it will be suitable for use in this process. For example, a sample of the material may be heated until a gentle rubbing action enables particles to be dislodged from its surface. If the temperature at which this occurs is below the softening temperature of the material then the polyurethane is acceptable for use in the process. Alternatively the surface of the sample may be heated by frictional contact with a workpiece and the surface of the sample examined at regular intervals. If the sample is abraded to form a particulate or powdery deposit then it will be suitable for use in this process, but if the surface melts or softens then it will not.

For most applications elastomeric polyurethanes are preferred for greater toughness, and these may be generally flexible or resilient depending upon the application. Rigid polyurethanes may also be used in some circumstances, for example, in abrasive cones.

The physical properties of the polyurethane substrate will of course depend on the chosen application and should be arranged such that the polyurethane has a softening point higher than the intended working surface temperature of the composition. The softening point is herein defined as the softening point determined by the Kofler hot bar technique. Preferably the softening point of the polyurethane is greater than 150°C and most preferably in the range of from 200° to 300°C.

For abrasive compositions, and in particular for abrasive discs, rods, belts and other shapes, the polyurethane preferably has a tensile strength of 600 to 9,000 lb/in^2, an elongation to break of from 100 to 200% and a tear strength of from 50 to 200 pounds per linear inch (Graves Test).

The composition may comprise a variety of friction or abrasive particles, but particularly preferred are those having a low oil absorbency.

Preferably the friction or abrasive composition contains at least 10% by weight of abrasive particles, and most preferably from 15 to 90% by weight of abrasive particles based on the total weight of the composition.

Example: An abrasive disc suitable for use on a pistol drill operating at about 2,400 revolutions per minute is composed as follows:

Material	Parts by Weight
Polyol (obtained from Bostik Ltd. under Compound PM 117X)	70
4,4'-diphenylmethane diisocyanate	20
Abrasive 60 grit	180

The abrasive grit is tungsten carbide though this may be replaced by silicon carbide or glass. The polyol is defined in the usual way by the physical properties of the polyurethane which is produced.

In this case the polyurethane produced by the polyol has a hardness of 65 to 75 on the Shore A scale, a heat deflection temperature of 100°C, a tensile strength of 1,000 psi (ASTM D 638-61 T) and an elongation at break of 110%. The material was molded to produce a one-piece abrasive disc which was mounted on a pistol drill. The disc is used to abrade a 5 inch thick block of aluminum. After it has abraded the aluminum through, the disc is examined and no signs of clogging of the working surface of the disc are found, the disc being only slightly worn and being suitable for continued use.

Resin-Metal Composite Bonding Matrix

R.H. Sioui; U.S. Patent 4,042,347; August 16, 1977; assigned to Norton Company has discovered that grinding wheels which are particularly effective for the wet grinding of cemented carbide tools can be made by employing diamond grit bonded in a composite matrix of resin and metal so fabricated that all of the powder particles (both resin and metal) have been coalesced into solid continuous, or essentially continuous, phases. Neither the resin nor the metal then is a "filler" in the other, in the sense of a particulate powder, the powders having lost their identity as such in the application of heat and pressure.

In order to achieve the abovedescribed result, it has been found necessary to employ, as the metal part of the system, a combination of metal powders which is fabricable below 450°C, but which, after fabrication, results in an intermetallic compound or an alloy which melts above 500°C. The limit of 450°C is established by the recently available high-temperature resins, none of which are sufficiently heat stable to be fabricated at temperatures above 450°C. It will be possible to increase this limit as resins of increasingly greater thermal stability are developed.

The preferred systems require the presence of at least one elemental metal powder in the mix to be fabricated, selected from the group consisting of tin, bismuth, and indium. Tin will form intermetallic compounds (melting above 500°C) with silver, cobalt, copper, iron, manganese, nickel, tantalum, and titanium. Bismuth will form suitable intermetallics with manganese, nickel, and titanium, and indium will form suitable intermetallics with silver, copper, manganese and nickel. Bonds may also be formed by combinations of the above systems, the only requirement being that the mix to be fabricated include a metal powder which will melt during fabrication and react with other metal present to form a metal phase having a melting point above 500°C.

Under processing conditions employed in this process, it has been found that the elemental metals do not completely react with each other. Thus, when a copper-tin system is employed, the resulting product will include elemental tin, elemental copper, and the intermetallic Cu_3Sn and a lesser amount of other Cu-Sn intermetallics. The metal matrix in such a case is thus composed of three individual phases which form a single mechanically interconnecting or continuous matrix.

A preferred embodiment of the process employs resin bond type copper clad diamond, with a mixture of copper and tin powders as the precursor of the metal matrix. In this preferred embodiment, a portion of the elemental tin reacts with the copper cladding of the diamond to make the cladding a mechanically continuous part of the metal matrix of the bonding matrix. In cases where borazon (cubic boron nitride) is employed as the abrasive, in coated form, a nickel-coated abrasive grain is preferred.

The resin phase of the matrix may be any bonding resin which is infusible in its final form. Thus the precursor for the resin phase of the bond may be a thermosetting prepolymer such as a "B" stage phenolic powder, or may ba a coalescible powder of an infusible polymer such as a polyimide as in U.S. Patent 3,523,773. By infusible, is meant a resin which does not melt upon heating to 250°C. This term thus includes thermosetting resins and high-temperature essentially non-crosslinked polymers such as the polyimides and polyphenylene sulfides, which can be molded by application of heat and pressure to the resin in powdered form.

The bond of this process may also include conventional finely divided particulate fillers heretofore employed in grinding wheels such as aluminum oxide, silicon carbide, and boron carbide as abrasive fillers, MoS_2, polytetrafluoroethylene, graphite, hexagonal boron nitride, as lubricating fillers, and metal fillers that do not melt or coalesce to become part of the continuous metal matrix. Among these fillers, silicon carbide and graphite are preferred.

The operative ratio of metal to resin, by volume, for improved results against a standard commercial phenolic bonded diamond wheel of the same diamond content, is from 5:95 to 95:5, the preferred range is from 15:85 to 85:15, and the optimum range is from 55:45 to 75:25. The abrasive content can be as high as 65 volume percent, the preferred range is from 5 to 50% by volume, the optimum is from 10 to 30% by volume.

The metal powder, for forming the metal matrix of the bond, may contain from 10 to 80%, by weight of the low-melting metal, preferably from 30 to 50% by weight, of the metal powders.

Active Filler Placed on Abrasive Grain

Active fillers are those which, based on chemical and physical reactions, cause an increase in the service life of the abrasive grain and lead to a "cooler abrasion." (Temperatures which arise during the grinding are kept at a lower level.) One advantage is that the cutting surface is smoother; therefore improved surface quality results.

According to the process of *W. Oberbichler; U.S. Patent 4,021,208; May 3, 1977; assigned to Tyrolit-Schleifmittelwerk Swarovski KG, Austria*, porous abrasive grain whose pore space is completely or at least partially filled with active fillers, as well as ordinary, nonporous abrasive grains which are, according to the process, covered with a filler coating will be used. The extent of the filling largely depends on the manufacturing process. According to the process the coating is exclusively or at least primarily active in the abrasive process. This is an important contrast to hitherto known grain coatings, which have exclusively been employed to influence the adhesiveness of the grain without the bond and to improve heat transfer from the grain to the bond (e.g., covering with a ceramic such as an inorganic coating which contains grains and completely or partially inert substances, such as metallic oxide).

According to the process the abovementioned porous abrasive articles and the ordinary abrasive articles with a filler coating will be covered with a sealing coating, if hygroscopic, liquid or sublimating fillers are used.

Example: An abrasive disk according to the process may be composed of the following in percent by weight: Al_2O_3, normal corundum (24 mesh), 80%; cryolite melted on the grain as coating, 7%; phenolic resin, liquid-resole (e.g., Bakelite 433), 4%; and phenolic resin, pulverous (novolak hexa, e.g., Bakelite 222), 9%. The density of the disk is 2.45 g/cm^3.

In an embodiment of the process porous abrasive grain, e.g., sintered corundum, as well as smelter corundum may be used. The pore space of the abrasive grain is almost completely soaked with the filler. The abrasive grains together with the filler are, in turn, embedded as a whole in the bonding agent. This embodiment may be composed of the following in percent by weight: porous corundum Al_2O_3 with approximately 20% of the pore space filled up to around 80% with elementary sulfur, mesh 24, 87%; phenolic resin, liquid (Bakelite 433), 4%; and phenolic resin, pulverous (Bakelite 222), 9%. The density of the disk was 2.45 g/cm^3.

In this embodiment it is possible for it to act directly on the grit surface, thus contributing to an essential prolongation of the service life of the abrasive disk and to a cooler abrasion.

The data refer to a disk 150 x 25 x 20 cm. The respective blend was pressed into a form and cured in the oven at a temperature of 180°C under the usual conditions for abrasive disks bonded with phenolic resin.

The advantage of such an abrasive disk is obvious. The fillers may become active in the very place of the abrasive process, namely in the abrasive zone of the abrasive grain.

Pyrolytic Carbon Matrix

It is an object of the process of *D.H. Leeds; U.S. Patent 3,874,856; Apil 1, 1975; assigned to Ducommun Incorporated* to provide a method of making a product in which the multiplicity of abrasive or refractory particles is held in form and position in the product and in which the supporting network is capable of withstanding high temperatures and of efficiently dissipating heat, is capable of withstanding thermal shock, and which will conduct electricity through the product.

In the method the abrasive particles are first mixed with a suitable binder which will hold the abrasive in place during a carbonizing process and which will carbonize leaving a carbon skeletal network, or a skeletal network prior to the time the pyrolytic carbon matrix is being formed, with the result that the pyrolytic matrix replaces in part the original binder without the abrasive particles falling from their proper locations.

In making the composite product the abrasive particles and binders are premixed either in a dry mix or a wet mix. The mixture is then cast into a suitable shape by pouring it into a form.

The green product may be air dried at ambient or less than 200°F temperature in order to develop strength without rupturing bonds developing in the early stages of drying.

The next step to be performed is that of "burning out" or pyrolyzing the organic binder into a carbon and/or inorganic skeleton which will hold the abrasive or refractory particles in position during the infiltrating and graphitizing steps. This is accomplished by raising the temperature of the green product to a suitable temperature for a sufficient period of time to accomplish the pyrolyzing to form the skeletal inorganic/carbon network. This usually occurs at a temperature of 400° to 900°F.

The green product is placed in a suitable furnace for infiltrating with pyrolytic carbon in which furnace the temperature is raised over a period of from 1 hr to 5 hours, to a temperature (after first pyrolyzing the organic binder and forming the supportive inorganic/carbon network), at which carbonaceous gas such as methane will deposit an internal coating of pyrolytic carbon, or a coating of pyrolytic carbon and vitreous binder or a matrix of pyrolytic carbon and a nonvitreous binder.

The abrasive particles may be any of those commonly used for abrasive wheels, stones, hones, or the like. The size of the abrasive particles may vary in accordance with the requirements and are normally those commonly used in the grinding industry. The binders which may be used comprise glucose, dextrose, sucrose, maltose, starch, dextrin and gums, and mixtures such as dextrose and vitreous material. Also ethyl silicate may be used which will form an inorganic vitreous bond. Sodium silicate has been successfully used as an inorganic binder. Where an inorganic binder is used the inorganic material is mixed with the abrasive or refractory particles and liquid is added, if necessary, to form a slush.

In the infiltration step of the process, methane or propane or natural gas or other suitable carbonaceous gas or mixtures of carbonaceous gases may be used. The temperature at which the infiltrating is done is found to be highly satisfactory in the range of approximately 1,900° to approximately 2,300°F. The pressure maintained in the furnace during infiltration is from 1 to 765 mm Hg. Very good results are obtained in pressures between 10 to 50 mm Hg. The infiltration time is from a minimum (gas on to gas off) of approximately 10 to 60 hours to a maximum of approximately 350 to 400 hours. The rate of flow of gas through the furnace has been successfully practiced at from 8 to 2,100 ft^3/hr, depending on the furnace size and loading. In the practicing of the infiltration step better results are obtained where the gas is preheated, and air is excluded from the furnace.

The carbonaceous gas will deposit carbon both on the abrasive particles and on the network whether it be carbonaceous or glassy or vitreous or another inorganic product. The porosity of the finished product generally varies from less than one-fifth the particle size to larger than actual particle size.

Grains

ALUMINA

Dry, High Transmissive Energy Milling

The process of *B. Hill; U.S. Patent 4,205,964; June 3, 1980; assigned to The International Nickel Company, Inc.* involves dry, high transmissive energy milling of powdered ceramic materials to produce dense, composite, product powder particles, the product particles being characterized by an intimate intradispersion of initial constituent particles, a large internal interfacial surface within individual product powder particles, and a surface area less than that of the original individual particles when the particle size of the initial powder constituent is small, less than 2-3 microns.

A single ceramic material or a mixture of two or more thereof can be subjected to dry, high transmissive energy milling, whereupon composite particles are produced having individual constituents distributed substantially uniformly at close interparticle spacings within the product powder particles. Ceramic powder composites of two or more constituents are less brittle than the brittle components from which they are formed and are, unexpectedly, quite homogeneous, being intimately interdispersed through mechanical cold bonding in the solid state. Depending upon particle size of the initial materials and milling time, interparticle spacings in the product powder will usually be substantially less than 10 microns, e.g., 5 microns or 1 micron or even much less, e.g., 0.1 micron.

The high transmissive energy dry milled ceramic powders of the process are characterized by improved bulk or pack density (e.g., when poured into a container and packed by tapping container), by enhanced pressability to provide better green strength and by superior hot pressing characteristics in comparison with identical materials prepared by ordinary ball milling, the latter utilizing liquid media for particle fragmentation or a surfactant under dry conditions for a similar purpose, grinding being accomplished largely through gravitational action. For example, dry, high transmissive energy milled micron or submicron sized pure alumina bodies are readily hot pressed within 15 minutes to virtually 100% density at 1500°C, whereas conventionally ball milled alumina seldom reaches 100% density even when sintered at substantially higher temperatures, e.g., 1650°C. Furthermore, pressed and sintered or hot pressed alumina compacts are fine grained and are uniform in grain structure, with the deviation from the mean grain size seldom exceeding about 10%. Additionally, less shrinkage occurs during firing, thereby contributing dimensional accuracy to the finished part. Hot pressed bodies produced according to the process are devoid of porosity.

High transmissive energy dry milling may be accomplished in high-energy machines such as the Szegvari attritor, the high-speed laboratory shaker mill (Spex mill) or even in vibratory ball mills, but the ratio of ball-to-powder should be at least about 3:1, preferably at least 5:1.

It is important that the surface lattice structure of the ceramic powders be broken down, altered or otherwise disrupted by the dry, high transmissive energy bombardment. By so doing, a considerable number of surface defects are formed, it being deemed that such defects are necessary to enable interfacial bonding to occur, and particle growth of small initial powders to ultimately take place; otherwise, mere agglomeration will primarily take place (agglomeration has largely occasioned the use of liquid media and surfactants heretofore).

The onset of the mechanical cold bonding phenomenon can be referred to as the "threshold point" and is depicted at region **X** in the curves of Figure 4.1. It represents the approximate point in time during which the slope of the comminution vs bonding curve is changing from a comminution zone (powders are primarily comminuted or fragmented) through a neutral zone (tangent to the curve being about zero at which point neither comminution nor bonding significantly exceeds the other) and then into a bonding zone value (particle growth and reduced surface area predominates). Milling should be continued well beyond the "threshold point."

This mechanical cold bonding is depicted in a general sense in Figure 4.1, the curves representing alumina (average particle size of 0.3 micron), dry, high transmissive energy milled (using a Spex mill) at two different ball-to-powder ratios, 5:1 (Curve **A**) and 20:1 (Curve **B**). It will be observed that for approximately 15 minutes (Curve **A**) and for about 7-8 minutes (Curve **B**), respectively, the alumina was undergoing comminution; thereafter, however, the particles mechanically bonded, surface area decreasing.

The overall milling period, as will be understood by those skilled in the art, cannot be precisely defined for it will obviously depend on the particular ceramic powders being milled, the amount of energy transmitted from the bombarding media to the powder, including ball-to-powder ratio, etc. For commercial purposes, a milling period of at least 1 or 2 hours should be used. The upper end of the milling period need not exceed 10-15 hours.

Figure 4.1: Milling Time vs Surface Area of Alumina Powder

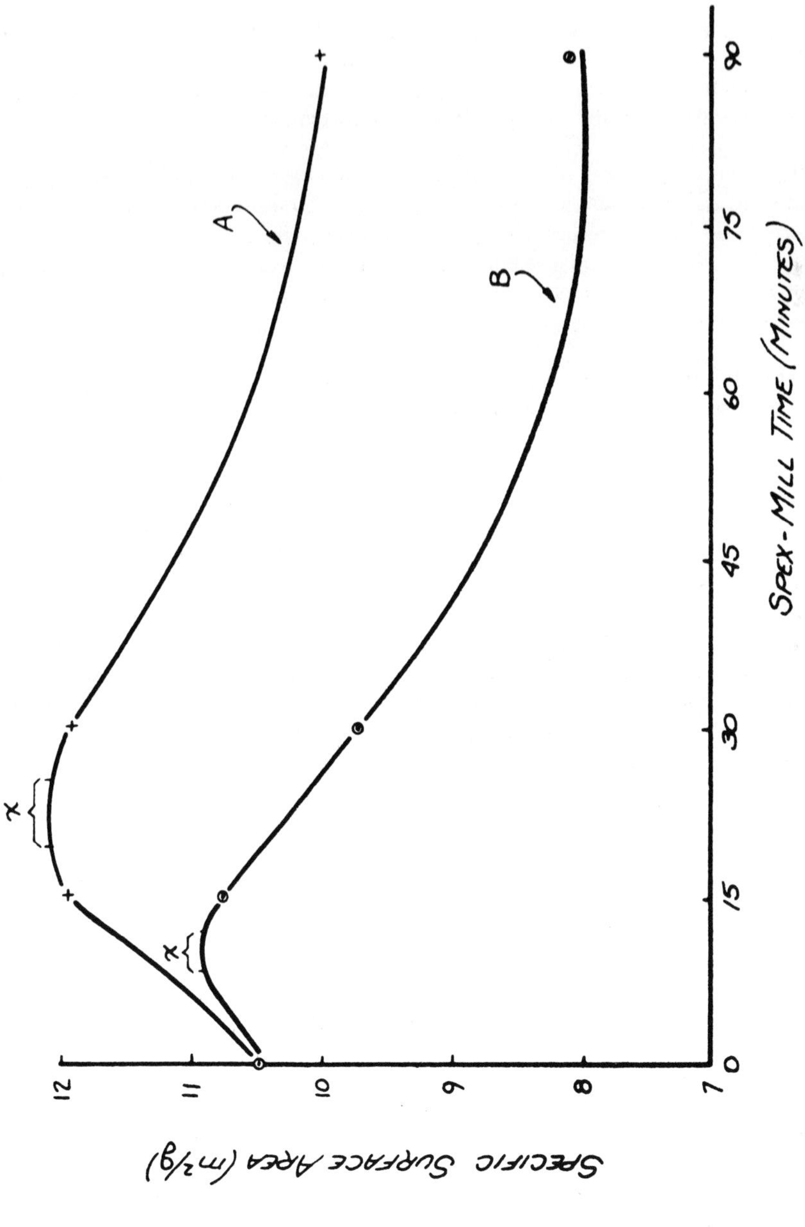

Source: U.S. Patent 4,205,964

With regard to the particle size range of powdered materials treatable with the process, they may vary over the range from about 0.01 to about 150 microns, preferably not exceeding about 0.05 to about 10 microns.

From Spent Catalyst

H. Zeiringer; U.S. Patent 4,142,871; March 6, 1979; assigned to Treibacher Chemische Werke AG, Austria found that high-quality abrasive materials may be obtained not only from clay or bauxite but also from spent catalytic material comprised of a metallic catalyst on an alumina carrier.

The process comprises the steps of melting the catalytic starting material with a reducing agent to obtain a melt consisting of a melt component including the alumina on an alloy residue, cooling the melt at a speed correlated with a desired crystallite size of the abrasive material to be obtained, and mechanically separating the melt component from the alloy residue before or after solidification. The melt component constitutes the abrasive material.

Preferably, the starting material is melted with at least one reducing agent, such as carbon, for instance coke, so that the melt component comprises at least a portion of the reducing agent. Aluminum may also be used as reducing agent. If the abrasive is to contain not only alumina but another component, an additive selected from the group consisting of zirconia, titania, chromic oxide, lime, magnesia, silica, zinc oxide and a rare earth oxide may be added to the alumina melt component. It is preferred to roast the catalytic starting material before it is melted.

The starting material may be melted in an electric arc furnace in the presence of the reducing agent or agents. If it contains much sulfur, it is advantageous to roast the material first to remove the sulfur. Depending on the desired properties of the abrasive material, any of the above listed additives may be added in any desired amounts, particularly useful additions including about 5 to 60%, by weight, of zirconia and/or up to 10%, by weight, of titania.

The size of the corundum crystallites obtained in the process may be influenced within wide limits by the rate of cooling of the melt, the crystallite size of the abrasive determining its usefulness for various fields of utilization. Considerable differences in the crystallite sizes and the corresponding utility of the abrasive can be obtained by solidifying the melt very slowly in a block, on the one hand, and rapidly cooling by casting it over steel balls, on the other hand. The resultant sizes of the corundum crystallites may accordingly vary, in dependence on the method of cooling, from 1 mm to 0.001 mm.

The solid alumina product component of the process is an excellent abrasive after comminution and classification into various grain sizes, with or without heat treatment. Depending on the type and amount of additives as well as the crystallite size, these abrasives may be used for polishing steel, precision polishing under very light pressures or polishing wood.

Example: Ten thousand kilograms of a roasted spent MoCo catalyst comprised, by weight, of 3.5% Co, 7.7% of Mo, 0.05% of S, 0.62% of Si and the remainder an Al_2O_3 carrier was mixed with 500 kg of coke, and the mixture was melted in a three-phase electric arc furnace. The melting vessel was a water-cooled, upwardly tapering sheet metal cylinder. After the vessel was filled with melt, electric current was cut off, and the melt was stored for 1 week during which it cooled slowly. After the solidified block was broken up, 1,206 kg of CoMo alloy was obtained. The remaining block of alumina weighed 8,239 kg and 443 kg of a poorly molten transition zone containing CoMo alloy was removed.

The alumina abrasive material had the following composition, by weight: 0.104% Fe_2O_3, 0.15% CoO, 0.06% MoO_3, 0.04% S, 0.05% C, 0.58% SiO_2, 0.05% MgO, and remainder Al_2O_3.

Coarse Crystals by Use of Foreign Salts

B. Schepers, H. Hellinghausen and A. Kramer; U.S. Patent 4,130,402; Dec. 19, 1978; assigned to Guilini Chemie GmbH, Germany found that aluminum oxides with an average primary crystal size of 6.1 to 25 μ, particularly 8 to 16 μ, can be produced by effecting the calcination of the aluminum hydroxide in the presence of at least one fluoride and vanadium salt.

Vanadium salt is here understood to mean the foreign salts which are precipitated from concentrated sodium aluminate liquor at many alumina plants, which salts contain, in addition to much sodium carbonate, fluctuating quantities of vanadium, phosphorus, fluorine and arsenic compounds. The foreign salts are known to accumulate in circulating sodium aluminate liquors of the Bayer process and are separated from the sodium aluminate liquors by the addition of CaO, either together with the red mud or by means of a special crystallization process, until the liquor drops below the saturation limit.

The composition of the so-called vanadium salt fluctuates considerably. As a result the quantities of the vanadium salt employed can be given only if the composition of the vanadium salt is known. The calcination of the process preferably is effected at temperatures of 1100° to 1300°C.

The aluminum hydroxide which customarily is used for the calcination to produce aluminum oxide has an Al_2O_3 of 65% by weight and contains the following quantities of trace elements by weight: $V_2O_5 < 0.005\%$, e.g. 0.0003%; $P_2O_5 < 0.005\%$, e.g. 0.0004%; $F < 0.01\%$, e.g. 0.004-0.008%; and $Na_2O < 0.6\%$, e.g. 0.3-0.5%.

At least one fluoride compound is added as a mineralizer to the starting aluminum hydroxide. The fluoride compound which is added as a mineralizer can be a fluoride such as AlF_3, CaF_2, or Na_3AlF_6, or mixtures thereof, and generally is added in an amount 0.001 to 0.1% by weight with respect to the Al_2O_3 content of the starting aluminum hydroxide.

In the practice of the process, and without considering the added 0.001 to 0.1% by weight of mineralizing fluorides with respect to Al_2O_3, the starting aluminum hydroxide according to the process has added to it so much of the vanadium-salt-containing foreign salt that the starting aluminum hydroxide (alumina hydrate) contains the following quantities of trace elements by weight: V_2O_5 > 0.005%, e.g. 0.005-0.03%; P_2O_5 > 0.005%, e.g. 0.005-0.03%; F > 0.01%, e.g. 0.01-0.03%; and Na_2O > 0.6%, e.g. 0.8-3%.

If an aluminum hydroxide of this composition is fed into a rotary drum furnace in the presence of the added fluoride compounds required according to the process which act as mineralizers, and particularly in the presence of 0.001 to 0.1% by weight of such fluoride compounds with respect to the Al_2O_3, the resulting growth in the average primary particle size during calcination at temperatures above 1100°C and with the otherwise customary furnace settings in dependence on time is such that the resulting average primary particle size of the aluminum oxide is greater than 7 μ. Average primary particle sizes of greater than 10 μ, 12 μ, or 16 μ can be achieved by the practice of the process, and at most the average primary particle size can be about 25 μ.

The chemical composition of the resulting aluminum oxides indicates the following weight contents: V_2O_5 > 0.005%, e.g. 0.005 to 0.05%; P_2O_5 > 0.005%, e.g. 0.005 to 0.05%; and Na_2O > 0.8%, e.g. 0.8 to 3%.

Example: An aluminum hydroxide doped with vanadium salt to contain 0.01% V_2O_5, 0.01% P_2O_5, 0.02% F and 1.2% Na_2O, and doped with an additional 0.02% AlF_3 with respect to the Al_2O_3, is charged into a rotary drum furnace which is operated for the manufacture of a known aluminum oxide. In the production of the known aluminum oxide, a mixture of gamma and alpha Al_2O_3 is produced with a specific surface of 5 to 30 m^2/g and an average primary particle size of about 1 μ. The calcination temperature is about 1200°C.

By using the doped aluminum hydroxide in accordance with the process as the starting material which is charged into the rotary drum furnace, already after 6 hours at about 1200°C, heavy crystal growth occurs and, after 18 to 24 hours, an aluminum oxide is obtained which has an average primary crystal size of 12 μ, as required for certain purposes in the grinding and polishing art.

Friable Sintered Grains

It is an object of the process of *J.F. Busch and S.M. Freeman; U.S. Patent 4,086,067; April 25, 1978; assigned to International Telephone and Telegraph Corporation* to produce sintered aluminous abrasive grains of such friability as will assure the grains will break down when in use as integral parts of grinding wheels to maintain clean cutting surfaces on the grinding wheels.

The foregoing and ancillary objects are accomplished by inducing pores into grains of sintered abrasive, during manufacture, where the pores are of sufficient size and number to allow the grains to chip and crack in a desired manner during grinding. This involves the introduction of sized pieces of coarse grit burn-out material such as, but not limited to pieces of paradichlorobenzene, crushed walnut shells, pieces of cork, pieces of plastic, or mustard seeds. Other organic resins or plastics such as polyethylene, polyesters, polystyrene, polyamides, phenolic and urea-formaldehyde resins and rubbers such as neoprene and butadiene-styrene may also be used.

Other combustible materials may also be used providing they can be prepared in the required size ranges, can be extruded when mixed in a paste with abrasive material and can be burned out of grains of abrasive material without producing undesirable effects on the abrasive.

By introducing burn-out material having an average diameter between about 20 and 80% of the cross-sectional width of the grain and in an amount between about 3.5 and 20 volume percent of the grain into the grain granules, a superior abrasive grain is produced after firing. When this abrasive grain is incorporated into a grinding wheel, such as would be used in the snagging of carbon steel billets or other grinding operations, a wheel results which cuts freely and resists loading. In addition, pressure levels approximately equal to those employed with fused abrasive grinding wheels need not be exceeded.

Also, the coarse grit burn-out material, in leaving pores behind, enables bonding to be more tenacious through increased contact surface areas permitting better interlocking of the abrasive grains and the bonding material.

In the production of abrasives for use in the practice of the process, a high alumina abrasive composition is first prepared. A typical composition would consist of about 93% aluminum oxide with varying amounts of other materials. A batch of this aluminous abrasive material is ground in a mill using conventional procedures until the particles produced have an average size of less than three microns. If wet milled, this material is then dried or dewatered to prepare it for controlled moisturizing in subsequent stages.

The batch is placed in a Muller Mixer and rewet by alternately adding water and a selected quantity of burn-out material of selected size until a body with sufficient plastic consistency for extrusion is obtained. The body of material is then extruded to produce elongated bodies having a cross-sectional size as defined below, is dried, granulated and fired above 2000°F. The burn-out material will consist generally of between about 3.5 and 20 volume percent of the batch material (with a permissible variation of about $\pm 0.5\%$) and is generally selected of an average size between about 20 and 80% of the cross-sectional width of the extruded material.

In preferred embodiments, extrusion has included the production of material ranging in cross-sectional size from a coarse mesh of about 8 to a fine mesh of about 20. The grains generally are cut to lengths between about $\frac{1}{8}$ and $\frac{3}{16}$ inch to produce grains retaining diameters substantially as defined above. Firing above 2000°F is maintained long enough to sinter the particles in the grains together and cause the burn-out material to burn. After the burn-out material is burned, pores are left in the grains having average diameters between about 20 and 80% of the cross-sectional diameter of the grains and occupying between about 3.5 and 20 volume percent of the grains, with small permissible additional variations, as indicated.

ALUMINA-ZIRCONIA

Immersion Process

H.F.G. Ueltz, M.A. Dashineau and J.J. Pino; U.S. Patents 4,194,887; March 25, 1980; 4,073,096; Feb. 14, 1978; both assigned to U.S. Industries, Inc. provide a process for the manufacture of abrasive material in which (1) an abrasive mix is brought to fusion as in an electric arc furnace, (2) a relatively cold substrate is dipped into the molten material whereby a layer of solid abrasive material is quickly frozen (or plated) on the substrate, (3) the plated substrate is withdrawn from the molten material and (4) the solidified abrasive material is broken away from the substrate and collected for further processing to produce abrasive grain.

Referring to Figure 4.2, a Higgins type furnace **1** is supplied with graphite electrodes **2** depending into the abrasive melt **3** therein. The abrasive material is kept in the molten state by passage of electric current from the electrodes through the said abrasive material in a manner well-known in the art. Passing overhead, transverse of furnace **1**, and between the electrodes **2**, is a trolley conveyor **4** including beam **5**, chain **6**, trolleys **7**, and clevis **8** which supports a plurality of heat sinks **9** as is shown. Movement of the chain, trolley, and clevis assembly transports heat sinks **9** serially down slope **10** so that the lower portions of heat sinks **9** are immersed in the melt.

Figure 4.2: Immersion Process

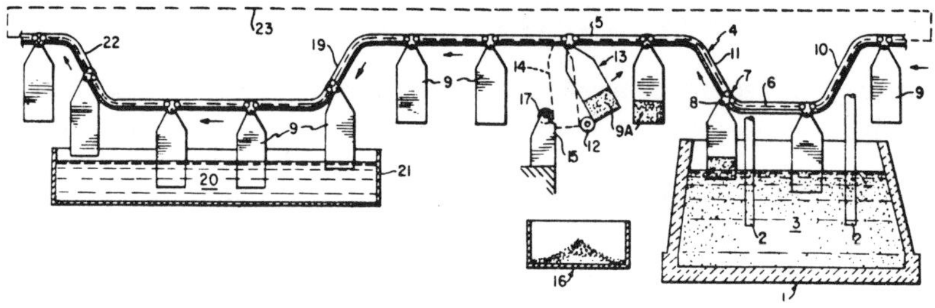

Source: U.S. Patent 4,194,887

Heat sinks **9** remain in the melt for a time established by the linear velocity of the chain, whereupon the heat sinks are withdrawn from the melt by an upward slope of the conveyor **11**. The time for which sinks **9** remain in the melt is sufficient to plate sinks **9** with the desired thickness of solidified abrasive material **9A** on the sinks.

Heat sinks **9** are serially transported so as to impinge upon a roller **12** which raises the heat sink to an inclined position (Reference sink is shown in inclined position **13**). As the heat sink rides over roller **12**, it falls abruptly to position **14** shown in dashed lines, where it forcibly strikes an immovable block **15**, or the like, thus knocking off the loosely-adherent plate-like abrasive material which falls into box **16** or alternately into a chute (not shown) for recovery and further processing into abrasive grain.

Heat sink **9**, when thus freed of abrasive material **9A**, is then transported so as to pass over roller **17** after which it resumes its normal depending position. At this point the heat sink is still hot. It is then transported down slope **19** so that the lower portion of the heat sink is immersed in water **20** preferably at a temperature between about 5° and 95°C in the case of the preferred steel heat sink, contained in tank **21** to cool the heat sink as it is moved along. Cooled heat sink **9** is raised from the water by slope **22** and further transported to the point of beginning. In this diagram, the continuity of the process is indicated by dotted line **23**. Not shown are conveyor drive, cornering mechanisms, and supporting structures, which may be of conventional design.

In most instances an abrasive deposit of between about 0.5 and 3.0 mm provides satisfactory abrasive grains from alumina-zirconia compositions. Preferably the thickness of the abrasive deposited on the heat sink substrate is approximately equivalent to one dimension of the abrasive grain size being sought in the final product.

Rapid Cooling in Molten Tin

In the process of *L. Coes, Jr.; U.S. Patent 4,140,494; Feb. 20, 1979; assigned to Norton Company* molten aluminous oxide abrasive, particularly molten mixtures of alumina and zirconia, are rapidly cooled and solidified by pouring into an agitated molten metal bath, having a boiling point above the temperature of the molten abrasive, such as molten tin, while being protected from oxidation. The solidified abrasive particles float to the top of the metal bath, are removed and separated from adhering metal, as by centrifuging, and crushed to abrasive grit-sized particles.

The cooling rate in the molten bath is more rapid than with the same melt in contact with a cold steel surface as in the prior art because of the higher heat transfer coefficient. That is there is no impedance of heat flow by the presence of oxide films on the steel or by lack of perfect contact.

The equipment for carrying out the above procedure is shown in Figure 4.3. In this is shown a conventional arc fusion furnace **1** in which the oxide or mixture of oxides are fused. The fused material is then poured into a bath of molten tin **2**. This is preferably done through a tube **8** which is kept filled with an inert gas, such as argon, introduced through inlet **9**.

Before the pouring of the abrasive the bath **2** is brought to slightly above the

Figure 4.3: Rapid Cooling in Molten Tin

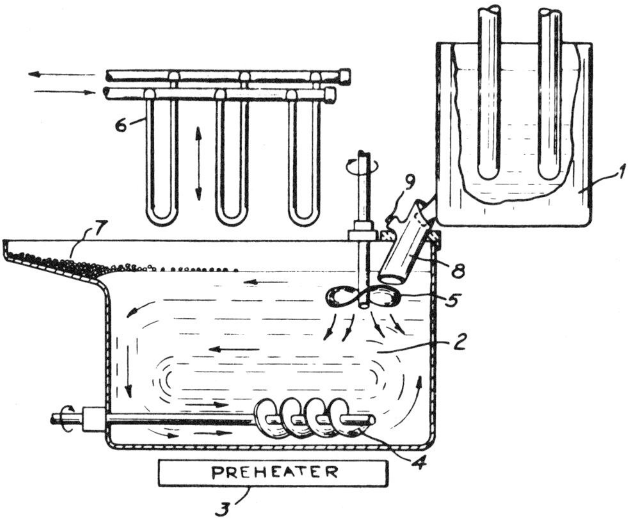

Source: U.S. Patent 4,140,494

melting point of tin by means of a preheater **3** which may be gas or oil fired, or electrically heated. The bath is kept in circulation by means of the impeller **4** and kept in turbulence in the pouring area by the agitator **5**. The speed of the agitator **5** is adjusted to give the required granularity to the product.

Once pouring has begun the bath temperature begins to rise because of the latent heat of the fused oxides. The bath temperature is controlled to somewhat above 235°C by means of the lowerable water cooled coils **6**.

The product floats to the surface, at a temperature below that at which oxygen can produce any deterioration, and collects in the area **7** from which it can readily be skimmed off. Any adhering tin can be removed by centrifuging the product at a temperature above about 235°C.

Rapidly Cooled Cofused Alloy

G.R. Watson; U.S. Patent 4,126,429; Nov. 21, 1978; assigned to Norton Co. produces a high purity cofused alumina-zirconia, containing from 22 to 28% zirconia and an amount of magnesia (in the range of 1 to 2% by weight) such that the magnesia to alumina weight ratio in the product is from 0.014 to 0.029. It is essential that the product be of high purity, and be rapidly cooled to a product in which the abrasive grits are polymicrocrystalline.

The rate of cooling of the product is important, and the methods of cooling of U.S. Patent 3,891,408 are employed. Sufficiently rapid cooling is achieved by pouring the melt into the interstices of a packed mass of carbon steel balls 1.5 inches in diameter or smaller, or between carbon steel plates, spaced $^3/_{16}$ inch apart or closer. The rapidity of cooling (other things being equal) determines the size of the crystal structure of the product.

Specifically, the cooling rate must be high enough to produce a solid product in which the rodlike (and sometimes platelike) zirconia crystallites in the alumina-zirconia eutectic have a rod-to-rod or plate-to-plate spacing of less than 4000 Å as measured at the eutectic colony centers by the random intercept technique. The product also contains colonies of spinel-zirconia eutectic, in which the rod spacing is finer than in the alumina-zirconia eutectic.

The solidified abrasive is impact crushed to produce strong abrasive grits in the desired size range (10 to 24 grit) when the abrasive is intended for use in grinding wheels intended for severe grinding applications such as heavy duty grinding. Where the abrasive is intended for use in coated abrasives (such as belts, or discs) the product may be jaw crushed and selectively screened (as by a slotted screen or by a Sutton table) to obtain elongated grits.

Other operations, such as mulling or pan milling may be performed on grits which are intended for heavy duty grinding.

Cooling Between Spaced Plates

In the process of *J.J. Scott; U.S. Patent 4,070,796; January 31, 1978; assigned to Norton Company* a molten abrasive composition is poured between a plurality of spaced metal or graphite plates spaced ½ inch or less apart, with the ratio of volume of plates to volume of space between plates being preferably at least about 3 to 1, depending upon the material being poured and the material of the plates. Lower ratios may be employed if supplemental cooling methods, such as internal water cooling or electrostatic cooling is employed or if reuse of the plates is not necessary.

The spaces in which the molten abrasive is poured should have a controlled maximum, and are conveniently uniform in thickness, but need not be planar or uniform in shape. A convenient and simple arrangement involves a plurality of flat plates mechanically held together in a frame and kept separated by spacers equal in thickness to the desired spacing between the plates.

Since the freezing of the molten abrasive proceeds from the face of each plate between which the material is poured, a line of separation is evident in the material, at its midpoint, where the solidification fronts meet. This is normally a plane of weakness in the material and thus requires that the plates be spaced apart the proper distance to maximize the yield of a specific desired grit size distribution.

The process is particularly suitable for the alloy type abrasives in which zirconia or other refractory oxides are cofused with alumina.

In a typical application, the spaced plates would be positioned under the pouring spout of a furnace and the interstices between the plates filled. The plates would then be allowed to cool sufficiently for safe handling and separated and the solidified abrasive removed and crushed by impact or rolls crushing or other conventional crushing technique, depending upon the particular grinding application for which the material is produced. A particular advantage of the process is the large yield of homogeneous finely crystalline usable abrasive produced upon crushing, as opposed to prior art cooling methods, since 95% or more of the surface of the abrasive in the mold is protected from the atmosphere by adjacent mold surface, and since the very rapid cooling results in extremely tough finely crystalline material.

Example: A fusion of alumina containing 25% by weight of zirconia was prepared in a casting furnace, by the conventional electric arc fusion technique well known in the art.

The product was cast between steel plates ¾ inch in thickness and spaced $^3/_{16}$ inch apart. The solidified product between the plates was collected, jaw crushed to ½ inch and finer lumps, then impact crushed. In the impact crushing operation all material passing through the standard 8 mesh Tyler Screen and retained on the 24 mesh was collected, coarser material was recycled, and finer grits were separated. The yield of 8 to 24 mesh material was 67%, as compared to a 45% yield by prior art casting methods, while maintaining the same grain strength.

Producing Material with Fine Crystal Structure

In the process of *W.Q. Richmond and P. Cichy; U.S. Patent 3,928,515; Dec. 23, 1975; assigned to The Carborundum Company,* an oxide refractory melt, e.g. alumina-zirconia mixtures, is rapidly cooled and solidified to produce a solidified melt having a fine crystal structure. A solidification chamber is first filled to a depth of at least about 15 cm with an initial supply of nonreactive, nonmelting particulate solid cooling media, of a material other than that of the oxide refractory material to be rapidly cooled and solidified.

The preferred cooling media are steel spheres, ranging from about 5 to about 60 mm in diameter. Additional cooling media are then added over the initial supply of cooling media while the melt of oxide refractory material is cast into the solidification chamber, over both the initial supply of cooling media and the additional cooling media, whereby the melt is rapidly cooled and solidified. The cooling media and solidified melt can then be recovered from the solidification chamber.

The process of U.S. Patent 3,928,515 is at best semicontinuous, since the solidification chamber, even if constructed to have very large dimensions, must still have some limit in size. Thus, when the solidification chamber becomes filled, the addition of cooling media and oxide refractory melt must be discontinued.

Furthermore, the solidification chamber, when filled, must somehow be emptied. While it is possible to recover the cooling media with a magnet (if magnetic cooling media are utilized), the solidified oxide refractory melt must, if the solidification chamber has a closed bottom, be poured from the solidification chamber.

Since the solidification chamber when filled with cooling media and solidified oxide refractory melt can be quite heavy, this poses problems. While these problems are not insurmountable, it is an object of the process of *P. Cichy; U.S. Patent 4,061,699; Dec. 6, 1977; assigned to The Carborundum Company* to provide for the semicontinuous operation of the above process, without the attendant problem of building up large amounts of material within the chamber which must then be turned to remove the contents of the chamber. It is an additional object to provide apparatus and process for continuously cooling and solidifying an oxide refractory melt.

The process is characterized in that the lower end of the solidification chamber comprises an opening through which the cooling media and solidified oxide refractory melt can be removed from the solidification chamber. The oxide refractory melt is cast into the solidification chamber, over the cooling media, and additional cooling media are added to the solidification chamber while the oxide refractory melt is cast into the solidification chamber.

According to one embodiment there is provided a substantially horizontal, detachable, preferably substantially conical, base which either temporarily completely, or continuously partially, blocks the opening in the lower end of the outer wall of the solidification chamber. By utilizing the base to open and close the lower end of the solidification chamber, the cooling media and solidified oxide refractory melt can be either periodically or continuously removed from the solidification chamber.

According to another embodiment, the apparatus solidification chamber consists essentially of a hollow cylindrical outer wall having a lower open end, this hollow cylindrical wall being inclined at an angle, whereby to control the continuous discharge of cooling media and solidified oxide refractoy melt from the lower open end of the solidification chamber.

Referring to Figure 4.3a an embodiment of the process adapted for batch or semicontinuous operation is shown. The solidification chamber for this embodiment comprises a substantially vertical hollow cylindrical outer wall **10**, and a substantially horizontal, detachable, preferably substantially conical, base **11**, capable of closing the opening in the lower end of the vertical hollow cylindrical outer wall. A "substantially horizontal" conical base, as used herein, has a vertical axis and a relatively blunt angle, but not so blunt that the cooling media and solidified melt will not run off the base of the cone.

Figure 4.3: Producing Material with Fine Crystal Structure

a.

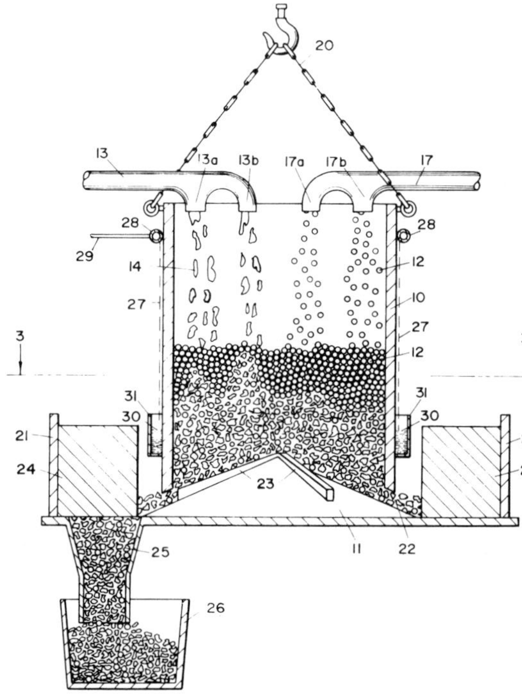

b.

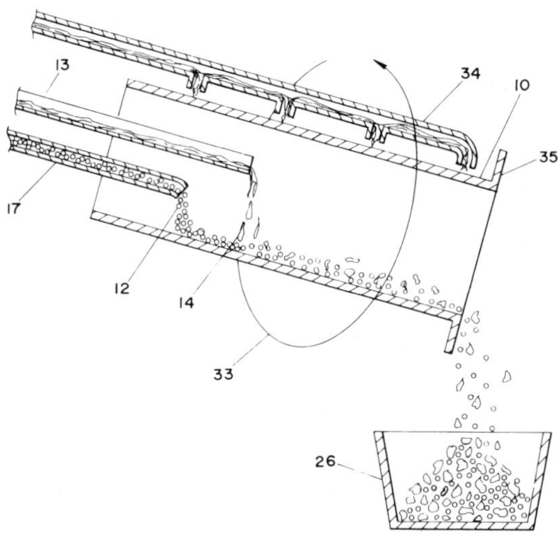

(a) Batch or semicontinuous process
(b) Continuous process

Source: U.S. Patent 4,061,699

The solidification chamber **10, 11** is provided with means **13** for casting oxide refractory melt **14** into the solidification chamber **10, 11** over the cooling media **12**, and also with means **17** for adding additional cooling media **12** to solidification chamber **10, 11** while the oxide refractory melt **14** is cast into the solidification chamber **10, 11**. The upper end of solidification chamber **10, 11**, is thus adapted for the addition of cooling media and oxide refractory melt. The substantially vertical hollow cylindrical outer wall is provided with an opening in its lower end through which cooling media and solidified oxide refractory melt can be removed. Means **13** for casting oxide refractory melt can be provided with a plurality of inlets **13a** and **13b** into solidification chamber **10, 11**. Means **17** for adding additional cooling media **12** can be provided with a plurality of inlets **17a** and **17b** into solidification chamber **10, 11**.

In operation, the solidification chamber **10, 11** is filled simultaneously with additional cooling media **12** and refractory oxide melt **14**, which melt **14** becomes rapidly solidified shortly after coming into contact with cooling media **12**. The solidification chamber **10, 11** eventually becomes filled, such that the addition of cooling media **12** and oxide refractory melt **14** must be discontinued. While cooling media **12** and oxide refractory melt **14** are being added to solidification chamber **10, 11**, it is preferred to rotate the solidification chamber **10, 11** (e.g., at a speed of about 3-10 rpm), to avoid the local buildup of high temperatures.

After the solidification chamber is substantially filled with solidified oxide refractory melt and cooling media, the addition of cooling media **12** and oxide refractory melt **14** is discontinued. The base **11** is then removed from the substantially vertical hollow cylindrical outer wall **10**. This is most conveniently accomplished by lifting outer wall **10**, for example by means **20**, so that the solidified oxide refractory melt and cooling media run down the substantially conical horizontal base **11** in all directions. If it is desired to prevent the cooling media **12** and solidified oxide refractory melt **14** from running down in all directions a retaining wall **21** can be provided around a portion of hollow cylindrical outer wall **10** in some portions of the apparatus. If such a retaining wall is utilized, it may be desirable to reshape the horizontal detachable base **11** so that the entire supply of cooling media and solidified oxide refractory melt leaves by the direction desired.

Figure 4.3b illustrates another embodiment, wherein no base is required to cover the open lower end of the solidification chamber. This is because the continuous discharge of cooling media and solidified oxide refractory melt from the lower end of hollow cylinder **10** is controlled by the angle of inclination of the cylinder **10** which constitutes the solidification chamber.

This apparatus is also provided with a source **13** of oxide refractory melt **14** and a source **17** of additional cooling media **12**. Hollow cylindrical outer wall **10** is preferably rotated, as shown by arrow **33**, both to prevent local overheating of the cooling media and the hollow outer cylindrical wall **10**, and also to assist in controlling the rate of discharge of cooling media and solidified oxide refractory melt into collecting bin **26**. If the apparatus is utilized in a continuous manner, it is preferable to provide means, such as source **34** of cooling water, for cooling hollow cylindrical outer wall **10**. The cooling water is collected in a suitable trough (not shown) and carried away. It is preferred to provide outer wall **10** with a flange **35** to prevent the cooling water from getting into the product, i.e., the cooling media and solidified oxide refractory melt.

In any of the embodiments, the direct output of the solidification chamber is a mixture of cooling media and solidified oxide refractory melt. If the cooling media are steel spheres, the cooling media can easily be removed of the solidified oxide refractory melt by means of an electromagnet. The remaining solidified oxide refractory melt is found to have an extremely fine crystal structure, for example, at least 90% by numerical count of the discontinuous crystals in a two-phase system, having a particle size smaller than about 40 microns. Material having such a fine crystal structure is highly desirable, since it can be crushed to produce an extremely tough abrasive grain for inclusion in abrasive articles such as grinding wheels and the like.

Casting into Molten Salt

According to the method of *B. Ilmaier and H. Zeiringer; U.S. Patent 4,059,417; Nov. 22, 1977; assigned to Treibacher Chemische Werke AG, Austria,* abrasive granules are produced on the basis of an oxide material consisting of alumina and a mixture thereof with zirconia, by melting the oxide material and rapidly cooling the melt, the melt being cast into a molten salt to cool the oxide material melt, the salt being separated from the cooled and solidified oxide material, and the abrasive granules thereafter being produced from the solidifed oxide material.

One of the major advantages of this method resides in the fact that the amount and/or the composition of the molten salt and the temperature thereof can be so chosen that the specific heat of the molten salt suffices to cool the entire cast melt rapidly to a temperature of about 1350° to 1550°C, i.e. a temperature range wherein crystal growth stops in zirconia-corundum eutectics. In this manner, a fine crystal structure is retained.

The indicated temperature range is between about 50° and 350°C, preferably 100° to 150°C or 250°C, below the solidification temperature of the fused oxide material. The rapid cooling to this temperature range proceeds within 10 seconds.

The cooling effect to obtain solidification of the fused oxide material is increased by selecting the evaporation point of the molten salt so that it partially is evaporated when the oxide material melt is cast thereinto, i.e. it is locally superheated so that the salt vapors rapidly circulate the oxide material melt in the bath. This circulation may be further enhanced by agitation.

The further cooling of the rapidly cooled oxide material below the range of about 1350° to 1550°C then proceeds more slowly to a temperature of about 800°C, preferably 500°C, at a maximum temperature of 40°C, preferably 20°C, per minute. This may be effected by cooling the material with iron spheres to obtain a material substantially free of tension and tension cracks. This material produces surprisingly high quality abrasive granules during the subsequent, generally conventional comminution treatment and greatly improved stability of the resultant abrasive granules in commercial operations.

Useful salts for cooling the fused oxide material include those having a relatively low melting point and a high boiling point. It will be commercially useful to use water-soluble salts to facilitate the complete separation of the salt from the abrasive material. Desirably, the molten salt will have a specific heat in the range of the molten state in excess of 0.1, preferably 0.2, cal/g/°C. Furthermore, the salt should be sufficiently inert to prevent chemical reactions with the abrasive oxide material. A preferred salt is molten calcium chloride, which has a melting point of 765°C and a boiling point of about 1600°C, its specific heat in the molten state between 0.2 and 0.3 cal/g/°C. It is highly water soluble. A mixture of calcium chloride with about 1 to 20%, by weight, of sodium chloride has similar characteristics.

Grind Sorting Process

In accordance with the process of *G.-E. Bockstiegel, M. Neidhardt, G. Rehfeld and W. Umlauf; U.S. Patent 4,049,397; Sept. 20, 1977; assigned to Dynamit Nobel AG, Germany,* it has been found that improved abrasives of increased grain toughness can be provided from known refractory materials in lump or block form if the known materials are initially subjected to a crushing process, and thereafter the resultant grains are subjected to a grind sorting process in such a manner that substantially only porous grits, grits containing flaws or coarsely crystalline grits are crushed and are separated from the rest of the material.

By the term "coarsely crystalline grits or grains," as used herein, there are meant those materials whose crystallites are larger than 70 to 80 microns. In the heretofore known processes for the achievement of abrasive grits having a finely crystalline internal structure, grits are generally produced whose crystalline sizes range between 0 and about 120 microns.

The process employs a material of such type and by the grind sorting operation, the more easily crushed grains are removed, including the coarsely crystalline grains, so that the end product will contain substantially only grains having crystallite sizes smaller than about 70 microns.

In the case of corundum or zirconium corundum, it is desirable to process a grit mixture of grain sizes from sieves No. 6 to about sieve No. 24. These grain sizes are those which are in greatest demand in practice for corundum and zirconium corundum abrasive. This is particularly true of zirconium corundum whose zirconium oxide content is 25% by weight, which on the basis of its special toughness and hardness characteristics is used for heavy operations in grain sizes corresponding to grain sizes Numbers 6 to 24.

In carrying out the process, a refractory abrasive composition having the following components, alumina 40 to 100 weight percent and zirconia 0 to 60 weight percent, and having particle size in the range of 6 to 24 mesh is subjected to ball milling employing a ball mill having a diameter between 80 and 200 cm and having a grinding medium therein of balls whose diameter is between 3 and 8 cm. The ball mill is revolved at a rate of between 8 and 40 rpm. The weight of the milling medium is generally between 1 and 4 times that of the weight of the charge and it is ball milled for between 0.5 and 4.0 hours. Under these conditions, the undesired grains are ground and those grains having the desired toughness are retained.

Tests show that good results are achieved when, in the case of a zirconium corundum having a zirconium content of 25% by weight, the material is treated in a tube mill which is 200 cm long and 90 cm in diameter for three hours at 21 rpm with 500 kg of grits and 1,000 kg of iron balls of 40 mm diameter in the mill.

Use of Hemispheric Molds

H. Sekigawa; U.S. Patent 3,977,132; August 31, 1976; assigned to The Japan Carlit Company, Ltd., Japan provides a mold suitable for manufacturing abrasive grains possessing high strength.

A mold set comprises two molds having hemispheric projections disposed regularly and closely to each other. The two molds may be brought into contact with each other or within 5 mm apart from each other in such a way that the projections of one mold surface fit into the cavities between the projections of the other. An abrasive composition which acquires needlelike crystal structure upon freezing is poured into the mold set, and subjected to cooling for freezing.

Thereafter, the frozen mass is separated from the molds, and crushed into abrasive grains. The abrasive composition employed in this process is capable of forming needlelike crystal structure upon freezing thereof. In particular, Al_2O_3-ZrO_2 alloy which forms dendritic structure (about 15 to 50 wt % of ZrO_2) is excellent as an abrasive composition.

The diameter of the hemispheric projections is in the range of from about 5 to 80 mm, and preferably, in the range of from 10 to 50 mm. If the diameter of the projections is smaller than 5 mm, the pouring operation of the molten abrasive composition into a mold set becomes very difficult, the mold set is damaged by the molten composition and the yields are very poor. On the other hand, if the diameter is larger than 80 mm, the spaces to be cooled are too large, whereby cooling of the molten abrasive composition will take longer resulting in weak abrasive products.

Referring to Figure 4.4, the two molds having projections **2** are brought into contact with each other or within 5 mm apart from each other in such a way that the projections **2** of one mold surface **1** fit into the cavities between the projections of the other, thereby forming small spaces **3**.

Figure 4.4: Use of Hemispheric Molds

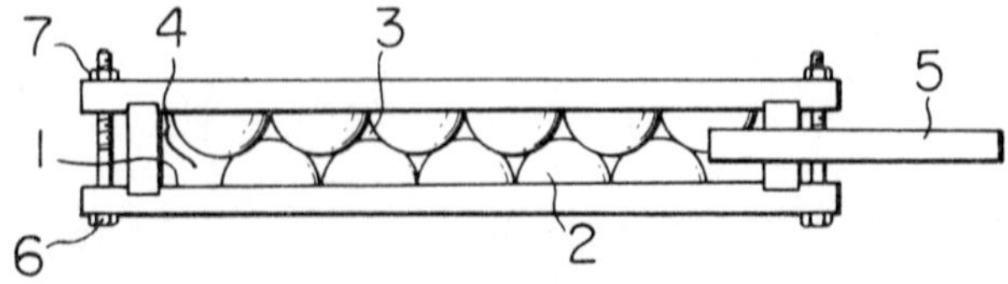

Source: U.S. Patent 3,977,132

Since there are relatively large spaces **4** at both sides and the bottom portion in the mold set, it is necessary to remove such relatively large spaces **4.** For this reason, iron bars are inserted into the spaces to form small spaces (not shown) in close proximity of within 5 mm to each other, and both sides and bottom portion are sealed up by means of bolt **6** and nut **7**. The mold set is then placed onto a thick, flat iron plate (not shown).

Cast iron may be used for the molds. A molten abrasive composition is poured into the small spaces **3** of the mold set as shown in Figure 4.4 from above the mold through an iron trough **5**, leaving a large space of more than 5 mm at the upper portion of the mold set. The molten abrasive composition is cooled from three directions at points on the curved surfaces of the projections, whereby the resulting frozen mass has fine, needlelike crystals which grow randomly in three dimensions. Thereafter, the frozen mass is removed from the mold set, and crushed and sifted by conventional methods into abrasive grains.

Use of Slivery Grits

Significantly improved grinding efficiency has been achieved according to the process of *R.A. Rowse and H.G. Lakhani; U.S. Patent 3,916,583; Nov. 4, 1975; assigned to Norton Company* by employing weak shaped alumina-zirconia alloy abrasive grits in thermoset or infusible resin bonded cut-off wheels.

U.S. Patent 3,181,939, discloses microcrystalline alumina-zirconia fused alloy abrasives containing from 10 to 60% zirconia. The finding of the process is that the more finely microcrystalline materials of this type, when produced in slivery or weak shaped grain significantly improve the grinding efficiency of thermoset resin bonded cut-off wheels in which they are incorporated.

Slivery or weak shaped grain of the desired size is produced by the rolls crushing of material of coarser grit size than required, followed by screening.

Best results are achieved with 100% of the abrasive of the alumina-zirconia alloy type. However up to 75% by volume of the abrasive may be replaced by blocky shaped fused alumina, and improved results achieved according to the process.

Such substitution may be advantageous to reduce the cost of the wheel.

Although operative in the range of 10 to 60% zirconia, the preferred abrasive for the process contains from 20 to 50% zirconia. The microcrystallinity of the abrasive should be such that the diameter of the primary alpha alumina crystals is less than 40 microns. Such abrasive can be produced by the methods disclosed in U.S. Patent 3,181,939, and British Patent 1,233,997 teaches one method of rapid cooling to achieve the preferred microcrystallinity.

Rolls crushing of the abrasive results in elongated grits which pack more loosely than strong shaped, more equidimensional grits. Useful abrasive grits are those selected from a population, for a 25% zirconia product having a specific gravity of 4.35, in which the 24 grit sized material (U.S. Dept. of Commerce Standards) has a loose packed (or bulk) density of not more than 130 pcf (2.1 g/cc).

For this process, a grain processing which results in a weight per cubic foot for the 24 grit fraction of not more than 135 pcf will be satisfactory. Although a precise lower limit has not been determined, lower densities, in the 24 grit fraction, than 115 pcf, would not be practical, for the 25% zirconia material. For material containing other than 25% zirconia or for 25% material having a specific gravity other than 4.35, the maximum and minimum bulk density should be corrected by multiplying 135 by the ratio of the specific gravity of the material to the specific gravity of the 25% material (4.35).

To achieve best results, the abrasive should be bonded with a phenol-aldehyde thermoset bond, with high bond content and high abrasive content, the abrasive ranging from 48 to 56 volume percent, and bond from 22 to 50% by volume, with any remainder being pores.

Rapid Cooling of Eutectic Mixtures

R.A. Rowse and G.R. Watson; U.S. Patent 3,891,408; June 24, 1975; assigned to Norton Company found that very rapid crystallization of eutectic and near eutectic molten mixtures of aluminum oxide and zirconium oxide, followed by crushing of the solidified melt, results in abrasive grits of very high strength combined with desirable microfracture properties.

The zirconium oxide in the material is in the form of rods (or platelets) which, on the average, are less than 3000 Å in diameter, and preferably at least 25%, by weight, of the zirconium oxide is in the tetragonal crystal form. The solidified melt is made up of cells or colonies typically 40 microns or less across their width. Groups of cells having identical orientation of microstructure form grains which typically include from 2 to 100 or more cells or colonies.

In crushing, the material fractures along grain boundaries and cell boundaries. Grinding improvement in excess of 100% of prior art standards is shown in tests of coated abrasive products employing the crushed abrasive material in typical applications and substantial improvement in bonded abrasive products.

Example: A mixture for fusion in the electric arc furnace was made up of 60 parts by weight of E-286 alumina (fused low soda alumina) with 44¼ parts of zirconia, and ½ part barley coal. The zirconia typically contained 2 to 3% hafnia and 12 of the 44¼ parts consisted of fused zirconia having the following typical analysis by weight:

	Percent
Al_2O_3	9.8*
SiO_2	5.24
Fe_2O_3	0.14
TiO_2	0.24
CaO	0.24
MgO	0.12
ZrO_2	84.2**

*by difference
**including hafnia

with the remaining 32¼ parts of zirconia having the following analysis by weight:

	Percent
SiO_2	0.56
Fe_2O_3	0.10
TiO_2	0.26
CaO	0.12
MgO	0.03
Al_2O_3	0.46
ZrO_2	98.5*

*including hafnia

A fusion was made in the conventional manner in an arc furnace arranged for pouring of the molten contents. The 4 ft^3 furnace employed two 4 inch diameter graphite electrodes, spaced 8" apart (center to center) and was operated at 85 volts and 175 kW. The average feed rate was 175 to 200 pounds an hour. The product was poured into a cast iron ingot mold filled with 1" diameter steel balls. A total of 1,712 pounds of material was cast in a series of pours. The average analysis of the composite product made from a number of similar runs was:

	Percent
Na_2O	0.04
SiO_2	0.24
Fe_2O_3	0.13
TiO_2	0.13
ZrO_2	39.85
Al_2O_3	59.61*

*by difference

The small amount of hafnia is reported at ZrO_2 above and throughout this patent. The average rod diameter of the zirconia in the center of the eutectic colonies was found, by a scanning electron microscope, to be less than 2000 Å, with a minimum diameter of below the resolution (300 Å) of the microscope, the larger diameter rods being found only in small amounts at the slowest cooled portions of the product (furthest removed from the ball cooling surface). The average rod spacing at the colony centers was typically less than 4000 Å and the platelet spacing was similar. The average platelet thickness was less than 2000 Å at the colony centers.

The zirconia was 31% in the tetragonal form, the remainder monoclinic. A cooling rate fast enough to cause at least 25% of the zirconia to be in the tetragonal form is necessary to produce the abrasives of this process. It has also been found that the crystalline form of Al_2O_3 known as delta-alumina is contained in samples which include tetragonal zirconia, the amount of delta-alumina increasing as the tetragonal zirconia content increases. In air quenched material containing 100% tetragonal zirconia it was found that essentially all of the alumina appears in the delta-form, as determined by standard powder x-ray diffraction; however the amount of delta-alumina is not believed to be proportional to the percent tetragonal zirconia since some analytical x-ray analyses of samples containing up to 40% tetragonal zirconia indicated that essentially all of the alumina was in the alpha form.

The product of the example was passed through a 20" x 6" jaw crusher to yield a product ½ inch and finer in size. The ½ inch product is then further crushed by rolls or impact to yield the desired grit sizes for use in bonded or coated abrasives. The abrasive yields superior results in medium and light duty applications. The abrasive grits were largely (51%) pseudohackly and predominantly (69%) all-eutectic.

OTHER ALUMINA-CONTAINING GRAINS

Aluminum Oxide-Titanium Oxide

T.B. Walker, R.J. Seider, and P. Cichy; U.S. Patents 4,157,898; June 12, 1979; and 4,111,668; September 5, 1978; both assigned to The Carborundum Co. provide a fused abrasive grain consisting essentially of titanium oxide containing from about 0.42 to about 0.84% titanium, by weight of the abrasive grain, the titanium being present as a reduced titanium oxide having an average oxidation state lower than in Ti_2O_3; from about 0.05 to about 0.3% by weight carbon; from about 0.02 to about 0.1% by weight Na_2O; from about 0 to about 0.1% by weight total calcium and silicon oxide; and alumina.

The abrasive has a gain on ignition in air at 1300°C, when of a size of about 147 microns and finer, of from about 0.4 to about 0.7% by weight before roasting. Such a grain, useful for example in coated abrasives, with or without a base layer of abrasive grain of diverse composition, and in bonded abrasives bonded with vitrified ceramic bonds or phenolic resin bonds, can be produced from high-purity titania, alumina containing soda as the only oxide impurity present in an amount in excess of 0.1% by weight, and carbon.

An important feature of the process is the amount of gain on ignition. The gain on ignition is a measure of the oxidation state of the titanium oxide and of the amount of carbon present. While various conditions could be used for determining the gain on ignition, a standard which has been used in determining the proper oxidation state of the grain for the process is the gain on ignition in air at 1300°C when the abrasive has been crushed and graded to a size of about 147 microns (100 mesh) and finer, and heated for 2 hours. This test is done before roasting the grain.

The preferred process for producing the fused abrasive grain comprises steps of charging an electric furnace; subjecting the mixture of ingredients within the electric furnace to the heat of an electric arc to melt the mixture; solidifying the melted mixture; crushing the solidified mixture; and roasting the crushed grain. The mixture to be charged to the electric furnace preferably comprises from about 0.7 to about 1.7 parts by weight of high-purity titania, such as "electronic grade" titania.

It is preferred that the titania contain at least 99% TiO_2, to avoid the introduction of unwanted impurities into the abrasive grain. A second raw material added to the furnace is preferably from about 98.6 to about 99.3 parts by weight of alumina containing Na_2O as the only oxide impurity present in a significant amount (i.e., in excess of about 0.1% by weight), for example tabular alumina having a maximum Na_2O content of 0.15%. The Na_2O content of the alumina raw material should be as low as convenient, as excess Na_2O must be reduced by reaction with carbon to leave less than about 0.1% by weight Na_2O in the finished product. Suitable alumina contains 0.3-0.6% Na_2O.

The third ingredient charged to the electric furnace of the process is carbon, such as calcined petroleum coke or graphite, in an amount from 1 to 10 times the theoretical amount necessary, based on the amount of TiO_2 and Na_2O added, to reduce the TiO_2 and Na_2O. As used herein, the amount necessary to reduce the TiO_2 and Na_2O is considered according to the following formulas:

(1) $$2TiO_2 + C \rightarrow Ti_2O_3 + CO, \text{ and}$$

(2) $$Na_2O + C \rightarrow 2Na + CO.$$

It is to be recognized, of course, that the TiO_2 will be reduced to an average titanium oxidation state less than in Ti_2O_3, and in particular it has been found that the average formula is more on the order of $TiO_{0.9}$. The reduction to the state Ti_2O_3 serves as a convenient benchmark in calculating the amount of carbon to add, however. The amount of carbon necessary to be added can be grossly estimated from the amount of TiO_2 and Na_2O present. By using carbon electrodes, however, additional carbon will be added to the reaction mixture from the carbon electrodes.

It is also desirable to use additional carbon past the theoretical amount necessary to reduce the TiO_2 and Na_2O to Ti_2O_3 and Na, respectively, in an amount of from about 3 to 10 times the calculated amount. Within this range, from about 4 to 7 times the theoretical amount necessary, based on the amount of TiO_2 and Na_2O added, to reduce the TiO_2 and Na_2O to Ti_2O_3 and Na, respectively, is preferred. The carbon can be in any convenient form, but finely divided calcined petroleum coke and graphite are preferred.

The mixture of titania, alumina and carbon is subjected within the electric furnace to the heat of an electric arc, the electric arc being a reducing arc passed from carbon electrodes to the mixture of titania, alumina, and carbon, for a time sufficient to melt the mixture. The term "reducing arc" is well known to those skilled in the electric furnace art, and is a relatively short arc produced by controlling the power input, phase voltage, electrode spacing, power factor, circuit configuration (single-phase or three-phase), and the like. As an example, the electric arc can be supplied by a single-phase power supply at a voltage of about 80 volts and a power input of from about 100 to about 120 kW.

After the mixture of titania, alumina and carbon is melted, it is solidified. It is highly preferred that this be accomplished by pouring the melted mixture into a cooling mold such as a water-cooled steel pan. A depth of melted mixture to be solidified of from about 2.5 to about 15 cm, more particularly from about 2.5 to about 6 cm, is preferred. An alternate cooling method is "ball casting," i.e., pouring the melted mixture into a solidification chamber containing steel spheres having a diameter ranging from about 5 to about 60 mm, more preferably of about 20 mm.

Regardless of the cooling apparatus used, it is desired to prevent reoxidation of the reduced titanium oxide. In order to prevent this reoxidation of the titanium, it is preferred that the length of the pouring stream from the electric furnace to the cooling mold be minimized, and the pour rate maximized, insofar as practical.

After the melted mixture is solidified, it is crushed to obtain abrasive grain and "roasted" by subjecting the crushed grain to an oxidizing atmosphere, preferably air, for a time of from about 5 minutes to about 64 hours or more at a temperature of from about 1250° to about 1450°C. The time of roasting more particularly is preferred to range from about 5 to about 20 minutes, optimum 10 minutes; and the temperature of roasting preferably ranges from about 1300° to about 1350°C, optimum 1300°C.

The product, as produced, should be bluish black in color. A reddish brown appearance indicates that the titanium is present in an oxidation state greater than desired. Such off-color material should be refurnaced to further reduce the titanium oxide present. The lumps exhibit a "glassy" conchoidal fracture which is suggestive of large crystal size. The titanium may be combined with carbon and oxygen to form a titanium oxycarbide, and although the chemical identity of the titanium compounds present is not critical, the titanium must be present in a form which has a very low solubility in aluminum oxide. The titanium compound is present as an easily recognizable second phase, which may also contain aluminum oxide or aluminum oxycarbide.

Freshly fractured crude abrasive (i.e., solidified melt, prior to crushing to obtain abrasive grain) has a "carbidic" odor, a condition well known to those skilled in the art as indicative or over-reduction of the aluminum oxide fusion. This odor is produced upon contact with water or water vapor (or even high-humidity air).

The rate of cooling the melted mixture to solidify the same is rapid, in order to ensure that crystals grow in highly-oriented columns. Equivalent "grain" diameters of between 0.25 and about 2.5 mm, with a median of about 1.2 mm, have been observed. This is unusually large for thin cast alumina.

The grains (crystalline columns within an abrasive granule) exhibit a cellular substructure. The individual cells which have an equivalent diameter of between 0.06 and 0.1 mm are clearly outlined in photomicrographs of thin sections cut perpendicular to the direction of solidification, by the titanium-rich second phase. These cells appear to be alumina dendrites. These dendrites appear to be a stack of rhombohedral crystallite units which share a common "c" axis. These columnar cells may, in some instances, be twinned stacks of rhombohedral units.

Alumina and Zirconia plus Oxide of V, Cr, Mn, or Co

J.A. Brothers, R.C. Doman and R.N. McNally; U.S. Patent 4,035,162; July 12, 1977; assigned to Corning Glass Works have discovered an abrasive grain that has substantially lesser tendency for chipping (or probability of fracture) and substantially lesser rates of attritious wear than the fused alumina-zirconia abrasives. These results were found by making fused mixtures of alumina, zirconia and at least one oxide of metal selected from vanadium, chromium, manganese and cobalt. While oxides of these metals have a tendency to form solid solutions with the corundum phase of the abrasive grain, experiments indicate that it is not necessary for all of these metal oxides to be in such solid solution to provide the improved abrasive grain.

Depending upon the relative oxidation-reduction ambient environment during fusing of the material, the resultant fused abrasive may contain dispersed particles of metal (or alloy) corresponding to the metals of the oxides, aluminates of such metal oxide combined with alumina, and/or the noted solid solution. In all these various situations, the resultant grain is found to be of the improved character indicated herein.

The fused abrasive grain consists of: (a) the crystalline phases of corundum and zirconia, and (b) analytically by weight, about 44 to 91% (preferably 44 to 80%) Al_2O_3, about 5 to 50% (preferably 17.6 to 50%) ZrO_2 and 0.1 to 12% (preferably 0.2 to 5%) R_2O_3 wherein R is at least one metal selected from vanadium, chromium, manganese and cobalt, and the percentage of R_2O_3 constitutes the entire R metal content calculated as R_2O_3.

Production of the grain is accomplished by conventional techniques well known in the art for manufacturing fused abrasives. The electrodes of electric arc melting furnaces are suitably adjusted during melting to either establish a "kissing arc" (i.e., lower tip of electrodes essentially just about touching surface of the melt) or be submerged for mainly resistance heating via current passing through the melt.

If it is desired to have a more oxidized melt and grain (i.e., no free metal particles therein), the reduced condition thereof caused by the common carbon or graphite electrodes can be simply reversed by the known practice (see U.S. Patent 960,712) of lancing the melt with oxidizing gas just prior to pouring of the melt. Solidification of a melt is accomplished in a reasonably rapid manner, according to the well known procedures (e.g., by pouring into molds of graphite or cast iron to form slabs or "pancake" castings or by pouring onto a plurality of discrete packed cooling media), so as to produce the uniform and very fine crystallization characteristic of the best quality fused abrasive grain.

Generally, the crystals are smaller than 300 microns, and the optimum average crystallite size in the fused grain is commonly about 40-60 microns. The solidified product is crushed as desired by conventional procedures. Common grain mesh or grit sizes range from about less than 6 to 250 mesh (U.S. Standard Sieve), depending upon the type of abrasive article in which the abrasive grains are to be employed. For heavy-duty grinding wheels, sizes between 8 and 28 mesh are usually employed. The finer sizes are mainly used in finishing wheels and coated abrasive articles.

Ceria-Alumina-Zirconia

In the process of *R.E. Barks; U.S. Patent 3,916,585; Nov. 4, 1975; assigned to Norton Company* very strong grits are obtained which are highly resistant to impact and have a low destructability factor in the rough grinding of metals such as stainless steel and carbon steels.

The broadest compositional range of the sintered abrasive of this process is bounded by the curve **AEBC** on Figure 4.5. The preferred composition for the abrasives of this process is indicated by the area enclosed by the parallelogram **DEBF**. Although high purity raw materials are preferred, materials of over 98% purity are expensive. Up to 5%, total, by weight of impurities or sintering aids, such as oxides of iron, silicon, titanium, magnesium, calcium, chromium, cobalt, vanadium, nickel, and manganese can be tolerated or even preferred in the abrasive composition.

The preparation of the abrasive grits involves (1) mixing the raw materials, (2) forming the mixture into abrasive grit sized particles or shapes, and (3) drying and firing the preformed shaped grits.

The raw materials should be well mixed and in a finely divided condition, the particles of aluminum oxide, cerium oxide and zirconium oxide being between 0.5 and 10 microns in size or finer, preferably 4 microns or finer in average size. The firing operation should not cause crystal growth above 5 microns, average size.

Figure 4.5: Ceria-Alumina-Zirconia Ternary Diagram

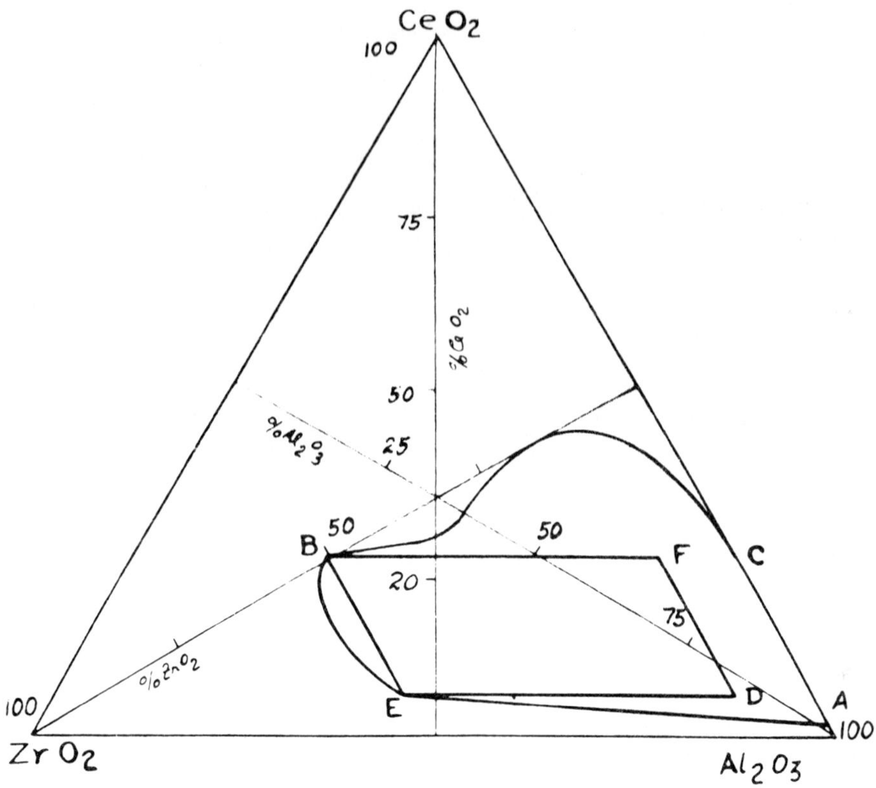

Source: U.S. Patent 3,916,585

Several different forming methods can be used, such as pressing into bricks which are then granulated into abrasive sized particles, extrusion, pouring into thin layers which are dried and then broken-up, molding etc. It is preferred to prepare slurries containing deflocculents for use in casting into molds of the desired size, without any need for pressing, as when drier mixes are used. Suitable deflocculents are ferric ammonium citrate (U.S. Patent 2,637,360) or commercial deflocculents such as Darvan C or 7.

Firing can be at temperatures of from 1350° to 1600°C, for from 1 to 24 hours. The fired abrasive grain can be employed in resin bonded grinding wheels adapted for rotation at high surface speeds and high pressures.

Isostatic Hot-Pressing

L. Coes, Jr.; U.S. Patent 3,909,991; October 7, 1975; assigned to Norton Co. has found that superior abrasives of the sintered type can be fabricated by the isostatic hot-pressing of preformed, unfired particles of base refractory material. The improved abrasive properties result from the fact that by virtue of the simultaneous application of heat and pressure the resultant abrasive particles have essentially an average crystal size not greater than the average particle size of the base refractory material used and have a specific gravity essentially equal to the true specific gravity of the base refractory material. Such abrasives are possible only through the simultaneous application of heat and isostatic pressing.

In order to discourage crystal growth the temperature must be below the fusion temperature of the refractory material yet must be above its sintering temperature; this alone however does not prevent crystal growth. However the application of isostatic pressure simultaneously with heating facilitates almost instantaneous compression of the particles to essentially theoretical density with the time at temperature being so short that no evident increase in average crystal size occurs.

In the preferred practice of this process a base refractory material, e.g., natural or synthetic bauxite, spinel, alpha-alumina, zirconia, and mixes of these, is first comminuted to an average particle size of less than one micron, the comminution being accomplished by any of the many known methods, although wet ball milling is preferred. The slurry of water and comminuted base refractory material is then freed of excess water by allowing the refractory material to settle and decanting the water and/or by subjecting the material to a filter-pressing process. The remaining water is then removed by evaporation preferably by subjecting the partially dehydrated material to heat.

The abrasive to be produced may be of the random shaped granule type or the preshaped type both of which are formed by conventional methods. With both the random shaped granules and the preshaped abrasive products the next step is the mixing of the unfired abrasive shapes with the isostatic pressing medium. The pressing medium can be any powdered material which will not sinter, melt, decompose, or react with the base refractory material making up the unfired abrasive particles. Graphite or boron nitride is preferred. The amount of pressing medium used relative to the amount of abrasive particles, can be varied widely. The optimum amount is approximately 50% by weight of each.

When the unfired abrasive particles and the pressing medium have been blended together, the resultant mixture is then loaded into a series of mold setups, the mold being composed preferably of graphite however, any material stable at the temperatures at which the pressing is to be carried out, may be effectively employed. The filled molds are then moved into an induction press or the like and rapidly heated to the sintering temperature of the refractory material involved, at which point pressure is then exerted on the mold plates resulting in sintering and densification of the particulate material suspended in the pressing medium.

The mold setup is then ejected from the pressing station and rapidly cooled. The pressing temperature will vary with the particular base refractory material being used but for the materials of interest here a temperature range of from 1200° to 1600° is the effective range. The pressure also can vary widely, it being affected to some degree by the amount of pressing medium involved and the temperature, but is most affected by the thickness of the pressing medium-abrasive particle mixture. For most practical thicknesses of pressing medium-abrasive particle mixtures a pressure of not less than 2,000 psi is most desirable.

The parameters, of these two variables, which facilitate the most rapid heating and pressing are the most desirable. If for example, the pressed thickness of the mixture would be on the order of 1" then the heat-up time would be about 10 minutes and the time during which pressure would be applied would be on the order of a minute or slightly less.

When the mold setups are removed from the pressing zone and cooled somewhat, the content thereof is removed and subjected to a coarse pressing operation such as jaw crushing, roll crushing, or the like being careful however to set the jaws or the rolls sufficiently wide enough apart to avoid crushing of the fired abrasive particles. The rough crushed mixture is then screened to remove the readily freed pressing medium while the fired abrasive particles, having some pressing medium adhering to the surface thereof, are further treated to remove this remaining material. This can be accomplished mechanically or by calcining the material in the presence of air if the pressing medium used was graphite. After mechanical cleaning or calcining, the abrasive is finished and ready to be used.

Calcining Alumina-Zirconia-Ceria Grain

Improved fused zirconia alumina is described by *G.R. Watson; U.S. Patent 3,881,282; May 6, 1975; assigned to Norton Company*. The abrasive is superior to normal alumina-zirconia abrasive in two ways:

> Where processed into grain, the grain yield is much higher, which gives a lower comminution cost per pound of useable grain.
>
> The grain durability is much higher, while maintaining or improving the rate of metal removal, in snagging tests.

These improvements are realized by making an addition of from 1 to 7% of ceria to an alumina-zirconium fusion.

The product is then chill-cast to obtain very fine crystal size. In the preferred embodiment of the process the grain is strengthened by heating to 800°-1250°C for 2 hours.

Example: A small arc furnace of the casting type was employed. The starting mixture was: calcined Bayer alumina, 69 lb 3 oz; baddeleyite (ZrO_2), 23 lb 3 oz; and ceria, 7 lb 3 oz.

The furnace is normally operated at 85 volts 185 kW when fusing alumina-zirconia. With ceria in the furnace feed it was necessary to operate at 100 volts 185 kW to minimize the tendency of ceria to be reduced to a lower oxide state. Six taps were made producing 480 pounds of product with a light greenish tinge.

The material was then impact crushed and shaped under controlled comminution conditions. A sample of the crude product is accumulated during a conventional jaw crushing operation by taking a portion of each shovelful of -½ inch crushed product. This sample of crude product is further reduced in size by standard sample-room practice (jaw + rolls crushing) to yield a 14 grit Tyler screen sample and, after further pulverization a portion of this sample that is 140 mesh and finer is collected, both of these 2-oz size samples are analyzed, the former by microscopic examination, the latter by chemical analysis.

The remainder of the crude sample of the abrasive product that is about minus ½ inch in size is first weighed then passed through an impact crusher like that shown in U.S. Patent 2,644,644 that can be fed by means of a Syntron Vibrating Feeder at the rate of 40 lb/min. The paddle speed of the crusher is set at 1500 rpm with a paddle radius of about 22 inches so the tip of the impeller is running at about 8,600 ft/min.

After each pass through the impact crusher, the product is passed over a 2 x 6 inch Rotex screen fitted with 6 mesh and 24 mesh Tyler screens. This yields three products, namely, +6, -6+24 and -24. The +6 product is recycled through the impact crusher. After each pass through the crusher the product is rescreened and the +6 size recycled again. This process is continued until 95% of the initial feed has been crushed to -6 mesh. The remaining +6 material is rolls crushed to -6 mesh and added to the –6 mesh products.

The -6+24 product developed as above, is passed through the crusher three times in succession at 1,500 rpm. A representative sample is taken and the impact strength measured by the k value method described in detail in the patent. If the 8/9 and 12/14 fractions have k values that meet a specified value no further shaping is done. If the k values are still too high, the material is given one or more additional shapes until k value measurements are acceptable.

The final crushed and impact shaped product is again sampled and a screen analysis made to determine what grit sizes have been produced and the yield of these sizes.

The -6+24 crushed and impact shaped product is then placed in the three-foot pan mill (60-70 lb per batch) and dry-panned for from 1 to 2 hours, or until the grain has been rounded sufficiently so a desired weight per cubic foot specification is met. The WPCF specification is adjusted for density differences, of course. The panned product is then screened into desired grit sizes.

DIAMOND

Annealing Diamond to Convert Type Ib Nitrogen to Type Ia

Diamonds are generally classified into four main types: Ia, Ib, IIa, and IIb. These types are most easily distinguished by infrared and ultraviolet spectra and sometimes by electron paramagnetic resonance (EPR). Types Ia and Ib diamonds contain dissolved nitrogen; in Ia diamonds, most of the nitrogen is not EPR active and appears to be in aggregated form; in Ib diamonds most of the nitrogen is EPR active, and is atomically dispersed. Types IIa and IIb diamonds do not contain appreciable nitrogen. Each type of diamond has typical infrared and ultraviolet spectra with characteristic features.

The large majority of synthesized diamonds are type Ib, but type IIa diamonds can easily be made either by excluding nitrogen from the diamond growing media or by using appropriate nitrogen getters.

The large majority of natural diamonds examined are type Ia. Natural type Ia diamond crystals can have a variety of colors, with many being pale yellow to colorless. Such a diamond crystal can also be a combination of pale yellow and colorless areas as well as exhibit local variations in its characteristic color in different parts of the crystal. Ordinarily, it has a rounded dodecahedral or octahedral morphology.

There are two forms of type Ia diamond, an A band form and a B band form, and these forms are distinguishable by their infrared, visible and ultraviolet absorption spectra. Usually, however, these two forms are most easily differentiated by their infrared spectra wherein the A band form has its main absorption band coming at 1280 cm^{-1} and the B band form has its main absorption band coming at 1175 cm^{-1}. While each form of type Ia appears to be thermodynamically more stable than type Ib diamond, the process described below has thus far produced only the A band form which is hereinafter referred to broadly as type Ia.

Synthetic diamonds are substantially the same as natural diamonds but there are enough differences between them to distinguish between the natural and synthetic crystals. These differences are mainly in morphology, surface appearance, impurity inclusions and the nature of impurity imperfections such as the different forms of nitrogen. As found, natural diamond crystals most frequently have curved edges and convex faces. On the other hand, synthesized diamonds crystals, as grown, have sharp edges, flat and relatively smooth faces. Depending on the conditions of growth, synthetic type Ib crystals have octahedral or cubo-octahedral morphology, the latter sometimes having small (113) faces.

Also, depending on the conditions of growth, a synthetic type Ib crystal may have no definite morphology and can have a wide variety of shapes with some substantially distorted from the octahedral or cubo-octahedral regular shapes, as well as in extreme cases, highly irregular particles of no particular shape. Impurity inclusions in synthetic diamonds are metal catalysts whereas in natural diamonds they are a variety of minerals, and these impurity inclusions are detectable by several techniques such as electron diffraction analysis or x-ray analysis.

According to the process of *H.M. Strong, R.M. Chrenko and R.E.Tuft; U.S. Patent 4,174,380; Nov. 13, 1979; assigned to General Electric Company,* type Ib synthetic diamond is converted to type Ia. Briefly stated, the process comprises annealing type Ib synthetic diamond crystal at an annealing temperature ranging from about 1500° to about 2200°C under a pressure which prevents significant graphitization of the diamond during annealing to convert at least about 20% of the total amount of type Ib nitrogen present in the crystal to type Ia nitrogen.

In the process the amount of conversion of type Ib nitrogen to type Ia nitrogen is determinable by a number of conventional techniques. The most frequently used technique is one where it is revealed by the differences or changes in the absorption spectra of the Ib crystal taken before and after annealing.

Synthetic type Ib diamond crystal has a color depending on the amount of nitrogen dissolved in the crystal. The color of the crystal ranges from a green to a greenish-yellow to a yellow with the maximum or largest amount of dissolved nitrogen producing the greenish-yellow color. Likewise, the amount of nitrogen dissolved in the crystal determines the intensity of the yellow color which can range from a deep golden yellow to a pale yellow with the deep golden yellow indicating substantially more dissolved nitrogen than the pale yellow. In addition, the Ib diamond crystal can exhibit a mixture of greenish-yellow and/or yellow colors or shades, i.e. it can exhibit local variations in its characteristic color and intensity, which indicates regions of varying nitrogen content.

In the process there is no limitation on the size of the type Ib diamond crystal. Specifically, the minimum size of the crystal can be one micron or less and the maximum size is limited only by the capacity of the annealing equipment. For most applications, the Ib crystal size ranges from about 0.25 mm to about 6 mm. The size of the diamond crystal given herein is that measured along the longest edge dimension of the crystal.

This annealing process is carried out in high temperature-high pressure apparatus normally used for synthesizing diamonds by application of high temperatures and pressures to a suitable reaction mass or specimen.

The directly heated embodiment of the reaction vessel is preferred in the process, and a particularly preferred form is shown in Figure 4.6a. Specifically, this reaction vessel includes a hollow outer cylinder **3** made of nonconducting material such as pyrophyllite. Positioned concentrically within and adjacent pyrophyllite

Figure 4.6: Annealing Diamond

a.

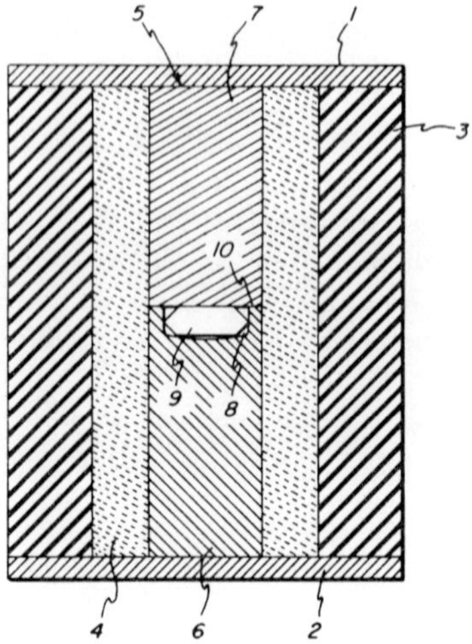

b.

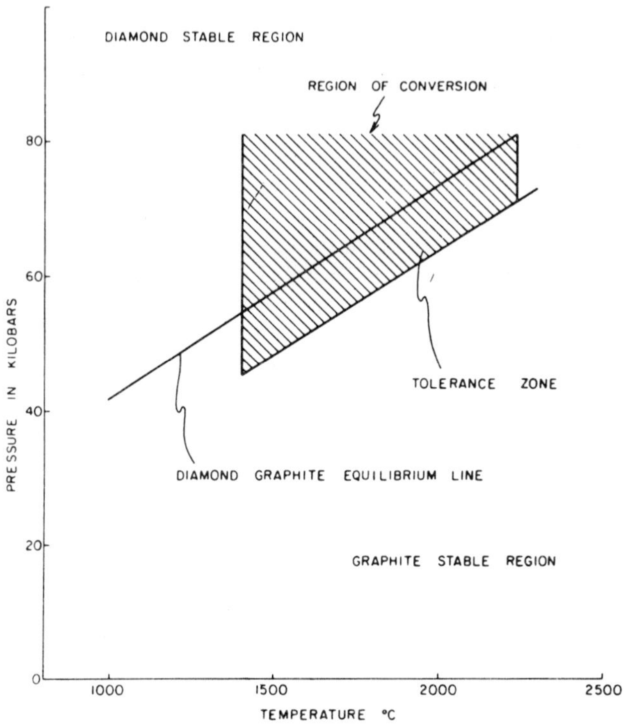

(a) Preferred reaction vessel

(b) Phase diagram of carbon showing region of conversion

Source. U.S. Patent 4,174,380

cylinder **3** is ceramic cylinder **4** preferably made of alumina. Charge element or insert assembly **5** is adapted to fit concentrically in ceramic cylinder **4** and is dimensioned for a close fit with cylinder **4**. Charge element **5** is comprised of graphite rod **7** and graphite rod **6** wherein the graphite is of spectroscopic purity.

Graphite rod **6** has hole **8** which is drilled to fit closely around diamond crystal **9**, i.e., the diamond to be annealed. Diamond crystal **9** should not project outside of hole **8** since such projection would prevent a close contiguous fit of rod **7** with rod **6**. Rod **7** should be in electrical contact with rod **6** at surface **10**. Preferably, the top surface of diamond **9** is flush with surface **10** of graphite rod **6**. Any space between diamond crystal **9** and hole **8** is preferably filled with an electrically conducting material, such as graphite powder of spectroscopic purity, to promote passage of the electric current and thereby promote heating of diamond **9**.

Electrically conducting circular metallic discs **1** and **2** close the ends of graphite rods **6**, **7**, and cylinders **3** and **4**. Discs **1** and **2** are preferably made of a metal such as nickel or tantalum and must be in electrical contact with graphite rods **7** and **6**, respectively. Since graphite rod **6** is electrically conducting and diamond crystal **9** is not electrically conducting, the highest temperatures are attained and maintained at the thinnest portions of graphite rod **6**, e.g., the area of graphite rod **6** surrounding diamond crystal **9**.

After assembly of the reaction vessel and introduction thereof into the high pressure-high temperature apparatus within the gasket/insulation assemblies preferably pressure is raised first and then the temperature. The rates of increase of pressure or temperature are not critical. When pressure and temperature are at a level in the region of conversion defined in Figure 4.6b, they are held at that level for a period of time sufficient to attain the desired conversion of at least about 20% of the diamond's Ib type nitrogen to type Ia nitrogen. When the desired conversion is attained, the electrical power which heats the diamond crystal is shut off and the crystal cools to about room temperature quickly, usually in about 1 minute. Generally, when the crystal has cooled to below 50°C, the pressure is then released, preferably at a rate of about 10 kilobars per minute to atmospheric pressure.

Annealing time, i.e. the period of time at annealing temperature and pressure, is determinable empirically and can range from about one minute to about 50 hours, and preferably up to about 20 hours.

The annealed diamond crystals produced by the process are useful as abrasives. The abrasive industry requires numerous types of abrasive materials to carry out various grinding or machining operations, the requirements of which are determined largely by the properties of the material being machined and, to some extent, the results desired. For certain operations in the abrasive industry, synthetic type Ib crystal has been satisfactory and for other operations natural type Ia crystal has been satisfactory.

However, as a result of the process, the abrasive industry has available a crystal which is a mixture of types Ib and Ia, the composition of which can be controlled to produce crystals with graded physical properties over a wide range to adjust the crystal to the particular abrasive use to which it is applied. Specifically, with increasing degrees of conversion of type Ib to type Ia, the crystal changes in abrasive properties, usually becoming harder and stronger. As a result, a mixed type Ib-Ia crystal can be produced having optimum properties for a particular abrasive use.

In related work by *H.M. Strong, R.M. Chrenko and R.E. Tuft; U.S. Patent 4,124,690; Nov. 7, 1978; assigned to General Electric Company* type Ib or mixed type Ib-Ia natural diamond is annealed to convert type Ib nitrogen to type Ia nitrogen. The process comprises annealing type Ib or mixed type Ib-Ia natural diamond crystal at an annealing temperature ranging from about 1500° to about 2200°C under a pressure which prevents significant graphitization of the diamond crystal during annealing to convert at least about 20% of the total amount of type Ib nitrogen present in the crystal to type Ia nitrogen.

In the process the diamond crystal can be wholly type Ib or mixed type Ib-Ia. The mixed type crystal can range in type Ib nitrogen content from about 99% to about 1% of the total amount of nitrogen present in the crystal.

Elongate Synthetic RD Diamond

At least three types of synthetic diamond may be produced. A first type is known as resin bond diamond (RD), a second type is metal bond diamond (MD) and a third type is saw diamond (SD). The RD type of diamond has an irregular shape and tends to fracture, so continually presenting fresh surfaces, when subjected to abrasive operations. MD and SD diamonds, on the other hand, generally present a blocky crystal of high impact resistance. These particles tend to wear by abrasion processes rather than by fracture.

According to the process of *A.R. Roy and A.B. Clarke; U.S. Patents 4,122,636; Oct. 31, 1978; and 4,036,937; July 19, 1977,* there is provided an elongate synthetic RD diamond particle having a long axis and a transverse short axis, the ratio of the long axis to the short axis being at least 3 to 1. The diamonds of the process are RD diamonds which means that they are friable and tend to fracture during abrasive operations so continually presenting fresh sharp cutting points. The particles may have a ratio of long axis to short axis of at least 5 to 1.

The length to width aspect is more pronounced with the larger particles. The particles are preferably in the range 60/170 U.S. mesh and more preferably either in the range 80/100 U.S. mesh or 100/120 U.S. mesh.

The particle of the process preferably has a ribbed structure down its long axis, the ribs of the structure being traces of $\{111\}$ crystallographic planes and the long axis providing the backbone for the structure and being in the $\langle 100 \rangle$

crystallographic direction. The particles of the process may have an enlarged end. The backbone is preferably of irregular shape.

The particles of the process, particularly those having enlarged ends, are generally not of uniform width. Consequently, in these cases the short axis of largest dimension is taken when determining the ratio of long axis to short axis. The long axis is generally uniquely defined, but when it is not the long axis of largest dimension is taken for the ratio determination.

The diamond particles of the process may be metal coated using techniques well known in the art. Such coated particles find application particularly in resin bond wheels.

The conditions necessary for diamond growth are well known in the art. Referring to Figure 4.7a, the diamond growing region is shown by the shaded area of the graph, i.e. the area above the line **A** and to the right of the line **B**. Line **A** defines the Berman-Simon line [*Zeit. F. Elektrochemie,* 59 355 (1955)] and line **B** is the eutectic melting temperature of the solvent metal. The position of line **B** will vary according to the type of solvent metal used.

In order to produce RD diamonds the conditions of temperature and pressure are maintained in the diamond growing region for a relatively short period, i.e. generally for a period of between 4 to 10 minutes. It is preferred that the temperature and pressure conditions be raised from ambient to the diamond growing region in a stepwise manner.

The carbonaceous material may be provided in the form of a sleeve around a core of the solvent metal. The sleeve may be a continuous sleeve or a number of segments which together define the sleeve. A carbonaceous disc may be provided at each end of the core. The ratio of the width of the sleeve to the diameter of the core is preferably in the range 1:20 to 1:5. In particular, the width of the sleeve may be about one-tenth the diameter of the core. The weight ratio of metal to carbonaceous material will generally be greater than 1.

Referring to Figures 4.7b and 4.7c, a manganese-cobalt alloy core **18** is placed inside a graphite sleeve **20**. Graphite discs **22** are placed on both ends of the core **18** and the whole is placed inside a pyrophyllite sleeve **24**.

The core **18** was made by mixing powdered manganese and cobalt in equal amounts by weight and compressing the mixture into the cylindrical form using a pressure of about 25 tons/in^2. The manganese powder was 80/100 U.S. mesh and the cobalt powder was minus 200 U.S. mesh. The compressed slug had a height:diameter ratio of about 1:1.

The graphite sleeve and discs were made from commercially available electrode grade graphite. The width of the sleeve was about $^1/_{10}$ the diameter of the core. The weight ratio of the graphite to the manganese-cobalt alloy was about 1:4. The graphite/metal/pyrophillite composite was placed in a high temperature/pressure apparatus of the type described in U.S. Patent 2,941,248 in the usual way.

Figure 4.7: Producing Synthetic RD Diamond

a.

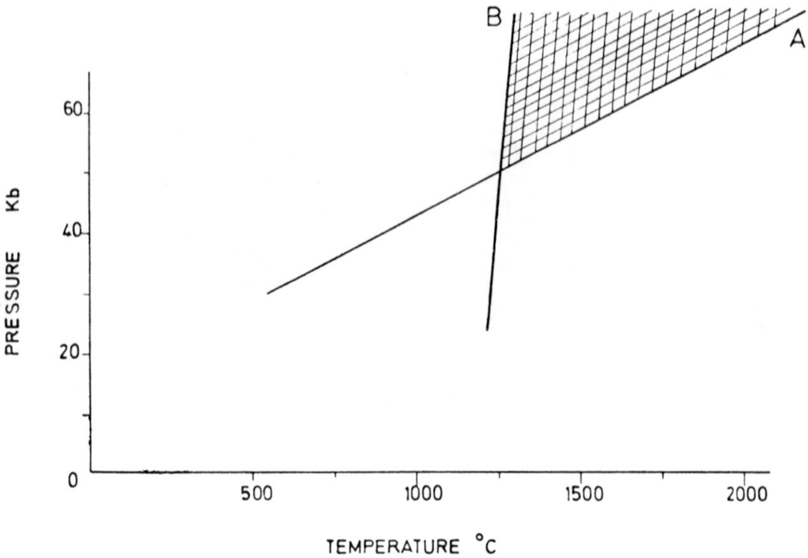

b.

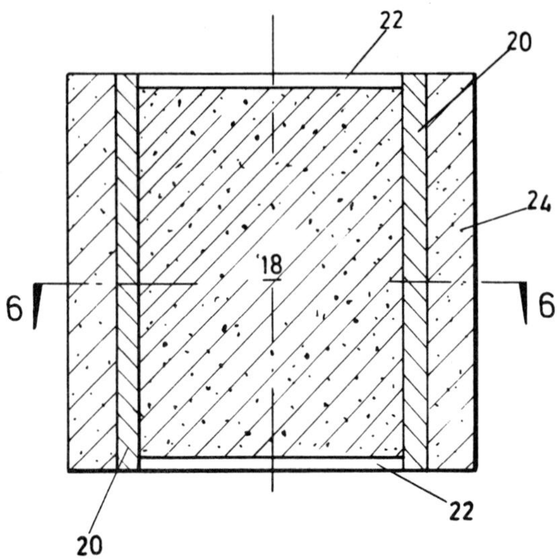

(continued)

Figure 4.7: (continued)

c.

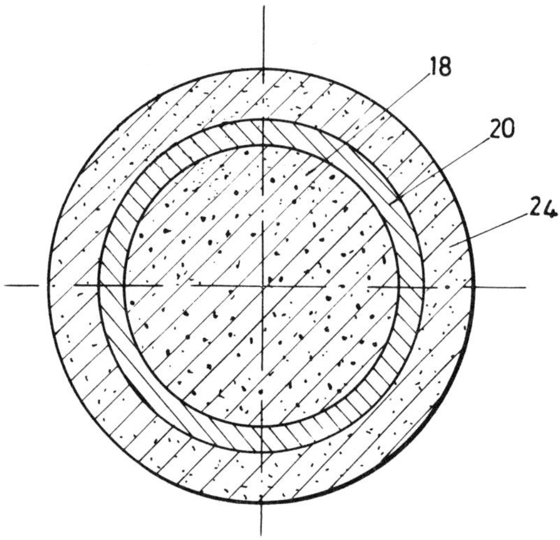

(a) Diamond growing region
(b) Sectional side view of reaction capsule
(c) Section along line **6–6** of Figure 4.7b.

Source: U.S. Patent 4,036,937

The temperature and pressure of the contents of the reaction zone were raised to the diamond growing region of the pressure temperature graph illustrated by Figure 4.7a in a stepwise manner, i.e., the pressure was raised a certain number of kilobars and then the temperature raised a certain number of degrees and so on until the desired conditions were obtained. A number of runs were performed using pressures in the range of from 55 to 65 kilobars and temperatures in the range of from 1300° to 1600°C. In each run the diamond growing temperature and pressure conditions were maintained for a period of between 4 to 10 minutes during which RD diamond growth occurred. Thereafter, the reaction capsule was quenched and the pressure allowed to return to ambient. The diamond content of the reaction capsule was recovered using conventional techniques. The recovered diamond was found to consist of a mixture of normal RD diamond and elongate RD diamond having a length to width ratio of 3:1 or more.

A number of other runs were performed in the same manner using an iron/manganese alloy and a cobalt/nickel/manganese alloy. In each case the diamond recovered from the reaction capsule contained a certain proportion of the elongate particles.

Eliminating Magnetism of Synthetic Diamond Grains

If an electrodeposited diamond tool is manufactured by treating in a plating bath a tool base on which conventional synthetic diamond grains are arranged, the plating metal will deposit not only on the tool base in spaces between the diamond grains but also on the surfaces of the diamond grains to reduce grinding or cutting ability of the resulting diamond tool due to the metal layer covering the grinding or cutting surface of each diamond grain and also reduce the strength for holding the diamond grains on the tool base, since at least a part of the plated metal is used not for holding the diamond grains on the tool base but for covering the diamond grains.

The drawbacks due to the magnetism make it difficult or rather impossible to use the conventional synthetic diamond grains for manufacturing a desired electrodeposited diamond tool, in which a major part of diamond grains is embedded in a relatively thick plating metal layer to be formed on the tool base and a remaining part only is exposed and protruded from a plating metal layer surface without any deposition of the plating metal thereon to form a sharp grinding or cutting surface of the tool product.

An object of the process of *H. Ishizuka, K. Suzuki and S. Tamon; U.S. Patent 4,119,500; October 10, 1978; assigned to Hiroshi Ishizuka, Japan* is to provide synthetic diamond grains having little or no magnetism and being suitable as a raw material for manufacturing electrodeposited diamond tools. This can be attained by treating in an inert atmosphere synthetic diamond grains at a temperature higher than the Curie temperature of the metal contained as an impurity in the raw material diamond grains, adjusting the friability of the thermally treated diamond grains and magnetically separating the resulting diamond grains to obtain demagnetized diamond grains and diamond grains of a relatively low magnetism.

As the raw material diamond grains, it is preferable to use the same having a bulk density of 1.75 to 1.90 g/cc, although the value may be changed depending on the size used. The magnetic separating step may be carried out prior to the friability adjusting step.

It is preferable to carry out the thermal treatment in vacuo or under nitrogen or the like inert gas stream in an electric furnace set at a temperature between about 800° and about 1300°C. The thermally treated diamond grains are cooled to room temperature in the electric furnace, from which the electric source has been disconnected or after having been discharged from the furnace. It has been found that the cooling speed does not have any influence on the magnetism of the resulting diamond grains but it requires about 30 minutes as the cooling time when the thermally treated diamond grains were left to stand.

The cooled diamond grains may have a thin graphite layer thereon and friability thereof has been reduced due to a difference in thermal expansion coefficients of the diamond and the metal impurity therein and a strain or distortion of the diamond grains per se during the thermal treatment. Therefore, the cooled diamond grains were first treated with an oxidizing agent to remove the graphite layer and then treated with a mill to adjust the friability thereof. As the oxidizing agent, chromic acid, perchloric acid or the like for dissolving graphite may be used. As the mill, a ball mill or the like conventional mill may be employed.

For using the resulting diamond grains on electrodeposited diamond tools, it is important that the grains have a friability of 30 to 55 weight %/90 sec.

Values of bulk densities and friabilities are those measured by methods as described in ANST-B74-1971 (American National Standard Test for Bulk Density of Diamond Abrasive Grains) and L.M. Zsolnay "Physical Evaluation of Diamond Abrasive Grain" *DWMI Technical Symposium,* Chicago, Sept. 1971, pages 8-1 to 8-4, respectively.

Nucleating Diamonds in Absence of External Container

At atmospheric pressure diamond represents an unstable form of carbon which will tend to transform to graphite if the temperature is sufficiently high, and the soft transformed graphite will be rapidly worn away. A very thin coating of nickel will smear over the surface of an active coated diamond grain and in so doing could alter the rate of wear by actually changing the equilibrium between the conversion of diamond to graphite and the conversion of graphite to carbon. Local conditions at the tip of the abrasive grain will be in the range of those found in the apparatus used to synthesize diamond from graphite in the presence of nickel.

In surface grinding extremely high energies are dissipated in a small area. It has been shown that the surface temperatures developed may be well above the melting point of the metal being ground. The temperature rise is of such short duration, however, that melting of the metal does not occur.

It is estimated that when grinding steel, surface temperatures on the order of 3000°C are obtained with grain-chip contact pressures as high as 2×10^6 psi (130 kiloatmospheres) which exceeds the range of operation of that employed in diamond synthesis. It is to be expected that under such extreme conditions the diamond nuclei produced will be extremely small, closely spaced and spontaneously formed.

In order to test the feasibility of generating diamond nuclei at the tip of an abrasive grain *R. Komanduri and M.C. Shaw; U.S. Patent 4,074,471; Feb. 21, 1978; assigned to Carnegie-Mellon University* performed the following experiment. It should be understood that the conventional abrasive grain can include tungsten carbide or diamond. Referring to Figure 4.8, a tungsten carbide sphere **10**, 1.5 mm in diameter, was coated with layers of nickel and graphite **12**, the nickel serving as a catalyst. The sphere was first cleaned by sputter etching and then three thin layers were deposited in the order nickel-carbon-nickel **12** using a sequential sputtering apparatus.

The coated tungsten carbide sphere was then cemented into a hollow head screw **14** which in turn was threaded radially into the surface of an 8 inch

(20 cm) diameter aluminum disc. The disc was mounted on the spindle of a surface grinding machine in place of the usual grinding wheel and used to cut a groove in a piece of hardened ball bearing steel (AISI 52100 steel of hardness R_c 61). The speed of the disc (V) was in the range of conventional grinding (30 m/sec) while the table speed (v) and depth of cut (d) were adjusted to provide individual chips that were comparable in size to those in ordinary surface grinding ($v = 2.54$ mm/sec and $d = 25\ \mu$). The length of groove produced was 12.5 mm which corresponds to 250 individual cuts.

The tungsten carbide sphere acted like an abrasive grain in the surface of a grinding wheel and produced individual chips that were projected into the atmosphere as sparks.

After cutting a small flat was found on the tip of the tungsten carbide sphere. When the worn surface of the composite coating was examined in a scanning electron microscope for evidence of diamond nuclei, many small nucleation sites about a micron in diameter were evident at the inner nickel layer. It has previously been reported that in the synthesis of diamonds actual transformation from carbon to diamond occurs in a very thin layer of nickel that separates the carbon from the diamond. The nuclei which were observed were produced in a thin nickel layer and are believed to be diamond nuclei generated under the extreme conditions of temperature and pressure which are generated during cutting.

An effort was made to confirm the presence of a diamond by x-ray diffraction. Tungsten carbide, titanium carbide, and nickel patterns were clearly evident. In addition, one spot (corresponding to the strongest diamond line) was also found.

Figure 4.8: Nucleating Diamonds in Absence of External Container

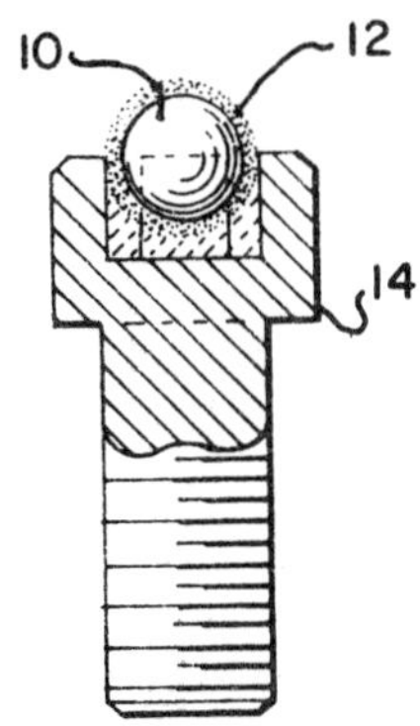

Source: U.S. Patent 4,074,471

COATED DIAMOND

Intermediate Metal Carbide Zone

M. Lee, L.E. Szala and L.E. Hibbs, Jr.; U.S. Patent 4,063,907; Dec. 20, 1977; assigned to General Electric Company produce diamond abrasive particles having a rough adherent covering. Their process comprises providing milling balls consisting of plastic or elastomer having a diameter ranging from about 1/16 inch to about 1/4 inch. Also provided are particles of a metal compound which is decomposible or reducible at atmospheric pressure at a temperature ranging from about 800° to 1400°C to produce metal and gaseous products of decomposition. The metal compound is selected from the group consisting of molybdenum sulfide, tungsten sulfide, titanium sulfide, niobium sulfide, tantalum sulfide, chromium chloride, zirconium sulfide, and mixtures thereof. Also provided are diamond particles ranging in size from about 10 to 2000 microns.

The milling balls and particles of metal compound and diamond are milled resulting in mechanically smearing a coating of the metal compound onto the surface of the diamond particles, coating the compound on at least 50 to about 100% of the surface area of the diamond particles. The milling balls have a shape and sufficient elasticity and resiliency to effect the mechanical smearing. The metal compound-coated diamond particles are recovered and fired in a reducing or inert atmosphere at a temperature ranging from at least about 800° to 1400°C thus decomposing or reducing the metal compound and producing the abrasive particles.

The abrasive particle consists essentially of a diamond particle having a rough granular adherent covering consisting of an outside surface coating ranging in composition from the metal to a carbide of the metal with all composition ratios of the metal and metal carbide falling within that range, and a layer of carbide of the metal intermediate the surface coating and the diamond particle chemically bonding the surface coating to the diamond particle. The covering is discontinuous or continuous.

Heat Treating in Nonoxidizing Atmosphere

M.G. Nicholas, P.M. Scott, B.I. Dewar; U.S.. Patent 4,062,660; Dec. 13, 1977; describe a method of bonding diamond to a metal capable of dissolving carbon but not of forming stable carbides under normal temperature and pressure conditions. The method includes the steps of intimately contacting the diamond with the metal to form a composite structure and heat treating the composite structure at a temperature above 500°C but below the melting temperature of the metal, both steps being carried out in a nonoxidizing atmosphere. This method has been found to result in interfacial bond strengths which are greater than those attributable to physical forces only. Suitable metals which have the properties defined above are nickel, cobalt and iron.

The nonoxidizing atmosphere may be provided by helium, argon, hydrogen, nitrogen or a vacuum on the order of 10^{-2} mm Hg or better.

In the case of diamond particles such as grit, the intimate contact may be achieved by coating the particles using standard deposition techniques. Vacuum deposition techniques are preferred because they produce very pure metal coatings. It has been found that the bonding strength is impaired if impurities are present. The grit may be of the RD, SD or MD type for application in resin bond or metal bond wheels, saws, compacts, crowns or single point tools.

Example 1: Pure nickel was swaged into an ingot of a desired diameter, e.g. 1.5 mm, and the ingot was then cut into desired lengths.

A length was placed on a diamond plaque which in turn rested on a graphite anvil in a chamber consisting of a quartz tube clamped between water cooled top and bottom brass plates. The chamber was evacuated by a rotary pump to 10^{-2} mm Hg or better pressure and maintained at this pressure during heating. A silica piston entered the vacuum chamber through a Wilson seal in the top plate and was used to apply pressure to the sample on the anvil. The pressure applied was sufficient to provide intimate contact between the nickel and the diamond, i.e. a pressure of between 3 and about 7.5 kg/mm^2. Induction heating was then used to raise the temperature of the chamber to 800°C which was maintained for about 2 hours. Excellent solid phase bonding of the nickel to diamond resulted with a bond strength of 25.8 kg/mm^2 being obtained. The bond strength, i.e. interfacial tensile strength, was measured in a standard manner using a shearing jig to which was attached a Hounsfield tensometer.

Example 2: 40-50 mesh SD diamond grit was coated with a layer of nickel using known vacuum deposition techniques. A layer amounting to 1 to 2% by weight of the uncoated particle was deposited. The coated particles were then heat treated at a temperature of 800°C for 2 hours in a vacuum furnace (10^{-2} mm Hg pressure). A layer of nickel was then deposited on the treated grit using known electroless deposition techniques. The outer layer amounted to 20% by weight of the original coated grit. For purposes of comparison, nickel coated grit was prepared as above, omitting the heat treatment step of the process. The two types of coated grit were incorporated in saws and sawing tests carried out. It was found that the saw containing the grit coated in accordance with the process showed less wear than the saw containing the prior art coated grit.

In a similar process *M.G. Nicholas, P.M. Scott and B.I. Dewar; U.S. Patent 3,924,031; Dec. 2, 1975; assigned to De Beers Industrial Diamond Division Limited, South Africa* provide a method of producing metal coated diamond particles in which the coatings are strongly bonded to the diamond particles. A layer of an alloy consisting of a major proportion of a metal such as copper, nickel or iron and a minor proportion of a carbide-forming metal such as titanium, chromium or vanadium is applied to the uncoated particles. Then the coated particles are heat treated to between 500°C and a temperature just below the melting point of the alloy long enough for a bonding carbide layer to form at the alloy/diamond interface, the steps being carried out in a nonoxidizing atmosphere. The bonding is achieved by means of a carbide layer, which is preferably continuous, at the alloy/diamond interface.

Composite Coating of Titanium and Metal Layer

According to the process of *R.J. Caveney; U.S. Patent 3,929,432; December 30, 1975; assigned to De Beers Industrial Diamond Division Limited, South Africa* there is provided a diamond particle of the MD or SD type having a composite coating comprising a thin continuous layer of titanium adjacent the diamond surface and a metal layer on the titanium layer, the metal of the metal layer being capable of forming an alloy with titanium under the influence of heat and the interface between the diamond and the titanium being essentially free of chemical bond formation.

The thin layer of titanium must be such as to provide a continuous layer around the diamond particle. The actual amount of titanium in the layer will vary according to the size of particle, but will, in general, be less than 3 wt % of the uncoated diamond. The titanium may be applied to the diamond particle by methods well known in the art, particularly by vacuum deposition.

The metal of the metal layer must be capable of forming alloys with titanium under the influence of heat. Examples of such metals are iron, nickel, cobalt, copper, and alloys thereof, e.g., iron-nickel alloys.

The metal layer may be applied to the titanium-coated diamond by methods known to the art. Suitable methods of application are electroplating, electroless plating, and vacuum deposition.

Nickel is the preferred metal for the metal layer. The nickel is preferably electroplated on to the Ti-coated diamond using nickel sulfamate as a source of nickel ions.

The metal of the metal layer is preferably present in an amount of from 10 to 100 wt % of the uncoated diamond.

Outer Nonmetallic Protective Layer

A.R. Roy and J.D. Van Schalkwyk; U.S. Patent 3,923,476; December 2, 1975 provide a method for producing a discrete coated abrasive particle in which the coating comprises an inner metal layer and an outer nonmetallic protective layer. The protective layer will be substantially inert to normal environmental conditions and will bond readily to the resin matrix of resin bond wheels. Preferably, the protective layer is a solid resin. Suitable resins are phenol-formaldehyde, epoxy, polyimide, alkyd, nonsaturated polyester and polyamide resins.

With this protective layer it has been found the coated particles can be stored for many months without significant deterioration; such particles also bond better to the resin matrix of wheels.

The metal of the inner layer will be any metal suitable in the field of metal-clad abrasive particles. The preferred material is nickel in the case of wet-grinding applications and copper in the case of dry-grinding applications.

The metal layer will generally be present in an amount of from 20 to 80 wt % of the coated particle. The protective layer will generally be present in an amount of from 1 to 3 wt % of the coated particle—this provides the particle with a thin continuous protective layer.

Preferably, the metal-coated particles are treated prior to the protective layer deposition substantially to remove any metal oxides from the exposed metal surface. The deposition should occur as soon after the treatment as possible so as to avoid, as far as possible, any reoxidation of the metal. This treatment will, in general, also remove other impurities such as greases and results in improved bonding between the metal layer and the protective layer.

The treatment is preferably a washing of the particles with a cleaning solution, preferably an alkaline cleaning solution. A suitable cleaning solution is a 10 to 30 wt % sodium carbonate cleaning solution. It has been found that elevated temperatures improve the cleaning powers of the solutions. Thus, it is preferable that the cleaning solutions be at or about their boiling points for the treatment.

Example: 35 carats of 80/100 RDA diamond grit coated with a 55 wt % layer of nickel (by weight being by weight of the coated particle) was refluxed in a 20% Imperial Zenith Cleaner aqueous solution. Imperial Zenith Cleaner is a commercially available alkaline cleaner having the following composition:

Na_2SiO_3, %	9.76
H_2O, %	24.33
Na_2CO_3, %	65.01
Impurities	Balance

The nickel-coated grit was refluxed for half an hour and then removed and washed with distilled water. The particles were dried with analytical reagent acetone and immediately thereafter placed in a solution consisting of 40% phenol-formaldehyde dissolved in 60% tetrahydrofuran, percentages being by weight. The particles and solution were agitated by stirring and the resin allowed to deposit on the metal-coated particles and set. The amount of solution was so chosen that the total quantity of resin for the 35 carats of diamond amounted to 0.24 g. With this amount of resin, the particles each had a thin continuous layer of resin amounting to about 3 wt % of the coated particle. The particles were recovered as discrete coated particles.

Diamond with Metallic Coating of Specific Thickness

A problem with grinding wheels containing diamond grits as abrasive has been that the abrasive is easily damaged during or after its embedment in the wheel matrix. Another problem has been to ensure that these grits will remain fixed when subjected to high loads and/or high temperatures.

These and other associated problems are solved by the use of diamonds prepared according to the process of *O. Lindstrom and E. Lundblad; U.S. Patent 3,904,391; September 9, 1975; assigned to Allmanna Svenska Elektriska A/B, Sweden*. The diamond grits are coated with a layer of metallic material with a thickness of between 0.5 μ and 155 μ, preferably between 5 μ and 20 μ. In a preferred form, the diamond grits, for example, synthetic diamonds having a grit size of 420 μ at the most, are coated with a layer, the thickness of which is between 10 μ and 15 μ. The layer consists essentially of one or more of the metals nickel, cobalt, silver, copper, molybdenum, titanium, aluminum, manganese, cadmium, tin, lead, zinc, chromium, gold, tungsten, indium, iron, zirconium, platinum group metals such as platinum, rhodium, palladium, or alloys or mixtures containing at least one of these metals, for example, tin-nickel. Such a layer, which can be applied in one of several ways, forms a continuous mechanically strong shell around the diamond grit. This shell is able to keep together splinters which may be formed in the diamond grit. Furthermore, the bonding surface towards the resin bond is enlarged, and the heat developed during the grinding will be transferred from the grits over a larger mass, thus preventing extreme temperature peaks in the grits and also thermally induced aging effects in the resin bond. It should be noted that the layer must be at least 0.5 μ thick, which means many times greater than a molecular layer, in order to obtain the advantages dealt with here.

Grinding wheels containing such diamond grits have, in comparison with those containing unprepared diamond grits, a better grinding performance and a longer life. It has been shown that grinding wheels containing diamond grits prepared according to the process will give an essentially larger grinding ratio than those containing untreated diamond grits. Moreover, an even higher grinding ratio is obtained within the layer thickness range 0.5 μ to 25 μ. It has also been shown that there is an optimum region between 5 μ and 20 μ. Layers with a thickness of between 10 μ and 15 μ have been shown to be particularly favorable for wet grinding.

Example 1: Nickel is reduced on diamond grits of 140 mesh to 170 mesh size in an autoclave under hydrogen at a pressure of 20 atm from an aqueous solution of a nickel salt containing 230 g/ℓ $NiSO_4 \cdot 6H_2O$, 300 g/ℓ $(NH_4)_2SO_4$ and 130 ml concentrated aqueous ammonia/ℓ, the diamonds being dispersed in the solution. The autoclave is equipped with a stirrer and the reduction takes place at a temperature of 160° to 170°C. The reduction is started by adding small amounts of a catalyst such as anthraquinone. The treatment is repeated until the thickness of the layer has reached the desired value, say 100 μ.

Example 2: 25 carats of diamond grits of 40 mesh to 60 mesh are cleaned in a mixture of equal parts of concentrated nitric acid and hydrochloric acid and rinsed thoroughly in distilled water, whereupon they are activated for 10 minutes each in the following two aqueous solutions: 5 g $SnCl_2$ + 10 ml HCl/ℓ and 0.2 g $PdCl_2$ + 1 ml HCl/ℓ. The activated diamonds are then dipped at a temperature of 70°C in an aqueous solution containing 30 g/ℓ $NiSO_4 \cdot 7H_2O$, 10 g/ℓ $NaOCOCH_3 \cdot 3H_2O$ and 10 g/ℓ $NaH_2PO_2 \cdot H_2O$. The acidity is adjusted to pH 5 with CH_3COOH.

Example 3: Synthetic diamond grits of 80 mesh to 100 mesh are cleaned by boiling in concentrated HNO_3 and rinsed in distilled water. The diamond grits are then silvered by reduction of silver from an ammoniacal solution of silver nitrate and a reducing agent, for example, according to Brashear, *Handbook of Chemistry and Physics*, 44th edition. The thickness of the layer obtained is less than 0.25 μ. The object of the silvering is to make the diamond grits electrically conducting. The thickness of the layer can range from molecular layers to some tenths of a micron. 75 carats of the silvered diamond grits are transferred to a rotatable cylindrical drum, containing an aqueous solution of $NiSO_4 \cdot 6H_2O$ (140 g/ℓ), NH_4Cl (30 g/ℓ), and H_3BO_3 (30 g/ℓ). A nickel anode is placed above the diamond grits, while the cathode is connected to the drum containing the diamond grits. The current is switched on and the current density is adjusted so that a value of about 20 mA/cm^2 area of the top layer of the diamond grits facing the anode is obtained. The drum is rotated at a speed of 1 to 2 rpm. The electrolysis is continued for about 24 hours, whereupon the diamond grits are removed, rinsed and dried. The thickness of the layer is by then 25 μ.

In related work *O. Lindstrom and E. Lundblad; U.S. Patent 3,957,461; May 18, 1976; assigned to Allmanna Svenska Elektriska A/B, Sweden* provide a grinding element having a grinding section comprising diamond particles individually embedded in a resin binding medium with improved performance characteristics. The improved characteristics are obtained by utilizing a metal coating on the diamond particles in the grinding section, which metal coating has a thickness of between 0.5 and 25 microns, preferably 5 to 20 microns, to impart to the section superior grinding properties to those which can be obtained with a like section wherein the diamond particles are not so coated.

Encapsulation with Metal Envelope of Allotriomorphic Dendrites

In the process of *A.G. Wilder and H.C. Bridwell; U.S. Patent 3,871,840; Mar. 18, 1975; assigned to Christensen Diamond Products Company*, abrasive particles are improved in function by encapsulating them with a metallic envelope; preferably the envelope is made of a pure metal in dendritic crystalline form. Desirably the abrasive substrate is placed in contraction by the envelope which is heat shrunk onto the abrasive substrate. The preferred method is to deposit the metal on the substrate at an elevated temperature by contacting a vapor of the metallic compound with the substrate particle under reducing conditions. The preferred primary abrasive is a diamond, and it is preferably etched before coating.

The deposits of the process are substantially pure metal envelopes, substantially free of intergranular inclusions.

The metallic envelope of the abrasive in the abrader structure is composed of crystal grains which are dendrites starting at and extending from the substrate surface, creating a superposition of grain growth interrupted by other grain skeleton deposit on top thereof. The grains thus deposited form a mechanically interlocked grain structure giving to the metal sheath high tensile strength. Such deposits are termed "allotriomorphic."

To obtain a compressive force on the substrate, a metal is chosen for the envelope having a substantially greater coefficient of expansion than the substrate. Thus, when the metal is deposited on the substrate at elevated deposition temperature, the metal sheath will contract on cooling more than the substrate, putting the substrate under compression. Since the linear coefficient of thermal expansion of suitable abrasives are in the range of about 1 to about 5×10^{-6} in/in/°F, metal sheaths are selected having a higher coefficient of expansion than the substrate, for example, metals having linear coefficients of expansion of about 2×10^{-6} to about 10^{-5} in/in/°F.

Example 1: To etch the diamonds, they are immersed in a molten bath of an alkali metal nitrate or alkaline earth nitrate at a temperature below the decomposition temperature; thus, in using potassium nitrate, temperature would range from 630+°F and under 750°F; sodium nitrate, about 580°F and under about 700°F; barium nitrate, at or above 1100°F and below its decomposition temperature. It is preferred to employ potassium nitrate at about 630°F for about an hour. The bath is contained in a nitrogen or other inert gas atmosphere. At the completion of the heating process, the molten bath is cooled and the cooled bath is then leached with water to dissolve the salt, leaving the etched diamonds which may then be separated and dried. The degree of etching depends upon the immersion time and a suitable time will be about an hour under which conditions the particles will lose from about ½ to 1% of their weight. The surface of the diamonds is roughened and pitted and forms a desirable and improved substrate base.

Figure 4.9: Diamond Coating Apparatus

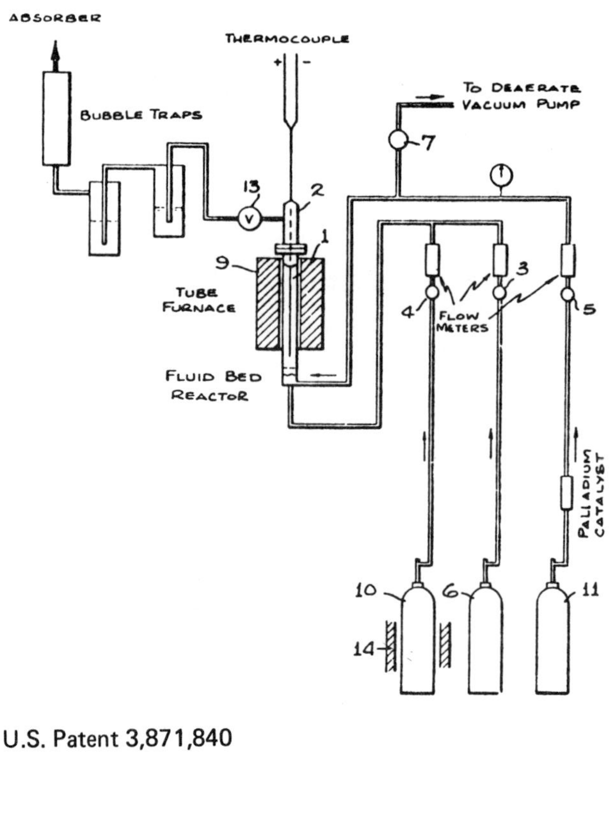

Source: U.S. Patent 3,871,840

Example 2: Diamonds, either synthetic or natural, preferably etched as above, of mesh size suitable for fluidizing are introduced into the reactor **1** (Figure 4.9). The actual mesh size employed depends upon the service to which the abrader is to be placed. Preferably, 40 to 100 mesh material, for example, -40+50 mesh is used. In depositing tungsten, it is preferred to employ tungsten hexafluoride, which is contained and vaporized in **10**. It is volatile at atmospheric temperatures and need not be heated. In the reactor employed after the system has been deaerated and back filled, hydrogen flow is established at a low flow rate of about 100 ml/min and the temperatures in the reactor **1** having been adjusted to 1150°F, as measured by the thermocouples, the hydrogen flow is increased to about 1,250 to 1,350 ml/min, and the flow of the tungsten fluoride vapor to about 150 ml/min and the argon gas is adjusted to about 285 ml/min, all as measured by the flow meters as indicated in the figure, the hydrogen being in stoichiometric excess over the tungsten hexafluoride.

The thickness of the coat of the tungsten on the diamond depends on the duration of the treatment and suitably for the 40 to 50 mesh diamonds described above, the coat will be 1-mil thick in about 1 hour. Suitable thickness deposit will run from about 0.1 to about 1.5 mils thick.

CUBIC BORON NITRIDE

Rough Adherent Coating

M. Lee, L.E. Szala and L.E. Hibbs, Jr.; U.S. Patent 4,011,064; March 8, 1977; assigned to General Electric Company provide a process for coating cubic boron nitride particles. The process comprises providing milling balls made of plastic or elastomer having a diameter ranging from about one-sixteenth inch to about one-fourth inch and providing particles of a metal compound which is decomposable or reducible at atmospheric pressure at a temperature ranging from about 800° to 1400°C to produce metal and gaseous products of decomposition. Cubic boron nitride particles are also provided ranging in size from about 10 microns to 500 microns. The milling balls and particles of metal compound and cubic boron nitride are milled, mechanically smearing a coating of the metal compound onto the surface of the cubic boron nitride particles, coating the compound on at least 50% to about 100% of the surface area of the cubic boron nitride particles.

The metal compound-coated cubic boron nitride particles are fired in a reducing or inert atmosphere at a temperature ranging from at least about 800° to 1400°C, decomposing or reducing the metal compound and producing abrasive particles. The abrasive particle consists of a cubic boron nitride particle having a rough granular adherent covering composed of an outside surface coating ranging in composition from metal to a nitride and/or boride of the metal with all composition ratios, i.e., mixtures, of the metal and metal nitride and/or boride falling within the range and layer of nitride and/or boride of the metal intermediate the surface coating and cubic boron nitride chemically bonding the surface coating to the cubic boron nitride. The granular covering is discontinuous or continuous and noninterconnecting or interconnecting and covers from at least about 50% to about 100% of the surface area of the cubic boron nitride particle.

Example 1: A clean polyethylene bottle, about 2⅛ inches high by ⅞ inch inside diameter was charged two-thirds full with polytetrafluoroethylene (Teflon) ⅛-inch-diameter spheres, i.e., balls. 1.239 g of powdered tungsten sulfide (WS_2), having a particle size of less than 40 microns were also added to the bottle which was then closed and rolled, i.e., milled, on a laboratory mill, at 60 rpm for 3 hours. Examination of the resulting balls showed them to be uniformly coated with about 0.892 g of WS_2, the remainder being coated on the inside walls of the bottle. The coated balls, along with 1.116 g of 100/120 mesh, i.e., about 125 microns to 149 microns in particle size, cubic boron nitride were placed in another identical clean bottle and rolled for 19 hours at 60 rpm on the laboratory mill. The balls and coated cubic boron nitride particles were then separated by suitable wire mesh sieves. The compound-coated cubic boron nitride particles weighed 1.415 g, indicating a WS_2 coating weight of 0.299 g. About 100% of the surface area of the cubic boron nitride particles was coated with WS_2.

Example 2: The WS_2-coated cubic boron nitride particles of Example 1 were fired in a ceramic tube furnace in an atmosphere of hydrogen at a temperature of 1040°C for 1 hour. The resulting fired particles had a very adherent substantially continuous rough granular coating or covering. The granular structure of the covering was flaky in character, of substantially uniform fine size. X-ray diffraction analysis of the fired coated particles determined that the covering or coating on the cubic boron nitride particles was tungsten metal. Also, the intermediate metal boride and/or nitride bonding layer, although not shown by x-ray diffraction analysis, would have been detectable by transmission microscopy.

Catalyst Systems for CBN Synthesis

Aluminum alloys of cobalt, nickel and manganese have been found by *R.H. Wentorf, Jr. and W.A. Rocco; U.S. Patent 3,918,219; November 11, 1975; assigned to General Electric Company* to be useful catalysts for the conversion of hexagonal BN to the CBN form at high pressure and high temperature. The habit and size of the CBN crystals developed depends on whether hexagonal BN or a mixture of hexagonal BN and CBN crystals (or sintered carbide grains) is employed with the catalyst metal.

One preferred form of a high-pressure, high-temperature apparatus in which the process may be practiced is the subject of U.S. Patent 2,941,248 and was described in U.S. Patent 4,188,194 in the preceding chapter of this book. It is shown in Figure 4.10a with a preferred form of reaction vessel **14**.

Reaction vessel **14** includes a hollow salt cylinder **19**. Cylinder **19** may be of other material, such as talc, which (1) is not converted during high-pressure/high-temperature operation to a stronger, stiffer state (as by phase transformation and/or compaction) and (2) is substantially free of volume discontinuities occurring under the application of high temperatures and pressures, as occurs, for example, with pyrophyllite and porous alumina. Materials meeting the criteria set forth in U.S. Patent 3,030,662 are useful for preparing cylinder **19**.

Positioned concentrically within and adjacent cylinder **19** is a graphite electrical resistance heater tube **20**. Within graphite heater tube **20** there is in turn concentrically positioned the cylindrical salt liner **21**. The ends of liner **21** are fitted with salt plugs **22, 22'**, disposed at the top and bottom, respectively. Liner **21** may have a cylindrical hollow core to receive one large charge assembly containing subassemblies or the liner may consist of a series of mold assemblies arranged in a stack for the preparation of a plurality of composite tool inserts.

Figure 4.10: Apparatus for Use in CBN Synthesis

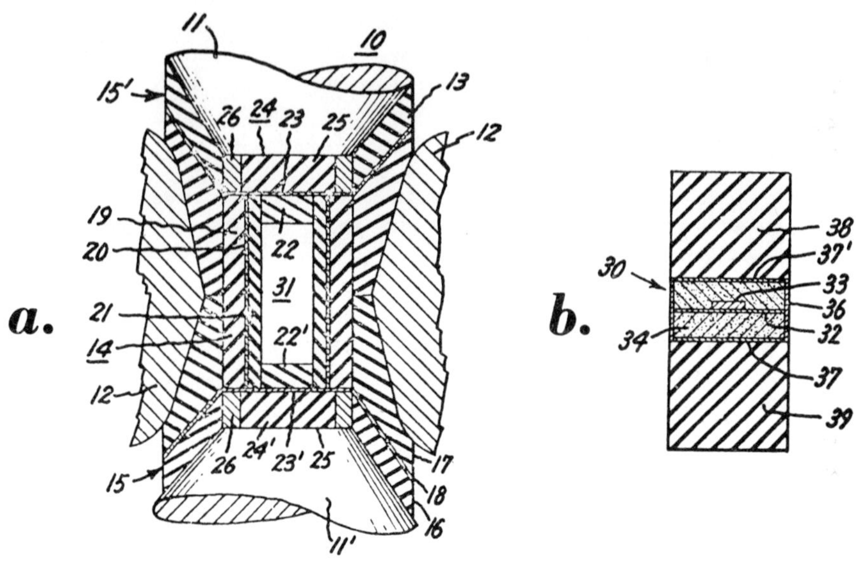

Source: U.S. Patent 3,918,219

Electrically conductive metal end discs **23** and **23'** are utilized at each end of cylinder **19** to provide electrical connection to graphite heater tube **20**. Adjacent each disc **23, 23'** is an end cap assembly **24** and **24'** each of which comprises a pyrophyllite plug or disc **25** surrounded by an electrical conducting ring **26**.

An exemplary charge assembly **30** for the conversion of hexagonal BN to CBN is shown schematically in Figure 4.10b. In the arrangement of materials shown the aluminum alloy forms in situ during the high-pressure, high-temperature conditions generated by means of the apparatus **10** to establish CBN-stable conditions. Charge assembly **30** fits within volume **31** of the apparatus of Figure 4.10a.

Example: The arrangement shown in Figure 4.10b was used wherein disc **32** was of cobalt metal. The hexagonal BN was baked for an hour at 900°C in ammonia before being loaded into the assembly. This was done to minimize oxygen in the system. After exposure to 55 kb (kilobars) and 1500°C for 60 min, heating was stopped and pressure was reduced to ambient. CBN crystals had been formed clustered around the region in which the cobalt and aluminum had alloyed. These crystals were identified as cubic boron nitride by x-ray diffraction. The crystals were brown in color measuring from about 10 to 100 micrometers in largest dimension. The habit of crystal growth was generally columnar and the crystals were often intergrown.

Containing Phosphorus

In the process of *J.D. Birle; U.S. Patent 3,881,890; May 6, 1975; assigned to General Electric Company* a tough cubic boron nitride crystal having greater-than-usual impact resistance is prepared by effecting the catalytic transformation of soft hexagonal boron nitride into cubic boron nitride in the presence of sufficient phosphorus or phosphorus-containing compound to substantially saturate the abrasive crystal with about 0.042 wt % of phosphorus.

Following usual procedure the phosphorus-containing charge is placed in the reaction chamber of a high-pressure high-temperature apparatus. The temperature is then raised to at least 1200°C and the pressure to a level where cubic boron nitride is stable at the particular temperature to which the charge is subjected. This is usually a pressure in excess of 40 kb. The temperature is then allowed to fall by cutting off the heating circuit and then the pressure is reduced. The charge is then withdrawn from the apparatus and the cubic boron nitride particles separated from the charge matrix.

Not all phosphorus-containing formulations yield the same results.

A preferred composition consists by weight of about 80% soft hexagonal boron nitride, 10% lithium nitride, 3% lithium hydride, 3% lithium hydroxide, 2% phosphorus and 2% ammonium chloride.

OTHER BORON-CONTAINING PARTICLES

Wurtzite Structure by Shock-Wave Treatment

F.R. Corrigan; U.S. Patents 4,201,757; May 6, 1980 and 4,231,980; November 4, 1980; both assigned to General Electric Company provides a way to produce high-density boron nitride particles which are larger than 50 microns in diameter and generally are several hundred microns in diameter. This is achieved by subjecting recrystallized pyrolytic boron nitride such as that disclosed in U.S. Patent 3,578,403 to dynamic shock-wave treatment which creates pressure and temperature conditions in which the stable form of boron nitride is high-density boron nitride. Most of the high-density boron nitride thus formed is wurtzite boron nitride.

In Figure 4.11a the boron nitride phase diagram in the pressure-temperature plane is marked into three regions outside of the liquid phase. In region I low-density boron nitride is stable and both wurtzite and zinc blende boron nitride are metastable. In region II wurtzite and zinc blende boron nitride are stable and low-density boron nitride is metastable. Region III is divided into area A and area B. In area A the pressure and temperature conditions will promote the formation of wurtzite boron nitride. In area B, these conditions will promote the formation of zinc blende boron nitride. The boundary between region II and region III is shown as an area because the precise point at which conversion takes place has not been determined for line accuracy. Likewise the dividing line between area A and area B is shown dashed because its position is not known accurately.

In the shock conversion of low-density boron nitride to high-density boron nitride the low-density material is subjected to a shock wave of sufficient amplitude to shock-compress the material into the direct conversion region III of Figure 4.11a.

Figure 4.11: Wurtzite Structure by Shock-Wave Treatment

a.

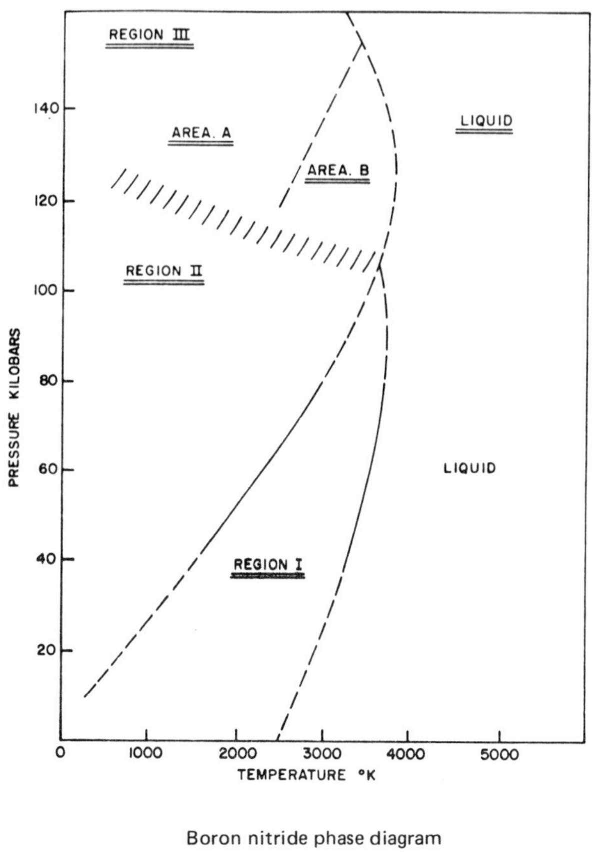

Boron nitride phase diagram

(continued)

Figure 4.11: (continued)

b.

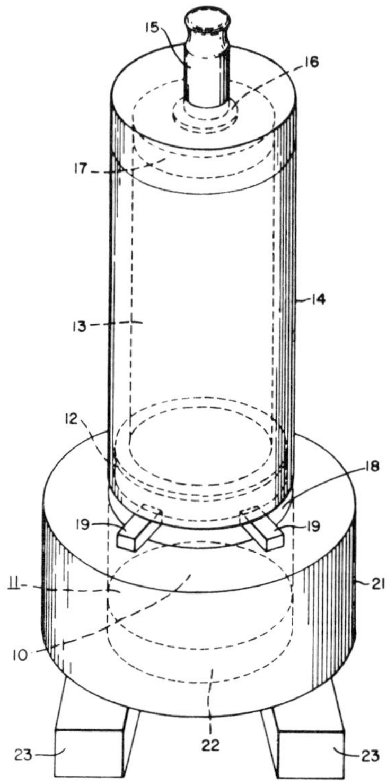

Shock-wave apparatus

Source: U.S. Patent 4,201,757

In the shock conversion process the high pressure applied is transient in nature, the duration being generally of the order of a microsecond. Because of the extremely short duration of the shock wave pressure pulse, there will be insufficient time for diffusion rate controlled chemical reaction mechanisms to be effective in the shock transformation process. However, there is ample time for diffusionless transition to occur under shock pressure conditions. A simple C-axis compression is probably the mechanism for transformation of low-density boron nitride to high-density.

Figure 4.11b shows one type of apparatus in which the process is carried out. In this apparatus a shock wave is introduced into a sample **10** positioned in a sample holder assembly **11** by impacting the top surface of the assembly **11** with a flyer plate **12** driven by a high-explosive driving charge **13** positioned in a copper tube **14**. The high-explosive charge **13** is fired by means of an electrically controlled blasting cap **15** which initiates a sensitive sheet explosive **16** which in turn fires a booster charge **17** which in turn fires the driving charge **13**. Detonation of the charge **13** accelerates the flyer plate **12** which impacts the front surface of the sample holder assembly **11** resulting in a shock wave being introduced into the sample container front wall, which then propagates through the container wall into the boron nitride sample **10**. An air gap **18** is provided between the explosive portion of the assembly and the sample holder **11** by three aluminum standoffs **19**. The sample holder assembly **11** is surrounded by a lead member **21** which aids in container retention and sample retrieval. The side of the sample holder assembly **11** away from the explosive charge is backed with a cylindrical momentum trap **22**. For convenience the entire assembly illustrated in Figure 4.11b may be placed on wooden blocks **23** which serve as a stand.

Typical dimensions for an assembly such as is illustrated in Figure 4.11b would have a booster charge **17** of about 0.75 inch thickness, a driving charge **13** of about 5.375-inch thickness, an air gap **18** of about one-quarter inch, a sample holder assembly **11** of 2 to 2.5 inches long and 2.5 to 3.5 inches in diameter, a steel flyer **12** one-sixteenth- to one-half-inch thick and 2.5 to 3.5 inches in diameter, a lead doughnut-shaped surround **21** of about two inches in wall thickness, and a one-half-inch-thick momentum trap **22**.

The pressure level and pressure duration attained by the firing of the explosive charge can be controlled by varying the quantity and geometry of the charge and the thickness and impedance of the flyer plate to produce pressures above 100 kb, and even above 500 kb. Typical pressures of the process are about 475 kb in the steel container. Under such conditions, almost 100% of the low-density boron nitride is converted to high-density boron nitride. The high-density boron nitride is mostly wurtzite boron nitride but there may also be zinc blende boron nitride in the mixture.

Parameters which must be considered in providing high-density boron nitride having optimum properties include the weight and velocity of the flyer and the thickness of the low-density boron nitride sample. High pressures promote the formation of larger-sized particles. Flyer velocities of about 7,000 ft/sec will produce particles as large as 1 mm in longest dimension.

Converting Hexagonal Boron Carbonitride to Tetrahedral Form

A hard material prepared according to the process of *R.J. Wedlake and A.L. Penny; U.S. Patent 4,187,083; February 5, 1980* consists essentially of $B_xC_yN_z$, wherein x, y and z can have any value greater than or equal to 1, in tetrahedral form.

In the hard material, the various ratios x:y, x:z, and y:z will generally not exceed 2:1 and vice versa. In one example of the material x is equal to z and not equal to y.

A typical and preferred example of the material is that in which x, y and z are all equal to 1. This is the material BCN in tetrahedral form.

The hard material of the process, particularly the preferred material BCN, has a hardness comparable to that of diamond and cubic boron nitride. The material will scratch cubic boron nitride and the soft direction of diamond and has a Knoop hardness between 4,050 and 9,000 kg/mm^2. The material is an insulator in its pure state.

In one form, the method involves subjecting boron carbonitride in hexagonal (graphitic) or amorphous form, in the presence of a solvent for the boron carbonitride to elevated conditions of pressure and temperature, the pressure exceeding 50 kb and the temperature exceeding 1300°C. The pressure used will generally not exceed 100 kb and the temperature, 2000°C. These elevated conditions of temperature and pressure are maintained for a period sufficient to produce the material. In general, this period will be about 10 to 60 minutes.

The solvent for the boron carbonitride must be capable of wetting and dissolving the substance under the conditions of elevated temperature and pressure which are used. A suitable solvent is a mixture or alloy of a metal of Group VIII of the Periodic Table and aluminum. The Group VIII metal is preferably cobalt, nickel or iron. The ratio of Group VIII metal to aluminum will typically be in the range 15:1 to 1:15 on a weight basis.

The synthesis of the hard material is carried out in a suitable high temperature/pressure apparatus such as the so-called "belt" apparatus which forms the subject of U.S. Patent 2,941,248.

Crystalline Titanium Diboride in Metal Carbide Matrix

The process of *H. Valdsaar; U.S. Patent 4,132,534; January 2, 1979; assigned to E.I. Du Pont de Nemours and Company* is directed to abrasive particles comprising a composite system wherein one component is a matrix of one or more carbide phases composed of titanium carbide, zirconium carbide and tantalum carbide and the other component is crystalline titanium diboride. The titanium diboride component ranges from 2 to 35 mol % of the carbide-boride system. The carbide component ranges from 98 to 65 mol % of the carbide-boride system and itself contains from 1 to 36 mol % of tantalum carbide and from 18 to 54 mol % of zirconium carbide, both based on total mols of carbide, the remainder being titanium carbide.

The particles are prepared by thoroughly mixing tantalum diboride, titanium carbide, and zirconium carbide, the TiC being present in an amount in excess of the stoichiometric amount required to react with the total amount of TaB_2 according to the equation $TiC + TaB_2 \rightarrow TiB_2 + TaC$; hot-pressing the mixture from 1 to 60 minutes at a temperature from 1900° to 2200°C at a pressure of at least 2,000 psi to form a hot-pressed compact. The compact has a room temperature density of 5.1 to 8.3 g/cc. The resulting hot-pressed compact is crushed to form abrasive particles.

Alternatively, the abrasive particles can be prepared as described above except that a mixture of TiC, ZrB_2 and TaC or a mixture of TiC, ZrC, TaC and TiB_2 in the appropriate amounts to give the desired product is heated under pressure where the ingredients combine to form a hot-pressed compact.

The abrasive particles of this process have a particle size of from -20 to +400 mesh (Standard U.S. Sieve) or 841 to 37 μm. When the abrasive particles are used in abrasive tools, such as grinding wheels, sizes from -40 to +200 mesh (420 to 74 μm) are preferred. For applications where the abrasive particles are used in a lapping composition, finer particles are used, for example, sizes from -325 (44 μm) to +400 (37 μm) are preferred. The $Knoop_{1000}$ hardness of these abrasive particles, as measured on the cold compact prior to crushing, ranges upward from 1,600 to 2,000 kg/mm^2.

Crystallization of Amorphous Boron

A boron abrasive according to the process of *D.K. Kuehl; U.S. Patent 3,976,444; August 24, 1976; assigned to United Technologies Corporation* is produced by crystallization of amorphous boron in the presence of and/or admixture with heavy metal borides. Boron filaments may be crystallized by being subjected to elevated temperatures, less than the melting point of boron, prior to shearing into short lengths and grinding into powder. Alternatively, boron in bulk form may be heated above the melting point and seeded with a material which will cause the formation of crystalline borides, the resulting product when cooled including solidified borides and crystalline boron which may be ground to powder.

Considering first the embodiment which employs boron filaments as the raw material, it will be understood that such filaments are produced using a tungsten wire substrate and the known BCl_3-H_2 process. In this process, for example, as disclosed in U.S. Patent 3,549,424, a complex flow of BCl_3 and H_2 gas is established along a vertically oriented, multistage, glass reactor which encloses the tungsten substrate, the substrate being heated resistively. The chemical reaction within the tube results in the deposition of amorphous boron on the filament, the production of HCl and the conversion of the substrate to tungsten boride.

In accordance with the process boron filaments produced in the manner briefly described above are heated, in a gaseous atmosphere, to a temperature above 1300°C but less than the approximately 2300°C melting temperature of the boron to crystallize the boron. The crystallization is performed in a gaseous atmosphere, for example, in H_2 or an inert atmosphere such as argon or nitrogen, and the duration of the heating step will be inversely proportional to the temperature. If it is desired to obtain β-tetragonal crystalline boron the temperature will be in the range of 1300° to 1480°C. If the β-rhombohedral form of crystallized boron is to be formed the temperature will be in the range of 1480° to 2300°C.

After the amorphous filamentary boron has been converted into crystalline form the boron filaments will be chopped into short lengths and thereafter ground or milled to powder. This grinding may be performed within a ball mill.

In accordance with a second embodiment, the molten boron may be seeded with 1 to 10% tungsten by weight to produce tungsten boride mixed with crystalline boron. The tungsten boride and crystalline boron are permitted to solidify and thereafter reduced to powder by grinding or milling. Since borides have a higher melting point than boron, the tungsten borides will solidify first and will thus be evenly distributed throughout the crystallized boron. Other materials suitable for seeding molten boron include molybdenum, tantalum and similar metals which form crystalline borides. In the practice of the process elemental boron is placed in a boron nitride crucible and heated to a temperature above 2300°C with an induction coil using a graphite or titanium nitride susceptor. The seed crystals of the selected material are dropped into the molten boron and the heat is reduced. The solidified borides and β-rhombohedral boron which remain after cooling, when ground to powder, have enhanced abrasive qualities when compared to amorphous boron.

CARBIDES

Sintered Chromium Carbide

In accordance with the process of *T.J. Smith; U.S. Patent 4,173,457; Nov. 6, 1979; assigned to Alloys, Incorporated* sintered or cemented chromium carbide particles are used to replace cemented tungsten carbide particles as a hardfacing material.

Sintered chromium carbide particles are provided in which the particle size is within the range of 10 to 60 mesh and preferably 14 to 30 mesh. When a mesh size is specified herein, it is with reference to the U.S. Standard Screen size.

Each of the particles is made up of grains of chromium carbide having an average grain size within the range of 0.5 to 25 μ, and preferably in the range of 0.5 to 15 μ. These grains of chromium carbide are cemented with a binder of which a major portion is nickel. The broad range of composition of the carbide particles can be from 3 to 25% binder, the balance being chromium carbide. Particles having compositions in this range have been tested and found to have a hardness of 87.8 to 88.8 Rockwell A and a density of from 6.8 to 7.0 g/cm^3. A typical composition in this range is 10% nickel with the balance being chromium carbide. This composition was found to have a Rockwell A hardness of 88.1 and a density of 6.9 g/cm^3.

A somewhat more preferred composition for the particles is from 3 to 25% nichrome, with the balance being chromium carbide. Compositions in this range have been found to have a hardness from 88.0 to 89.0 Rockwell A and a density of 6.8 to 7.0 g/cm^3. A typical composition within this range is 12% nichrome and the balance chromium carbide, which has been found to have a hardness of 88.6 Rockwell A and a density of 6.9 g/cc. The term nichrome as used herein denotes an alloy of 75 to 95% nickel, with the balance being chromium.

When the chromium carbide particles have only chromium carbide and binder present during sintering, the chromium carbide undergoes grain growth when using normal sintering techniques. For example, when chromium carbide powders having an average particle size of 0.5 to 5.0 μ are used, the grains will normally grow in size to an average within the range of 5 to 25 μ. It has been found that when certain grain growth inhibitors are added in minor amounts, the average grain size can be maintained in the range of 0.5 to 15 μ. This results in improved hardness. Thus, one or more of, for instance, titanium carbide, molybdenum carbide, vanadium carbide or tantalum carbide can be added in minor amounts effective to retard grain growth of the chromium carbide. A preferred range for titanium carbide is from 2 to 12% based on the total weight of the particles. The preferred range for molybdenum carbide is from 0.2 to 5%. Therefore, a preferred composition will contain from 3 to 16% of nichrome binder, 2 to 12% of titanium carbide, 0.2 to 5% molybdenum carbide, with the balance being chromium carbide.

It has been found that particles of this composition have a hardness within the range of 90.6 to 91.6 Rockwell A and a density of 6.8 to 7.0 g/cc. A typical composition in this range is 1.2% molybdenum carbide, 4.6% titanium carbide, 6.5% nichrome with the balance being chromium carbide. This typical composition was found to have a Rockwell A hardness of 91.4 and a density of 6.9 g/cc.

The manufacturing techniques employed to make the chromium carbide particles of this process can be any one of those which have heretofore been used to manufacture tungsten carbide particles for hardfacing material.

Titanium Carbide, Zirconium Carbide and Titanium Diboride

The process of *H. Valdsaar; U.S. Patent 4,076,506; February 28, 1978; assigned to E.I. Du Pont de Nemours and Company* is directed to abrasive particles consisting essentially of titanium diboride (TiB_2) crystals in a carbide matrix of titanium carbide (TiC) and zirconium carbide (ZrC). The carbide matrix constitutes approximately 69 to 94%, and the titanium diboride constitutes approximately 31 to 6% of the total weight of the titanium and zirconium carbides and borides present. At least a part of the carbide matrix is in the form of one or more solid solutions of zirconium carbide and titanium carbide and contains from 10 to 46% of zirconium carbide, based on the total weight of titanium and zirconium carbides and borides present. In addition to the titanium and zirconium carbides and borides, the abrasive particles may contain other materials in amounts up to 20% of the total weight of the abrasive particles.

In the preferred abrasive particles of this process, the carbide matrix constitutes from 75 to 90% and titanium diboride constitutes from 25 to 10% of the titanium and zirconium carbides and borides present, while the zirconium carbide content is from 15 to 40%, based on the total weight of titanium and zirconium carbides and borides present.

The abrasive particles of this process are prepared by thoroughly mixing zirconium diboride (ZrB_2) and TiC, the TiC being present in an amount in excess of the stoichiometric amount required to react with the total amount of ZrB_2 according to the equation $TiC + ZrB_2 \rightarrow TiB_2 + ZrC$; hot-pressing the mixture from 1 to 60 minutes at a temperature from 1900° to 2400°C at a pressure of at least 2,000 psi to form a hot-pressed compact having a room temperature density of 4.8 to 5.3 g/cc; preferably the components will be hot-pressed for from 5 to 15 minutes at a temperature of from 1900° to 2100°C and at a pressure of at least 4,000 psi; and then crushing the resulting hot-pressed compact to form abrasive particles.

Alternatively, the abrasive particles of this process can be prepared as described above except that a mixture of TiC, ZrC and TiB_2 in the appropriate amounts to give the desired product is heated under pressure where they combine to form the hot-pressed compact.

The abrasive particles of this process have a particle size of from -20 to +400 mesh (Standard U.S. Sieve) or 841 to 37 μm. When the abrasive particles are used in abrasive tools, such as grinding wheels, sizes from -40 to +200 mesh (420 to 74 μm) are preferred. For applications where the abrasive particles are used in a lapping composition, finer particles are used, for example, sizes from -325 (44 μm) to +400 (37 μm) are preferred. The $Knoop_{1000}$ hardness of these abrasive particles, as measured on the cold compact prior to crushing, ranges upward from 1,600 to 2,100 kg/mm^2.

Example: 170 g of commercial titanium carbide and 80 g of commercial zirconium diboride powders having particle diameters up to 100 μm are thoroughly mixed and milled with 100 ml of tetrachloroethylene in a conventional porcelain ball mill having a 4.875-inch inside diameter and 5.85-inch outside diameter with ⅜ inch diameter alumina balls for 32 hours at 60 rpm. The ground powder mixture is air dried to remove the tetrachloroethylene. 40 g of the powder mixture is placed in a one-inch-inside-diameter cylindrical graphite mold surrounded with graphite powder and is pressed at 2000°C and 4,000 psi for 5 to 15 minutes under nitrogen. Heat-up time to 2000°C is 25 to 30 minutes. A pressure of 2,000 psi is applied at the start of heating and then increased to 4,000 psi when the mold reaches a temperature of about 1900°C and is held at this point during the rest of the heat-up and through the temperature hold period. The mold is then allowed to cool and the cold compact is removed.

To render the composition thus formed useful as a component of an abrasive tool, the cold compact is size-reduced in a conventional jaw crusher, followed by size reduction by hand in a steel mortar. The crushed product is then screened into four fractions of abrasive particles having the following U.S. Standard Sieve sizes: +20, -20+40, -40+80 and -80 (+840 μ, -840+420 μ, -420+177 μ, -177 μ).

This procedure is repeated twice and the resulting abrasive particles are composited. The density of the particles is found to be 5.07 g/cc by air pycnometer. $Knoop_{500}$ microhardness is 1985.

Titanium Carbide, Titania and Alumina

K. Ogawa, M. Miyahara and M. Furukawa; U.S. Patent 4,063,908; December 20, 1977; assigned to Nippon Tungsten Co., Ltd., Japan provide powder mixtures which are to be used in manufacturing ceramic cutting tool materials containing 19 to 70% powdered titanium carbide by weight, 1 to 10% powdered titania by weight and the remainder powdered alumina.

In addition to the above constituents a suitable amount of conventional grain growth inhibitor such as magnesia or chromium oxide may be optionally added to the abovedescribed powder mixture to prevent grain growth in the sintering body.

Example 1: *Simple Mixing Procedure* — 89.8% of a powder of commercial alumina with a purity of above 99.5% and mean particle size of 1 μ was mixed with 10% of powdered titania and 0.2% of magnesia and the mixture was subsequently thoroughly blended. Then the thus-prepared mixture was preliminarily roasted at a temperature of 1300°C for 2 hours.

A 100-g portion of commercial powdered titanium carbide with a total carbon content of 19.1% and a mean particle size of 2 μ was mixed with 100 g of the abovementioned powder mixture. The whole mixture was ball-milled wet to grind and blend to obtain a homogeneous composition. Then the ball-milled charge was dried and was compacted in a graphite mold of 50 x 50 x 5.5 mm, placed in a high-frequency coil furnace and hot-pressed at a temperature of 1600°C and a pressure of 200 kg/cm^2 for 60 minutes. Finally the pressure was released and the mixture was cooled. The thus-sintered body measuring 50 x 50 x 5.5 mm obtained thereby was cut with a diamond blade into small test pieces which were then ground with a diamond-faced grinding wheel yielding four small sectional test pieces measuring 24 x 8 x 5 mm. These test pieces were tested for the following physical and mechanical properties: 99.2% relative density (specific gravity to the theoretical specific gravity), 94.3 HRA hardness, and 78.5 kg/mm^2 bending rupture strength.

Example 2: *Modified Procedure Using Partially Carburized Titania* — A 140 g portion of commercial powdered alumina having a mean particle size of 1 μ and a purity of above 99.5% was mixed with 60 g of partially carburized titania powder which had been prepared by partially carburizing commercial powdered titania of a purity greater than 99% and which had a combined carbon content of 16%, an oxygen content of 4.2% and a mean particle size of 1.5 μ.

The resultant mixture was blended thoroughly and sintered by hot-pressing as in Example 1 to yield sintered bodies from which test pieces were prepared and were subjected to examination. The following evaluation values were obtained: 98.7% relative density (specific gravity to the theoretical specific gravity), 94.4 HRA hardness, and 77.0 kg/mm^2 bending rupture strength.

Example 3: *Modified Procedure Using Partially Oxidized Titanium Carbide* – A 140-g portion of the same commercial powdered alumina as used in Example 2 was mixed with 60 g of a partially oxidized titanium carbide powder which had been previously prepared by partially chemically oxidizing the same commercial titanium carbide powder as in Example 1 using an aqueous solution of hydrogen peroxide to an extent sufficient to result in an oxygen content of 4% in the partially oxidized titanium carbide powder. Then the mixture was thoroughly blended and sintered by hot-pressing as in Example 1 to yield sintered bodies. However, in this case the sintering temperature was 1580°C.

Test pieces were prepared and the evaluation of the experiment was conducted as indicated in Example 1; the following values were obtained: 99.7% relative density (specific gravity to the theoretical density), 94.4 HRA hardness, and 79.3 kg/mm^2 bending rupture strength.

Reclaiming Uncemented Tungsten Carbide

M.B. MacInnis and C.D. Vanderpool; U.S. Patent 3,947,555; March 30, 1976; assigned to GTE Sylvania Incorporated provide a simple process for recovering refractory metal values from uncemented refractory metal carbides.

The process comprises heating uncemented refractory metal carbide powders contaminated with an iron group metal in a boiling solution of glacial acetic acid for at least about 30 minutes, the weight of the powders divided by the weight of the acid being a numerical value less than about 5. Thereafter, the solids are removed from the acid, washed with sufficient water to remove the residual acid and dried and are satisfactorily purified to enable them to be used as raw materials in a refractory metal carbide process.

Example: About 20 parts by weight of tungsten carbide containing about 9% cobalt and 1.6% iron is charged into about 200 parts by weight of glacial acetic acid. The material is heated to the boiling point of about 118°C for about 1 hour. The solid materials are removed, washed and dried and analyzed and are found to contain 1.7% cobalt and 0.08% iron thus indicating that iron is very effectively removed and the cobalt content is appreciably lower. The material can be milled with sufficient additional cobalt to yield a satisfactory material for sintering to the cemented carbide state thereby enabling essentially 100% recovery of the tungsten values contained in the material.

Substantially similar results are achieved when other refractory metals are present in the cemented carbides. Additionally, carbides which are contaminated with nickel can be recovered in the above manner. Increasing the time to about 2 hours lowers the cobalt content to a level below about 1%. Beyond 24 hours no further decrease is detected.

OTHER GRAIN PREPARATION METHODS

Additive to Produce Porous Cake

An object of the process of *J.V. Naidich, I.A. Lavrinenko, V.A. Kondratsky, J.N. Semenov, B.M. Emelyanov and L.P. Kogosov; U.S. Patent 4,024,675; May 24, 1977* is to produce an abrasive cake having a porous structure.

This is achieved in a method of producing aggregated abrasive grains wherein a mixture of abrasive and metal powders is sintered in the presence of a liquid phase and the cake thus produced is subjected to crushing. According to the process the mixture of powdered materials to be sintered is in a loose or unbound bulk state. The required degree of wettability of elementary grains with metal is achieved and the cohesion of individual elementary grains to each other is improved by effecting the sintering operation in the presence of an additive, which is titanium and/or zirconium and/or chromium, and/or vanadium, and/or silicon. Owing to the sintered powders being in a loose bulk state and the resulting pore formation, the above additives, when introduced into the mixture, form a thin film on the surface of the abrasive particles, the film being chemically bound to the surface thereof and conducive to the curing of various defects on the surface of abrasive grains, thus improving the strength of the elementary grains 1.5 to 2-fold and strongly binding the grains to each other.

Specifically, the additive is introduced in amounts of 5 to 10 wt % of the basic metal powder.

The abrasive powder and metal powder can be mixed in the same proportions as generally used for similar methods. The preferred proportion is within the range of from 4:1 to 1:5 by volume, for the abrasive powder and the metal powder respectively. The binder metal or alloy is conventionally selected depending on the nature of the abrasive material.

The additive and the amount thereof is selected depending on the nature of the abrasive material. Thus, for example, a preferred additive in case of diamond powder is titanium, whereas a zirconium additive is preferred for a mixture of diamond and boron carbide powders.

Example 1: A diamond powder having a particle size of 100 to 80 μ and powdered alloy comprising 72 wt % of silver and 28 wt % of copper were used for sintering. The diamond powder and the alloy powder were taken in the proportion of 3 to 1, respectively. The sintering was carried out in the presence of a titanium additive in an amount of 7 wt % in excess of 100 wt % of the basic metal powder, at a temperature of 900°C under vacuum. The aggregate grains obtained after crushing the cake are characterized by a strongly developed structure and strong adhesion of the primary (elementary) grains to each other. The grains thus produced were used for the manufacture of grinding discs on an organic binder and subjected to testing by way of grinding hard alloys. The tests were carried out at a disc speed of 25 m/sec and longitudinal feed of 2 m/min and cross-feed of 0.2 mm per double pass. Specific diamond consumption was 0.06 to 0.08 mg/g.

Example 2: An abrasive powder of cubic boron nitride having a grain size in the range of 125 to 100 μ, and a powdered alloy comprised of 80% copper and 20% tin were used for sintering. The ratio of the abrasive powder and the alloy powder was 1.2 to 1, respectively. The sintering was carried out in the presence of a zirconium additive in an amount of 10% in excess of 100% of the weight of the basic metal powder at a temperature of 1050°C, under vacuum not lower than 2.10^{-5} mm Hg, for a period fo 25 minutes.

The aggregated grains thus produced are characterized by a developed surface and high strength. The grains were used for the manufacture of grinding wheels on an organic binder. The grinding wheels were tested in machining specimens of steel having 0.7 to 0.95 carbon content, at a wheel speed of 30 m/sec, longitudinal feed of 1 m/min and cross-feed of 0.02 mm per double pass, the specific consumption of cubic boron nitride being 0.2 to 0.26 mg/g.

Silicon-Base Alloy Coating

P.P. Otopkov, A.V. Nozhkina, V.I. Kostikov, J.I. Andropov, A.Y. Pesina, A.I. Senchakov and A.V. Kabanov; U.S. Patent 4,184,853; January 22, 1980 provide an abrasive material the grains whereof have a coating of a silicon-base alloy. The alloy comprises, separately or in combination, copper, silver, gold, aluminum, and transition metals of Group IV to VIII of the Periodic Table, wherein, according to the process, the alloy also comprises gallium, indium, thallium, germanium, tin, lead, phosphorus, antimony, tellurium and sulfur, separately or in combination, in a quantity of from 2 to 80 wt % of the total weight of the coating.

It was found that introduction into the coating of an element chosen from the group consisting of gallium, indium, thallium, germanium, tin, lead, phosphorus, antimony, tellurium and sulfur (fusible elements), and taken separately or in a combination in a quantity of from 2 to 80 wt % of the total weight of the coating, makes it possible to strengthen the abrasive material, as this leads to a greater coating plasticity and, consequently, to a higher abrasive strength, and also results in a lower melting temperature of the coating material which, in turn, increases the durability of the coating on the abrasive.

The abrasive material, according to the process, is most effective in processing hard-to-treat materials, for example, diamond. By properly selecting alloy components one can substantially improve the characteristics of the material. The appropriate thickness of a coating varies from 0.01 to 1.000 μ.

Chemically Removing Iron from Grits

W. Falkenhain and E. Feder; U.S. Patent 4,118,459; October 3, 1978; assigned to Dynamit Nobel AG, Germany provide a process for removing of iron from grits with aqueous hydrochloric acid. A mixture of grits and aqueous hydrochloric acid is agitated for removal of the iron from the grits, for the provision of a first suspension of the grits in an aqueous hydrochloric acid medium in which the iron is dissolved. Water is added to the first suspension to form a second suspension, and the second suspension is centrifuged and thereby separated into a specifically lighter fraction rich in hydrochloric acid and a specifically heavier fraction rich in grits. Water is added to the specifically heavier fraction to form a third suspension, and the third suspension is centrifuged to produce a specifically lighter fraction rich in hydrochloric acid and a specifically heavier fraction rich in grits.

The specifically heavier fraction produced in the second centrifuging step can be filtered to provide a filter cake containing the grits, and the filter cake can be dried to provide dried grits substantially free of iron.

In a preferred manner of carrying out the process, the first step, wherein the aforesaid first suspension is produced, is performed batchwise in at least two removal vessels synchronized in operation to provide a continuous supply of the first suspension, whereby the first suspension can be continuously subjected to the steps following the production thereof.

In contrast to the conventional process, in this process the grits are generally not separated into their individual sizes prior to the removal of the iron. The suspension of grits may accordingly contain grain sizes from about 2 all the way up to 500 μ.

The removal of the iron can be performed with an approximately 20% solution of hydrochloric acid, cold or heated if desired to as much as 70°C. It is preferred to give the iron removal tank, which may be made for example of acid-resistant plastic such as hard PVC, a funnel shape, and to discharge the suspension at the bottom. The agitation can be effected with any desired type of stirring unit. Indirect agitation, by means of compressed air, for example, is advantageous on account of the danger of corrosion. The compressed air in this case may be introduced into the tank from the bottom, or the hydrochloric acid suspension may be stirred by removing part of the suspension through a gap and feeding it back into the tank by compressed air or by means of an air lift pump.

The grits are generally sprinkled steadily or in portions into the iron removal tank and after being stirred for about two hours, or less if the hydrochloric acid is heated, they will be free of iron.

Spinel and Aluminum-Base Metal Cermet

L.E. Wilson; U.S. Patent 3,940,276; February 24, 1976; assigned to Corning Glass Works describes a cermet composition of matter which is a dense, coherent, substantially homogeneous material composed essentially of, on a weight basis, at least 60% magnesium aluminate spinel crystals, 0 to 10% alpha-alumina crystals, 0 to 2% mullite crystals, 0 to 5% magnesium aluminum boride crystals and 10 to 30% aluminum-base metal phase. In especially advantageous form, this cermet is a fine-grained material wherein at least 90% by volume of the crystals of spinel, alumina and mullite are of a size less than 1 μ. The average size of such submicron size crystals is desirably not more then 0.5 μ. The metal phase is finely interspersed throughout the mass of crystals of spinel, alumina and mullite. The amounts of alpha-alumina and mullite crystals are severely restricted in order to avoid undesirable abrasive chipping and wear characteristics. The substantial strength and toughness of the cermet appear to be mainly provided by the substantially homogeneous composite of predominately fine-grained spinel crystals thoroughly and finely interspersed with the aluminum-base metal phase.

The tendency to lesser abrasive chipping and wear is further enhanced by the presence of boron in the composition, which sometimes forms observable discrete crystals of boride scattered in an irregular random manner throughout the cermet (commonly in apparently preexisting small void spaces within or on the surface of the cermet mass). Such boride appears to be $MgAlB_{14}$ and in the form of platelets of submicron thickness, but with their longest dimension not infrequently exceeding 1 μ. However, the cermet often does not appear to contain any or enough boride crystals (even under electron microscopy examination) to account for all of the boron detectable in the cermet by chemical analysis; nevertheless, all such detectable boron appears to provide the noted enhancement of abrasive characteristics in this reaction cermet.

On an analytical basis, the cermet broadly comprises (by weight) 2.5 to 33% MgO, 39 to 76% Al_2O_3, 0 to 22% B_2O_3 and the balance being essentially aluminum-base metal, the latter forming a metal phase that does not exceed 26% by weight of the cermet. Most preferably, the cermet is composed wholly of (by weight) at least 75% magnesium aluminate spinel crystals, 0 to 2% magnesium aluminum boride crystals and 10 to 26% aluminum-base metal phase. For best abrasive characteristics, the analytical boron content (expressed as B_2O_3) is 3 to 10% by weight of the cermet.

The method of making the cermet comprises providing a mass of melted and solidified material consisting of, analytically by weight, at least 65% of a base composition and 0 to 35% B_2O_3. The base composition consists of 5 to less than 40% MgO, 5 to 43% Al_2O_3 and 35 to 85% SiO_2. The mass is contacted with molten aluminum-base metal maintained at temperature in the range of about 650° to 850°C when the MgO content of the base composition is at least 8% and in the range of about 650° to 750°C when the MgO content of the base composition is less than 8%. The mass is maintained in contact with the molten metal for a time sufficient to effect reaction between the mass and the molten metal and to transform a substantial portion of the mass into the cermet composition.

Particles of Uniform Size and Shape

The process of *W.S. Blanding and J.A. Brothers; U.S. Patent 3,859,407; Jan. 7, 1975; assigned to Corning Glass Works* is concerned with the problem of repeatedly forming particles having virtually identical size and shape.

A plurality of interconnected particles of like size and shape in ribbon form are provided from plastically formable material. The ribbon of connected particles, after rigidification is subsequently separable into uniform individual particles.

Referring to Figure 4.12, a pair of cooperating forming rollers **10, 12** which may be water-cooled if desired, and having engraved or recessed patterned surfaces **14, 16,** roll-forms a flow of plastically formable material **G**, such as green ceramic batch or molten glass or basalts, into a continuous ribbon **R** of interconnected particles **P**. The rollers are rotatable about parallel longitudinal axes toward one another as shown by the arrows, so that the flowable material **G** supplied thereto is roll-formed as it passes through the bite.

As shown, the ribbon may be transferred along conveyor rollers **18** to a pair of soft, off-set flexing rolls **20** which distort and flex the now rigidified ribbon **R** in various directions out of its natural plane and thereby liberate the connected particles **P** from the ribbon so they may fall as separate individual particles and be collected in a container **22**.

Figure 4.12: Apparatus for Forming Uniformly Sized and Shaped Particles

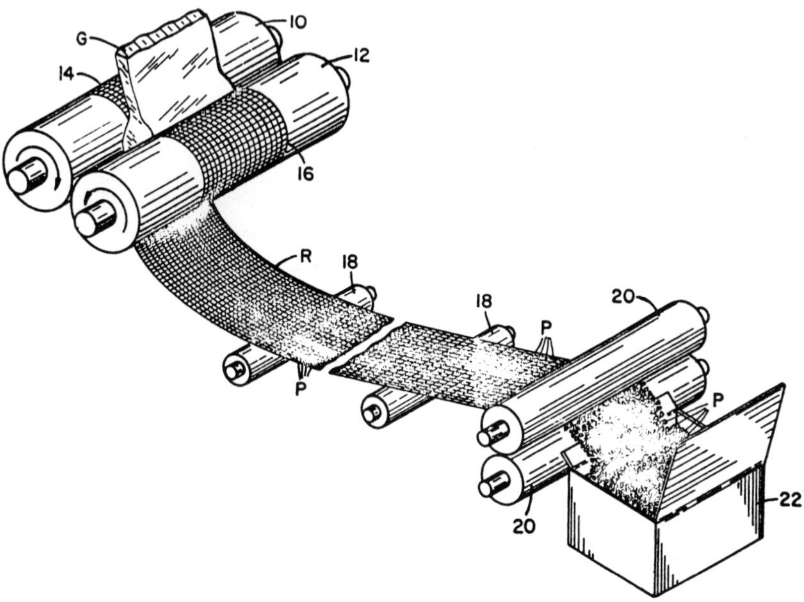

Source: U.S. Patent 3,859,407

The ribbon is shown as being discontinuous between conveyor rollers **18** to indicate that additional processing such as solidifying, drying, firing and metallizing may be imparted to the ribbon as desired prior to the liberating or separating of the various particles therefrom into their individual state. For example, the newly formed ribbon **R**, if of a siliceous vitreous material, may be subjected to a metallizing treatment such as by contacting it with molten aluminum as described in U.S. Patent 2,702,750, so as to form cermet particles which are especially suited for heavy-duty abrasive grain application.

The recessed or engraved patterned surface may be formed in either or both of the cooperating rollers **10, 12** as desired.

The fact that the process permits a highly controlled geometric shape of the particle is of particular significance when such particle is used as an abrasive grain. That is, improved abrasive performance can be achieved by controlling the shape to maximize grain sharpness and grain toughness. Also, the shape can be designed to supply anchors for the resin or other bond utilized to hold the grains on a grinding or snagging wheel and thereby increase the strength and life of such a wheel. Further, the shape may be optimized to increase the particle packing of the grains and thus increase the volume of grains per wheel. Finally, the process allows for very accurate control of the particle size distribution of the grains produced, and therefore permits the manufacture of highly reproducible abrasive products from such grains.

IMPACT AND TUMBLING CLEANING

Metallic Abrasive from Scarfer Spittings

According to the process of *H.W. Hitzrot, Jr.; U.S. Patent 4,190,422; Feb. 26, 1980; assigned to Bethlehem Steel Corporation* a steel plant waste material, scarfer spittings, is prepared for use as size-graded metallic abrasives suitable to clean the metallic and nonmetallic surfaces by machine or manual blasting techniques. Scarfer spittings are generally spherical particles, ranging in size from +50.8 mm to -100 sieve size. Scarfer spittings have a dual structure comprised of inner metallic cores encased within brittle outer shells comprised substantially of iron oxides.

To prepare the scarfer spittings for use as metallic abrasives, they are screened on a first screen to separate them from all large foreign matter which is inadvertently collected with them. Simultaneously, a size separation of the spittings is made, for example, at about 6.35 mm to separate substantially all the +6.35 mm spittings from the -6.35 mm spittings. The fraction containing the larger particles is recycled to the steel mill and is used as a source of iron. The fraction containing the smaller particles is charged into a grinding mill containing appropriate grinding media. The smaller fraction remains in the mill for a time to allow the grinding media to impinge upon the surface of the brittle shells and fracture the shells into relatively fine pieces substantially all of which fall away from the surfaces of the metallic cores. The grinding media do not affect the surfaces of the metallic cores; hence the cores remain substantially unchanged.

The mixture of metallic cores and the pieces of shells is screened in a second screening step using a 30 to 40 sieve size to separate some of the cores from substantially all the shells and simultaneously make a size separation of the cores. The +30 to 40 sieve fraction, which is comprised substantially of metallic cores, is stored. The -30 to 40 sieve fraction, which is comprised of some metallic cores and substantially all the shells, is screened in a third screening step on a sieve within the range of 100 to 120 sieve to separate substantially all the metallic cores which are larger than about 100 sieve from the metallic cores which are smaller than about 100 sieve and the broken shells. The cores in the +100 sieve fraction are stored with the +30 to 40 fraction from the second screening. The -100 sieve fraction is recycled to the steel mill.

The mixture of +30 to 40 sieve fraction and +100 sieve fraction cores is dried in a rotary drier prior to being screened on a series of sieves into a plurality of sizes to make size-graded metallic abrasives. The sized cores are used as metallic abrasives in blast cleaning equipment to machine or manually blast clean metallic and nonmetallic surfaces.

The metallic cores have a microstructure comprised of lathlike untempered martensite substantially devoid of retained austenite and intergranular and intragranular cracking, a grain size of about 3 to 4, and a hardness of about R_C 20 to 35 and are characterized by having good impact resistance and extended service life when compared to commercially available steel shot and grit. A major portion of the cores have irregular or subangular surfaces while a minor portion are spherical and have smooth surfaces.

H.W. Hitzrot, Jr.; U.S. Patent 4,115,076; September 19, 1978; assigned to Bethlehem Steel Corporation prepares an abrasive from scarfer spittings which can be used as a substitute for mineral abrasives or sand to manually blast clean the surface of ferrous metals prior to painting.

The useful fraction is preferably -6+100 mesh in size; however, -100 mesh size particles are used in special applications. The particles are generally spherical in shape and contain 85% iron, the remainder being substantially oxygen and minor amounts of elements usually found in carbon and alloy steels, and are comprised of a metallic core containing 30% of the iron as metallic iron and a shell surrounding the core containing 70% of the iron as iron oxides. The abrasive is characterized by having a specific gravity of 5 to 6 and a bulk density of not less than 180 pcf (2.88 g/cm^3).

Wet Blasting

The process of *N. Easton; U.S. Patent 4,125,969; November 21, 1978; assigned to A. Long & Company Limited, England* involves subjection of a work surface to a stream of carrier liquid and particulate of abrasive material from jet nozzle means from which particulate abrasive material and carrier liquid are expelled through an abrasive outlet and a plurality of liquid outlets respectively, the work surface being the first surface contacted by the stream downstream of the jet nozzle means.

The use of conventional wet abrasion blasting for certain operations, for example, the cleaning of tanks of oil tankers for which such a process would appear suitable has in fact proven unsatisfactory. Jets of water alone may not be sufficiently abrasive and jets of sand and water leave the problem of disposal of the sand residue. The process is particularly suitable for use in blast cleaning operations where at least a portion of the particulate abrasive material is soluble in the carrier liquid.

For the soluble abrasive material to remain sufficiently abrasive at the work surface it is necessary to minimize the dissolution of the particles in the carrier liquid between the jet nozzle means and the work surface, i.e., to minimize the amount of intermixing of the soluble abrasive material and the carrier liquid, and also to minimize, or avoid entirely, contact of the abrasive material with other surfaces between the jet nozzle means and the work surface.

This may involve dispensing with the stand-off barrel altogether, or the use of a shield means extending from the jet nozzle means, the shield means being so shaped and disposed as to be wholly outside the stream of abrasive and water downstream of the jet nozzle means.

The soluble particulate abrasive material may comprise abrasive particles which are soluble per se in the carrier liquid or particles which are substantially indissoluble per se in the carrier liquid but which break down into subparticles partially or completely soluble in the carrier liquid. This latter property may be achieved by physical or chemical treatment of the surfaces of particles of soluble abrasive material.

An abrasive material which is soluble per se and which has been found to be useful is powdered sodium silicate, either alone or in admixture with common salt. When a mixture of these materials is used the sodium silicate, which may be present in the neutral or alkaline commercially available forms, may comprise 5 to 95% by weight of the mixture and the salt may comprise from 5 to 95% by weight of the mixture.

A further suitable soluble crystalline particulate material may, for example, be treated to encourage the decomposition of its surface, followed by baking the particles to a hard glazed outer surface to render them slower to start dissolving. Encapsulation of each particle of soluble abrasive material with a brittle or friable skin of an acceptable insoluble material may also be appropriate. The insoluble material may be one which may be dried or baked over each particle.

Thus, a quite readily soluble particulate material may be maintained in an abrasive configuration for sufficient time to enable the desired cleaning of the work surface to be effected. Particles thus treated may also become easier to handle.

The residues from the abrasion jet, once the soluble components have dissolved in the carrier following the primary treatment, i.e., the abrasion cleaning step, may be such as can effect further, secondary, useful treatment of the work surface. For example, the jet could include a rust-inhibiting composition for protection of ferrous working surfaces, an etching agent such as phosphoric acid, which may be of assistance to future operations such as further surface coating and painting to be carried out on the work surface, or surface protection compositions. These materials or their immediate precursors may be present either in the soluble abrasive material, the insoluble portion (if present) of the abrasive particles, or indeed could be in the carrier itself.

Supply of the particulate abrasive material to the jet nozzle means may be conventional techniques of mechanical feed, e.g., by Archimedian screw conveyor, or vibration or other agitation of the particulate material. However, it is also contemplated that in some circumstances special techniques may be adopted. One such technique is the provision of a pressurized feed of the particulate material (either by air pressure or by supplying the particulate material from a pressurized container) into the carrier which can, among other things, prevent the carrier liquid feeding back up the feed tube to the store reservoir of the particulate material.

Glass-Bonded Finishing Chips

G.W. Balz; U.S. Patent 4,110,085; August 29, 1978; assigned to Roto-Finish Company describes abrasive finishing chips suitable for use as abrasive media in finishing processes and apparatus for the finishing of parts and workpieces. The chips comprise abrasive grains dispersed in a matrix comprising at least 50% by weight of ground glass bonded together with a substantially uniform bond and of a shape suitable for employment as an abrasive medium chip. The process for the production of such chips comprises the steps of providing ground glass, causing abrasive grains to be dispersed therein, subjecting the mixture to shape-forming procedure, heating the same to the sintering point of the ground glass, that is, at a temperature between the softening point and the working (fluid) point of the glass but below 1950°F, and allowing the same to cool.

Example: Common soda lime glass scrap, comprising mainly old bottle glass, is collected and reduced in size to particles of approximately 80 mesh size (U.S. standard). Grains of silica of approximately 50 μ in diameter on their greatest diameter are added, along with water, and water glass, to provide a mixture comprising 58% of ground glass, 32% of abrasive grains, 8% water, and 2% water glass, all by weight of the total mixture. The ratio of glass to abrasive is 1.8:1.

The mixture is then extruded in the form of cylindrical bars of ¼-inch diameter, which are cut into chips of 1-inch lengths. The chips are dried at 200°F for 15 hours and then fired at 1650°F for a period of one hour and subsequently allowed to cool. The grains of the abrasive become dispersed throughout the ground glass and the glass becomes substantially uniformly bonded to itself and to the abrasive grains throughout the chip.

In other operations, the mixture is cast into molds or dropped onto a flat plate, in either case resulting in the formation of raw or green chips having the approximate form of cones of the desired height and width, e.g., one inch in height and an approximate base diameter of one inch. In still another operation, immediately after extrusion, the cylindrical bar upon extrusion is cut into approximately one-inch-long cylinders having oblique ends by means of an approximately 45° angle shear. In each case, after the shape-forming operation, the abrasive, now in the form of chips, is subjected to drying in an oven at 200°F for a period of six hours, and is then fired in a gas-fired furnace for one hour at a temperature of 1650°F.

After drying and firing, and thereafter allowing to cool, the abrasive chips are collected and employed in the finishing of parts, for example, aluminum, zinc, steel, or plastic workpieces, in a Spiratron vibratory finishing machine and are found satisfactory for such purposes. The finishing chips provide a completely adequate cut rate and do not exhibit any observable unacceptable wear characteristics. In their performance and depreciation, they are entirely comparable to the best high-temperature high-cost ceramic-bonded abrasive media commercially available in the field.

Zircon Sand

Zircon sand is a naturally occurring form of the mineral zircon, the zirconium orthosilicate, $ZrSiO_4$. The known deposits of it have been found principally in certain beach areas in the states of Florida and Georgia and in Australia, Greenland, Uruguay, Brazil and Canada. The grains of this sand are generally quite fine in size and are extraordinarily hard and tough, having a Moh's hardness of about 7.5 which is about 15 to 45% greater than the Moh's hardness measurements of the glass beads most commonly used for impact treatment. Although they are not spherical, zircons and grains have a characteristic rounded oblong configuration with smoothly rounded convex edges and end contours and remarkably few sharp edges. This form enables them to act in some respects like spherical granules when they are hurled at a high velocity against surfaces to be treated. In addition, they possess a very high density, having a specific gravity which is greater even than the specific gravity of 4.49 possessed by specialty beads of leaded glass and is nearly twice the specific gravity of 3.48 possessed by the glass beads made of (standard soda lime) crown glass most commonly used for impact treatment.

It has been found by *J.W. Ayers; U.S. Patent 4,035,962; July 19, 1977* that the fine zircon sand grains can be employed with extraordinary effectiveness for peening, brightening, texturizing and/or removing adhesions from hard surfaces such as metal surfaces by bombardment of the grains against the workpieces. For a given blasting velocity these grains exert a far greater impacting force than do similarly sized crown glass beads, while their rounded contours accompanied by very few sharp edges enable this force to be exerted without objectionable cutting or tearing of the impacted surface. Also, due to their great hardness and structural density, the impacting zircon sand grains transmit their kinetic energy efficiently to the workpiece with relatively little dissipation of the energy through elastic deformation of the granules, and their resistance to breakdown or fracture and wear is so great that they can be kept in efficient use for extraordinarily long periods of service.

By virtue of those combined properties, the treatment of metal surfaces by high velocity bombardment of them with extremely fine sizes of zircon sand grains, such as sizes in the -200 to +325 mesh range, offers exceptional advantages over the use of correspondingly sized glass beads or of any other known nonmetallic impacting material.

The fine zircon sand grains provide still other advantages over the use of glass beads by virtue of their properties of thermal stability and chemical inertness. They remain solid at temperatures up to 1500°C. They have such a low coefficient of thermal expansion, which is less than that of fused quartz, that they can be used effectively for impact treatments at any atmospheric temperature, including subfreezing temperatures, and even at highly elevated temperatures; whereas glass beads soften at corresponding elevated temperatures and become embrittled at low temperatures. Also, they are highly inert to the chemical influences of water, of organic liquids and solvents, and of most acids and alkalis, so that they can be used without impairment under great varieties of working conditions. Still further, the use of them does not present the health hazards to workmen which are involved in the use of glass beads.

For the preparation of the material, a mass of zircon sand concentrated away from other minerals naturally occurring with the sand may be washed with water to remove water-soluble foreign matter and then, before being screened to the required size range or ranges, may be roasted at a temperature sufficiently high to burn off organic foreign matter remaining on or among the zircon sand grains. The roasting temperature should be in the range of 300° to 800°C.

J.W. Ayers; U.S. Patent 3,939,613; February 24, 1976 states that for peening very hard steel surfaces, the zircon sand grains may have sieve sizes (U.S. standard) ranging from -60 or -70 mesh to +100 mesh.

Olivine Grit

Olivine, also known as chrysolite or peridot, is the most common member of a group of silicates. Olivine is a magnesium-iron silicate, colored various shades of green (occasionally brown) with a hardness of 6.5 to 7 based on a diamond hardness scale of 10, and a specific gravity of 3.3. It is found in igneous rocks that are rich in magnesium and low in quartz. It is found both in small grains and in large, granular masses. The chemical composition of olivine is $(MgFe)_2SiO_4$.

In the process of *F.L. Cox, S. Crump and P. Hice; U.S. Patent 4,044,507; August 30, 1977; assigned to Silver Creek Minerals Corporation* surfaces are stripped, cleaned and treated by use of olivine grit driven against the surfaces by fluid pressure.

Referring to Figure 4.13, air pressure **11** may be supplied by any well-known means such as a compressor. This pressure is supplied to a container **13** through line **15** and valve **17**. Container **13** contains the olivine grit material **19** and includes an opening **21** which may be sealed by means of a door **23**. When door **23** is sealed and pressure is applied to container **13** through line **15**, the olivine material is forced under pressure through line **25** to blasting gun **27**.

Air pressure is also supplied directly to gun **27** by means of line **29** and valve **31**. Air pressure and olivine grit material are mixed in gun **27** and directed outwardly through nozzle **33** against the surface of the material **35** to be treated.

Figure 4.13: Treating with Olivine Grit

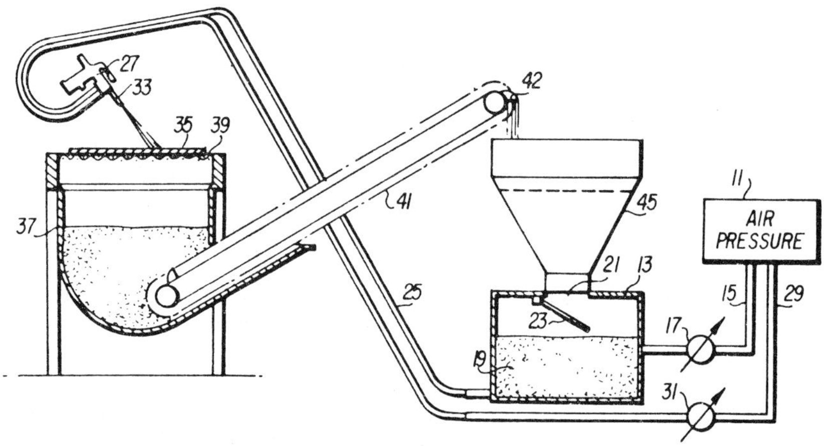

Source: U.S. Patent 4,044,507

A reclaim pan **37** is positioned below the working area, and a supporting screen **39** may be placed between the work area and pan **37**. A conveyor **41** having buckets **42** extends into the reclaim pan and may be driven by a motor (not shown). After the olivine material strikes the surface, it drops into pan **37** where the buckets of conveyor **41** scoop it up. The reclaimed material is deposited into a hopper **45** mounted above door **23**. When the material in the hopper **45** is to be used to replenish the material in container **13**, the pressure in line **15** is reduced and door **23** is opened. Prior to resuming operation, the door is closed and pressure in line **15** is raised to the desired level.

Coated Abrasives

COATINGS AND BONDING AGENTS

Microencapsulated Catalyst for Isocyanurate Formation

The known catalysts for trimerizing isocyanates can be divided into two general classes. First, those which rapidly trimerize isocyanate exothermically at room temperature, and, second, those which trimerize isocyanates only at elevated temperatures and with prolonged reaction times. The first group of catalysts lacks latency and therefore their use is generally limited to two-part compositions with the parts mixed together very shortly before use. The second group of catalysts generally have a short shelf life, for example, several weeks after mixing with the isocyanate starting materials, and requires long cure times, on the order of several hours at elevated temperatures.

An improved coated abrasive employing an isocyanurate resin binder is provided by *H.E. Rude; U.S. Patent 4,126,428; November 21, 1978; assigned to Minnesota Mining and Manufacturing Company.* Binding is accomplished by the use of a one-part, latent, room temperature stable isocyanate system trimerizable to polyisocyanurate. The trimerizable system, described in U.S. Patent 3,860,565, comprises an aromatic isocyanate having dispersed throughout the liquid, microcapsules of a catalyst for isocyanurate formation. The microcapsules have a permeable shell wall essentially of partially polymerized aromatic isocyanate which encapsulates the liquid catalyst at room temperature but, on activation of the isocyanate-catalyst by a suitable force such as heat, the liquid catalyst permeates the shell wall, mixes with the isocyanate, and cures the isocyanate to a fully cured thermoset isocyanurate resin.

The use of the aforementioned encapsulated isocyanate-catalyst system overcomes the problems associated with each of the two types of isocyanate catalysts described above. With this system, coated abrasive products can be made employing an isocyanurate resin binder. The unique characteristic of the system, which permits curing to be triggered upon application of the suitable force such as heat, provides control over the curing time of the isocyanate in the production of coated abrasive products.

The coated abrasive products according to the process are characterized by having at least a make coating or a size coating of isocyanurate binder. This binder layer contains a plurality of minute microcapsules which formerly contained catalyst (and which may contain residual catalyst) dispersed uniformly throughout the cured composition.

The conventional components going to form the coated abrasive product of this process will be selected from those typically used in this art.

The isocyanate trimerization catalysts used in the latent, one-part, triggerable isocyanate-catalyst capsules employed in making the binder of the coated abrasive products of this process comprise a viscous liquid catalyst solution, preferably a solution of an alkali metal salt of an organic acid, encapsulated in microcapsules. The microcapsules are prepared in situ, in a polyisocyanate material, by the addition of the liquid catalyst to the polyisocyanate under conditions of shear. During catalyst addition, the polyisocyanate is maintained at a temperature above the melting point of the polyisocyanate material but below about 50°C, and preferably below about 40°C, to avoid premature trimerization of the polyisocyanate.

The microcapsules prepared under these conditions have a shell wall which is essentially partially polymerized aromatic isocyanate. The wall is of such a nature that it isolates the viscous solution of alkali metal salt at temperatures below about 50°C. However, at temperatures above about 75°C, the viscosity of the alkali metal salt solution is reduced to less than about 5,000 cp, allowing the alkali metal salt solution to penetrate the shell wall, mix with the isocyanate material, and initiate polymerization of the aromatic polyisocyanate material. The capsules obtained are a distribution of sizes generally ranging from about 1 to 20 microns in diameter.

Suitable alkali metal salts for use in a liquid catalyst solution are those alkali metal salts of organic acids having a pK_a between about 2.5 and 10, preferably between about 3 and 6. The acids with a pK_a of about 3 to 6 provide catalysts that are easily synthesized, give good shelf stability when dispersed in isocyanate materials, yet cure rapidly when activated.

Generally speaking, the alkali metal salts used in this process are alkali metal salts of mono- or polycarboxylic acids and phenols having from 2 to about 8 carbon atoms and are preferably mono- and polycarboxylic acids having 2 to 5 carbon atoms.

The solubility of the alkali metal salt catalysts in polyols is such that the catalyst solution will generally contain more than 15% by weight salts, preferably about

50 to 75% by weight or more. The resulting solutions will generally have a viscosity of at least 5,000 cp or greater at temperatures up to about 50°C, preferably a viscosity greater than about 100,000 cp at temperatures up to about 40°C.

Solvents which can be used to dissolve the alkali metal salts, to form a viscous catalyst solution, are generally polar organic compounds and preferably have a boiling point sufficiently higher than water so that the solvent can be made essentially anhydrous, i.e., have a water content of 1% or less, by azeotropic distillation. Examples of such solvents include lower polyhydric alcohols having 2 to about 8 carbon atoms. The preferred solvents are the lower polyhydric alcohols and glycerol is the most preferred.

A preferred catalyst solution is an alkali metal salt of lactic acid in glycerol. This catalyst solution, when dispersed in polyisocyanate, especially "Mondur" MRS, gives consistently good catalyst-containing capsules which are efficient in releasing the catalyst and giving cured resins with good physical properties, including thermal stability, hydrolytic stability and useful dielectric properties.

Isocyanate-Catalyst Preparation: A variable speed laboratory type "Waring Blender" having a capacity of approximately 1 liter was charged with 17.46 parts of "Mondur" MRS (a polymethylene polyphenyl polyisocyanate having an isocyanate functionality of about 2.7). The blender was turned on and its speed increased to maximum followed by the rapid addition of 2.54 parts of a solution containing 70.86% by weight potassium lactate in glycerol, over a period of about one-fourth minute. A fine dispersion of liquid catalyst droplets was formed in the isocyanate.

The blender was stopped and the dispersion degassed in a vacuum desiccator. Upon standing at room temperature, an encapsulating shell formed around each droplet by reaction of the catalyst solution with isocyanate material. The resulting dispersion, an isocyanate-catalyst concentrate, contained approximately 9% by weight encapsulated potassium lactate which could be used as a resin or diluted with more polyisocyanate or isocyanate-terminated prepolymer composition to form useful compositions.

Two parts of the isocyanate-lactate catalyst concentrate was diluted with 8 parts of an isocyanate-terminated prepolymer made by reacting 75 parts of "Mondur" MRS with 25 parts of a 1,500 molecular weight triol, having an equivalent weight of 500 and being prepared from trimethylol propane, propylene oxide and potassium hydroxide, to obtain a coating mixture hereinafter identified as mixture A, having approximately an 80:20 isocyanate:polyol ratio. A thin film of the mixture A was cured by placing in an oven at 90°C for 2 minutes. The cured polyisocyanurate resin obtained had an average Knoop hardness number of 21.

Example: A 3 mil thick coextruded polyester backing film consisting of a 2.75 mil thick layer of polyethylene terephthalate and a 0.25 mil thick layer of a polyester comprised of 80 parts ethylene terephthalate and 20 parts ethylene isophthalate was coated on its copolyester surface with 3 to 4 mils (wet) of a slurry described below.

Slurry Components	Parts by Weight
Mixture A	10
Xylene	6
Al_2O_3 abrasive granules (9 microns)	30

The resultant coated composite was cured for one minute in a forced air oven heated at 250°F (120°C), producing a coated abrasive sheet which was evaluated against a standard commercial coated abrasive product known as "Imperial" Lapping Film (9 microns) (3M Company), employing a "Shuron" Random Motion Tester to polish a three-quarter inch (19 mm) diameter stainless steel rod under a load of twenty-five pounds (11 kg). The weight loss after 5½ minutes of abrading time was noted. Results are as follows:

	Weight Loss (mg)
Control	8.0
Example	17.1

Microballoons in Backsize on Polyester Backing

D.G. Adams; U.S. Patent 4,111,667; September 5, 1978; asssigned to Norton Company incorporates from 2 to 10% by weight, equivalent to 10 to 50% by volume, of hollow microspheres into an otherwise conventional backsize coating on the reverse (nonabrasive) side of a woven polyester backing.

By microbeads or microballoons is meant hollow spheres of resin or glass having a diameter in the range of between 5 and 125 microns, a typical average diameter being 40 microns. The preferred microspheres are cured phenolaldehyde spheres, but glass or thermoplastic resin spheres can also be used.

The incorporation of the beads into the backsize formulation enables a coating to be applied in no more than two steps which stiffens the fabric in an amount which would require 4 to 5 applications of conventional backsize and which is more readily sandblasted from the fabric along any given area where a lap joint, which requires removal of backsize from one end of the ends to be joined, is to be made in the coated abrasive product to form an endless belt. Such removal of conventional backsize coatings frequently requires such vigorous techniques that the tensile strength of the underlying polyester is reduced by 30 to 50%. This process is particularly adaptable to heavy duty use as in belt grinding with cofused alumina-zirconia abrasive of grits 60 and coarser. The advantages of this process, however, are also applicable to the use of more conventional fused aluminum oxide and silicon carbide abrasives. The preferred phenolic microballoons have a density of from 0.1 to 0.3 g/cc.

Irradiation-Cured Water-Resistant Abrasive

W. Hesse, J. Ritz and E. Teschner; U.S. Patent 4,047,903; September 13, 1977; assigned to Hoechst AG, Germany provides a process for preparing a waterproof abrasive. The process comprises forming a waterproof abrasive by coating a fibrous substrate with a plurality of successive layers including at least one base layer of a binder resin hardenable by irradiation, at least one intermediate layer of abrasive grains and at least one top layer of binding resin hardenable by irradiation and thereafter curing the resin layers by electron irradiation. At least one of the resin layers is selected from the group consisting of: (A) the reaction product of a polycarboxylic acid (A1) with an esterified epoxy resin (A2), prepared by reaction of an epoxy resin with a member selected from the group consisting of: (a) acrylic acid, (b) methacrylic acid, (c) a mixture of (a) and (b); and (B) the reaction product of (A2) first reacted with diketenes and then reacted with a chelate forming compound.

As to energetic radiation the binder coats are mainly cured by means of electron radiation having an energy of 175,000 to 1,000,000 preferably 200,000 to 600,000 volts. The dose of radiation depends on the properties of the resin which is used as binder. The present electron radiation sources ensure a sufficient processing speed; curing requires a dose of radiation of 0.5 to 30, preferably 1.0 to 12 Mrads. The processing speeds, i.e., the rates of feed of the coated material simultaneously cured, range from 10 to 120 m/min. The rates of curing and feed are variable according to width and length of the surface exposed to radiation. The dose of radiation can be varied according to the coating to be cured. Generally curing of the base coat and the top coat is carried out separately, whereby a lower dose of radiation is applied for the base coat than for the top coat. For curing the base coat the appropriate dose of radiation is for example 0.5 to 5 Mrads.

Example: *Production of Binder* — 400 parts of an epoxy resin of dimethylolpropane and epichlorohydrin with an epoxy number of 8.2 combined with 120 parts of acrylic acid, 0.66 part of hydroquinone and 1.35 parts of N,N-dimethylaniline are heated up to 100°C until after 7 hours the acid number of 0.4 is obtained. At that stage 133 parts of ethyl acrylate are added to form a solution of an unsaturated epoxy ester with a viscosity of 800 cp/20°C.

At room temperature 4 parts of dimethylaniline are added to this solution while stirring for one hour. Thereafter 31 parts of diketene are slowly added to the solution while stirring at 50°C. The temperature shall not exceed 60°C. After addition of diketene the batch is maintained at 50°C for further 30 minutes and subsequently cooled down to room temperature. The solution thus obtained has a solids content of 81.3% and a viscosity of 1,200 cp at 20°C.

100 parts of this solution comprising an epoxy resin esterified with acrylic acid and modified with acetoacetic ester groups are combined with 8.8 parts of a 50% solution of aluminum triisopropylate and anhydrous toluene while stirring for 15 minutes at room temperature. After 2 hours' rest the solution has a viscosity of 1,300 cp at 20°C which remains practically unchanged during the fol-

lowing 24 hours. An aluminum chelate complex of the modified epoxy resin has been formed.

Production of Abrasive Paper — A latex impregnated paper, normally used in the abrasives industry, of 120 g/m^2 is used as substrate. The abovementioned binder is applied to the latex coated paper at a wet coat thickness of 50 μ by means of a film application apparatus. Silicon carbide (grain FEPA No. 80) is sprinkled excessively on the wet binder film and the excess is removed. Subsequently the basic binder is cured with a radiation dose of 1.5 Mrads. The electron radiation source employed for curing is operated by an accelerating voltage of 400 kilovolts and an acceleration current of 60 mA. Operating speed is 47 m/min. 120 g/m^2 of the same binder is applied as top coat to the semiproduct by means of rubber rollers. The material is then cured with a dose of radiation of 3.0 Mrads, at a processing speed of 23.5 m/min.

Tests — The thus obtained abrasive paper samples are cut into round disks at an external diameter of 178 mm and an internal diameter of 22 mm and fastened on vulcanized fiber disks of the same size, thickness 0.8 mm, with contact adhesives. These test samples are attached to a conventional grinding machine with rubber plates which operates at 1,200 revolutions per minute. By means of a fixing device a flat-bar steel of 20 x 4 mm is radially moved inwards from the periphery of the disk in tangential order of 5 mm and pressed with a frictional contact of 25° in axial direction with an absolute force of 4 bars, the angle of attack being in the sense of rotation. During the whole grinding process of 10 minutes water of about 16°C is poured over the point of attack. After a grinding time of 10 minutes the abrasive coat is completely attrited. 11.1 grams were cut off from the test steel.

Urea-Formaldehyde, Phenolic Resin Blends

The process of *S.J. Supkis; U.S. Patent 4,038,046; July 26, 1977; assigned to Norton Company* relates to abrasive articles in which a thin layer or, more usually, a single layer of abrasive grit is adhesively bonded to a backing. Such products are coated abrasives, more commonly referred to as "sandpaper", and fibrous abrasive pads such as disclosed in U.S. Patent 2,958,593. Conventionally, animal glue or thermosetting resins such as phenol-formaldehyde resins and urea-formaldehyde resins have been used as either the maker coat (first bonding coat) or size coat (second bonding coat). The most common resinous bond for coated abrasives has been the phenolic resin, which is usually filled, in the maker coat, with finely particulate calcium carbonate.

Urea-formaldehyde resins, which are less expensive than phenolic resins, have been used to reduce the cost of such products, in the size coat. Since urea-formaldehyde precondensates employed to make coated abrasives require an acid catalyst, it has not been feasible to employ calcium carbonate filler with urea-formaldehyde resins.

It has been found that the cheaper urea-formaldehyde precondensates can be blended with liquid phenolic resin systems, the blend can be catalyzed by providing a basic environment (pH above 7), rather than the acid catalyst specified for unblended urea-formaldehyde prepolymers and a finished product at least equal in quality to that achieved when a straight phenolic resin is employed. Such blends can be employed in both the maker and size coats, and when employed in the size coat, reduce the loading tendency of the product when grinding certain materials, such as wood.

The blend of urea-formaldehyde liquid precondensate and phenolic resin precondensate may vary from 10 parts of urea-formaldehyde to 90 parts of phenolic solids, by weight to 90 parts of urea-formaldehyde to 10 parts of phenolic solids. The curing time and temperature is the same as conventionally used with phenolic resins, and calcium carbonate filler can be employed in the maker coat and, if desired, in the size coat.

The backing can be any conventional fibrous woven, or nonwoven material such as cotton or synthetic cloth, felted fibrous material such as paper or so-called vulcanized fiber, and open, lofty nonwoven fibrous material.

The abrasive may be any conventional abrasive suited for making coated abrasive products, such as fused alumina, cofused alumina-zirconia, silicon carbide, garnet, emery and flint.

Example: A standard greige cotton fabric in a 2 x 1 twill construction having a yarn count of 76 x 48 and weighing 7 oz/yd^2 was finished by applying a backsize of 51%, 58 mp glue to the backing which is dried by conventional means to a total dried weight of deposition of 2.0 lb/ream. After the web had been so dried a second coating was applied to the front side of the fabric containing a resol phenol-formaldehyde resin containing 76% active material (Varcum 2535, having F/P ratio of 2.08/1) 275 pounds, calcium carbonate 200 pounds and water 30 pounds, the second solution giving a viscosity of 2,100 cp at 80°F. The second solution was applied by inverted knife coating technique to a dry deposition of 6.3 lb/ream.

To this prepared backing a maker coat of adhesive was applied by roll coater, the maker coat containing a mixture of basic catalyzed formaldehyde resins; the first phenol-formaldehyde material had a formaldehyde to phenol ratio of 2.08 and the second phenol-formaldehyde resin had a formaldehyde to phenol ratio of approximately 0.95. These phenolic constituents which were basic (pH above 7) were mixed with a urea-formaldehyde condensate (Beetle 7238-20) and calcium carbonate filler. The exact weight ratios of these components were phenolic (2.08) 7.4 pounds; phenolic (0.95) 3.2 pounds; urea-formaldehyde resin 8.3 pounds; and calcium carbonate filler 21.4 pounds. This mixture was adjusted with water to a viscosity of 6,300 cp at 100°F. The pH was between 7 and 10.

The amount of maker adhesive applied to the backing member was 21 lb/ream. The backing member with the making coat of adhesive thereon was fed through a sandpaper making machine in which grit 50 aluminum oxide abrasive grain was applied in two steps. In the first step, 19.2 lb/ream were applied by a gravity method and in the second, 20.8 lb/ream were applied by an electro-coating method. After the abrasive grain had been applied to the web it continued into a festoon drying and curing oven where the coated abrasive web was cured for 25 minutes at 170°F, 25 minutes at 190°F, and 57 minutes at 225°F.

The fabric with the maker coat of adhesive plus abrasive grain which had been cured is given a second coat of adhesive which contains a mixture of phenol-formaldehyde resins similar to the maker adhesive but without the urea-formaldehyde component. The exact formulation of the second coat of adhesive which is applied over the abrasive grain contains phenol-formaldehyde condensate (F/P ratio 2.08) 28.7 pounds; phenol-formaldehyde condensate (F/P ratio 0.95) 12.3 pounds; calcium carbonate (average particle diameter 14.9 microns) 49.3 pounds, and water 9.7 pounds. The viscosity of this second coat of adhesive measures 550 cp at 100°F.

The second coat of adhesive is usually applied visually by a skilled operator so that points of abrasive grain are not completely covered by the second coat of adhesive. Usually the amount of ahdesive is 16 lb/ream. After the coated abrasive web has received a second coat of adhesive, it then proceeds into a sandpaper curing oven and is cured at 25 minutes at 125°F, 25 minutes at 135°F, 18 minutes at 190°F, 15 minutes at 225°F, and 8 hours at 230°F.

After the material is cooled to room temperature and emerges from the sandpaper making machine it then receives a number of finishing treatments to aid in curl correction of the product.

This product was tested on a bench backstand machine with a test designed to determine its shed resistance. The results show that a conventional coated abrasive belt lasted 3.87 minutes compared with belts made from the maker adhesive of this process, had an improved shed resistance over the control and lasted a period of 5.25 minutes.

This test is designed to show the shed resistance of the coated abrasive product and also to give an evaluation of the base adhesion between the maker adhesive and the backing member.

Instead of the straight phenolic size employed above, the maker (without abrasive, and with or without filler) may be diluted with water to give a viscosity of 550 cp at 100°F.

The phenolic resins useful in this process are A-stage, or resol phenolics, that is, alkaline catalyzed one-step fusible thermosetting resins.

Low Density Abrasive Using Isocyanurate Resin

U.S. Patent 2,958,593 describes a low density abrasive article comprising a uniform, lofty, open, nonwoven, three dimensional integral structure formed of many interlaced, randomly extended, flexible, durable, tough, resilient, organic fibers. The structure is unified throughout by firmly bonding fibers together at points where they cross and contact. Abrasive particles are distributed within the structure and firmly bonded to the fibers by an adhesive, which may perform the dual function of bonding web fibers together and bonding abrasive grains thereto. Alternatively, a first adhesive may be used for bonding the fibers to each other and a second adhesive for bonding the abrasive granules to the fibers. Despite great utility and broad commercial acceptance, most low density abrasive products are subject to tearing when the abrasive is snagged on a burr or other projection, largely because of the hard, brittle nature of the phenolic resins generally used in making low density abrasives.

Plasticizing a phenolic resin binder will lower its hardness and improve the tear resistance of a low density product made therewith, but as tear resistance increases, abrading effectiveness tends to decrease. Further, the phenolic systems not only have a tendency to degrade and weaken nylon fibers, but also show poor adhesion to polyester fibers, thereby limiting the potential of low density abrasives made from the phenolic systems using these resins.

Low density abrasive webs bonded with epoxy resins have excellent tear resistance, but commonly used, the cured resins do not have adequate heat resistance for severe abrasive application. In addition, catalyzed epoxy resins often have an excessively short pot life, making it commercially difficult to bond the fibers of a low density abrasive with an epoxy resin. Epoxy novolac systems as disclosed in U.S. Patent 3,510,283 have lower toxicity and better durability than other epoxy systems, but low density abrasive products bonded with the epoxy-novolacs tend to deposit melted fibers (known as smear) on the workpiece when exposed to the high stresses common in abrasive uses.

Another type of low density abrasive utilizes a flexible elastomeric binder, such as latex or a reclaimed rubber dispersion. The use of such binders gives a product which is useful in making scouring pads for hand use, but does not result in a low density abrasive useful in a wide variety of industrial applications where a hard binder is necessary to prevent smearing and maintain sufficient cut.

The low density abrasives made in accordance with the process of *M.R. Johnston; U.S. Patent 4,011,063; March 8, 1977; assigned to Minnesota Mining and Manufacturing Company* are structurally similar to products described in U.S. Patent 2,958,593, but employ an isocyanurate resin. Low density abrasives where the fibers are bonded to each other by a first resin and the abrasive granules are bonded to the structure by a second resin, at least one of the resins being an isocyanurate resin, are characterized by the same excellent conformability and abrading characteristics as the conventional products, but they possess significantly greater toughness and tear resistance than conventional products. The low density abrasives of this process have the additional desirable properties of bonding firmly to both polyester and nylon fibers without causing embrittlement or degradation.

Isocyanurate resins preferred in the practice of this process are the product of the catalyzed trimerization of polyisocyanate-terminated polyols known in the art.

A cured isocyanurate resin used in forming low density abrasive products must have a minimum hardness of at least 50 on the Shore D scale in order to maintain the necessary abrasive action. Those isocyanate-terminated prepolymers or polyisocyanate-polyhydroxyl systems which when cured yield, per 1,000 grams of resin, at least 0.5 gram equivalent of isocyanurate and a combined total of at least 1.3 gram equivalents of urethane plus isocyanurate, result in a resin having the necessary hardness to perform adequately under the abrasive conditions encountered by low density abrasive products. The preferred resins are those falling within the Shore D hardnesses range of 65 to 95.

The preferred isocyanurate resin is based on a prepolymer formed from polymethylene polyphenyl isocyanate (e.g., Mondur MRS), and polyoxypropylene glycol having a hydroxyl equivalent weight of about 385 (e.g., Pluracol P-710) heated to react all the hydroxyl with isocyanate, two hours at about 80°C being sufficient in most cases. After the reaction is complete and all the hydroxyl groups have reacted with the isocyanate to form an isocyanate-terminated prepolymer containing urethane linkages, an acceptably hard isocyanurate resin can be formed by any of several catalysts, such as borate esters, tertiary amines, or xanthates, which trimerize the remaining isocyanate to isocyanurate.

Also known as a catalyst which promotes trimerization of isocyanate is a carbamic acid ester (urethane) in combination with a tertiary amine, as disclosed in U.S. Patent 2,954,365. It was found that a trimerized isocyanurate resin, formed by the reaction of aromatic isocyanate-terminated prepolymers, catalyzed by a combination of N,N-dialkylalkanolamines, the tertiary amines having the structure $HO{-}R''{-}N(R)_2$ where R'' is lower alkylene having 2 to 4 carbon atoms and R is lower alkyl having 1 to 4 and preferably 1 to 2 carbon atoms, (e.g., N,N-dimethylethanolamine) and an epoxy is a particularly valuable resin for making low density abrasives.

The isocyanurate resins catalyzed by the amine epoxy catalyst systems have the desirable characteristic of good stability at ambient room temperature for up to a month; yet they cure quickly and completely in ten minutes to an hour at elevated temperatures approaching 135°C.

The coating weights in the example below are given in grains per 24 square inches of web face. Parts are by weight.

Example: A low density nonwoven web about ⅝ inch thick weighing 60 grains per 24 square inches was formed from 60 denier crimpset nylon fibers on a Rando Webber machine. The resulting web was needle tacked into 36 grains of 6,6 nylon scrim having 6 threads per inch in both the warp and fill directions (available from Burlington Industries as 61501/16/70) using a James Hunter Model 9-needle loom, forming a low density web weighing 96 grains per 24 square inches.

The resulting low density web was roll coat prebonded using about 130 dry grains of a solution comprising 210 parts polyester diol having a hydroxy equivalent weight of 1150 (Multron R-68), 4.07 parts glycerol, 4.07 parts trimethylolpropane, 210 parts of a 75% solids solution in ethyl acetate of the reaction product of three mols of tolylene diisocyanate per one mol of trimethylolpropane (Mondur CB75), 10 parts of red pigment (Setacyl Red R) and 31.0 parts of 1-acetoxy-2-ethoxyethane. The resin was cured to a nontacky condition by passing the coated web through a 90-foot convection oven maintained at about 170°C at a rate of approximately 12 feet per minute.

An isocyanate-terminated urethane prepolymer (78.4 parts Mondur MRS, 19.6 parts Pluracol P-710) was catalyzed by one part of N,N-dimethylethanolamine and one part of the diglycidyl ether of bisphenol A per 98 parts prepolymer to make an isocyanurate resin. The dried web was sprayed on the side opposite the nylon scrim with a slurry make coat comprising 3,000 parts grade 100 to 150 aluminum oxide, 1,000 parts of the isocyanurate resin, one part Setacyl Red R red pigment and 100 parts 1-acetoxy-2-ethoxyethane, such that a total of 225 dry grains of slurry per 24 square inches was applied. The resin on the web was cured by passing the web through a 180 foot convection oven having a temperature of 160°C at 12 feet per minute.

The resulting low density abrasive which was about 0.33 inch thick was slit to 3-inch widths and cut to 132-inch lengths. The ends of the resulting strip were ground to reduce their thickness and the ends adhesively joined using a 45° angle overlap splice forming an endless belt.

The endless belt was placed on a 14-inch diameter contact wheel mounted 51 inches (center-to-center) away from a 9-inch diameter idler roller. The contact wheel was driven at 1,750 rpm while a ½-inch diameter by 6-inch long steel bar, positioned parallel to the axis of the wheel and held firmly to prevent rotation was pushed against the belt with a total force of 30 pounds.

The steel rod and belt were weighed at one minute intervals to determine both incremental and cumulative cut and belt weight loss. The belt removed 282 grams of steel in 18 minutes and lost 149 grams. Grain efficiency, the ratio of stock removed over belt weight loss was 282/149 = 1.89. The ratio of stock removed on the last cycle to that removed on the first cycle, retention of cut, was 71.2%.

A control belt was made substituting base-catalyzed phenol-formaldehyde resin in the slurry make coat. The phenol-formaldehyde was applied by spraying as described above and subjecting the web to a curing step in a manner similar to that disclosed in U.S. Patent 2,958,593, particularly Example 3. This belt cut 226 grams in 18 minutes and lost 245 grams for a grain efficiency of 0.92, and the retention of cut was 46%.

Supersize Layer with Residual Unsaturation

The process of *S.J. Supkis; U.S. Patent 3,997,302; December 14, 1976; assigned to Norton Company* involves the provision at the exposed surface of a coated abrasive material of a supersize layer comprising an elastomeric material which in addition to supporting an active, solid, particulate filler material, also acts by itself to improve grinding.

Elastomeric materials which have been found to function in a dual manner, i.e., as a binder material as well as a grinding aid per se, are, in general, polymeric materials of monomers which, when polymerized, result in substantial residual unsaturation.

A grinding aid layer, as is provided herein, has been discovered to result in, in certain grinding applications, a noticeable reduction in frictional heat. (In grinding of titanium a noticeable reduction of sparking is achieved.) This is accompanied with a cooler cutting action which provides less potential for glazing over of the coated abrasive material thereby resulting in extended product life as well as more uniform cut over the life of the abrasive product.

The coated abrasive material of this process may be used to advantage in various dry grinding applications; however, it will be found particularly useful in light pressure operations, e.g., those involving off-hand sanding. These grinding operations include, among other things, the dry grinding of brazed welds and workpieces of hard to grind metals such as stainless steels and titanium. The ever increasing use of titanium metal in fabrication of various articles, e.g., in the aircraft industry, has made the need for a solution to safer, better, and faster grinding of this metal acute.

The supersize layer of the process contains polymers having substantial residual unsaturation. By this is meant a degree of unsaturation that is indicated by a weight percent of C=C functional groups of at least 10% or by an iodine number of at least 105. One class of such polymers includes polymeric materials of conjugated diene monomers, e.g., butadiene and its homologues.

The process, however, is not limited solely to homopolymers of these monomers and, in its preferred aspects, includes copolymers thereof with, for example, monomers copolymerizable therewith such as acrylonitrile and styrene. Other polymeric materials of this class and which can be desirably used are carboxylated butadiene-acrylonitrile and carboxylated butadiene-styrene copolymers. Where, however, a copolymer is utilized, the copolymer must contain no less than about 50% by weight (total) of the conjugated diolefin. Such a copolymer containing butadiene-styrene may have substantial unsaturation as represented by an iodine number of, e.g., about 160.

Fillers are also included. Those fillers which may be satisfactorily used are the metal fluorides, in particular, the complex alkali metal fluorides. Suitable such fluorides include, e.g., sodium aluminum hexafluoride, potassium aluminum hexafluoride, sodium silicofluoride, sodium fluoborate, potassium silicofluoride, and potassium fluoborate. Other known active fillers, however, may be used in place of those abovementioned, and in some instances to good advantage in combination therewith.

Example: A conventionally produced coated silicon carbide abrasive sheet material, 20 inches wide, (60X-R/R Durite Openkote) was coated on the abrasive surface with a supersize composition.

The supersize composition was prepared by adding to a latex composition a suitable amount of sodium fluoborate to provide a 1:1 ratio (by weight) sodium fluoborate to latex solids and to provide a 50% total solids content. The latex is Tylac 3502A, and has a weight percent of C=C functional groups of about 38%. The thus prepared composition was applied at room temperature, e.g., 75°F, to the abrasive surface by passing the above coated abrasive material (10.5 fpm) over a partially submerged coating roll (6 inches diameter) rotating in a vessel containing the above composition.

After application the coating composition was made smooth and uniformly thick and excess composition was removed by means of a conventional air knife operated at a pressure of 16.6 inches (water) utilizing compressed air at room temperature. The orifice of the air knife was set at an opening of 28 mils and was spaced at a distance of about ¾ inch from the coated surface. It was directed at an angle of 45° with respect to the coated web and against the direction of movement thereof.

A sufficient amount of the latex composition was provided on the front surface of the coated abrasive web to provide, on subsequent drying and curing, a weight of 7.4 pounds per sandpaper maker's ream of 480 sheets, 9 inches by 11 inches. The wet coating was air dried for 30 minutes at room temperature (100°F) after which it was heated for 10 minutes at 225°F to fuse the latex solids. On fusion, a layer of rubbery elastomeric material results.

The tips of the abrasive grains, on examination of the coated abrasive material, must be exposed sufficiently to be noticeable visually or when running one's fingers over the abrasive surface.

An endless abrasive belt (2½ inches by 60 inches) was manufactured from the thus coated abrasive material according to usual belt-making techniques. Grinding tests were then run on the abrasive belt under controlled conditions. The test performed was that conventionally known as a "backstand belt test."

The supersize layer resulted in not only improved cut but was discovered to have good adhesion with cured phenol-formaldehyde resin in the size coat (60X-R/R Durite) as evidenced by no flaking off of the supersize layer during belt usage.

On examination of the workpiece, the surface finish was seen to be finer than is usually provided with such abrasive grit.

As described by *M.T. Gladstone and S.J. Supkis; U.S. Patent 3,963,458; June 15, 1976; assigned to Norton Company* the supersize layer contains in addition to the ingredients described above a sufficient amount of antioxidant, at least 4% by weight, to limit oxidative degradation.

Without a sufficient amount of antioxidant, a noxious odor will be given off during usage of the coated abrasive material. The polymeric material used should be substantially free of any butadiene monomer.

This process is not limited to the use of any particular antioxidant. Any of those conventionally used for this purpose may be used; however, the minimum amount thereof in the supersize layer to achieve the desired results is deemed critical. Those antioxidants which will be found particularly suitable in the practice of this process include naphthylamines, alkylated phenols, polyalkyl polyphenols, and hydroquinones. Mixtures of antioxidants can also be mixed as desired.

Second Binder with Small Wax Particles

The object of the process of *H. Lukowski; U.S. Patent 3,926,585; December 16, 1975; assigned to Feldmuhle Anlagen- und Produktions- GmbH, Germany* is the provision of an improved grinding or polishing material of the type in which one of the major faces of a flexible base of sheet material carries a layer of binder material, and a multiplicity of abrasive grains are secured to the base by the binder material.

The improved grinding or polishing material includes an exposed covering layer covering the abrasive grains and essentially consisting of an intimate mixture of a second binder material and wax. The wax is present in the mixture in the form of particles having an average size of less than 500 microns, and the second binder material is solid and shape-retaining at least at 25°C when traveling at 10,000 ft/min in a circular path of 4 inch diameter. The weight ratio between the second binder material to the wax in the mixture is between 1:5 and 2:1, and preferably between 1:1 and 2:3.

The binder materials employed for securing the abrasive grains to the flexible base are conventional and include synthetic resin compositions, but also water-soluble glue, such as hide glue.

Binder materials combined with the wax particles in the covering layer may be synthetic resin compositions including phenolic, urea, epoxy, polyester, polyurethane, and alkyd resins, but also water-soluble glue.

A wax, for the purpose of this process, may be a mineral wax, an animal or vegetable wax, or a synthetic wax. Beeswax, carnauba wax, ouricuri wax, paraffin, ceresin, polyolefin waxes, and amide waxes are representative of suitable waxes.

The preferred amide waxes are derivatives of ethylenediamine in which two acyl radicals of fatty acids, particularly of stearic, palmitic, and oleic acid, are attached to at least one amine nitrogen, and more specifically the distearoylethylenediamines.

The effectiveness of a covering layer in the abrasive sheet material of this process is not impaired by the presence of a pulverulent, inorganic or organic filler which is softer than the abrasive grains employed if it does not amount to more than 60% of the weight of the covering layer. Calcium carbonate is typical of inexpensive filler materials readily available. The weight of the covering layer may vary between 5 and 1,000 g/m^2 and is selected according to the mesh or grit of the abrasive grains.

Example: An endless belt of cotton twill having two major faces 50 mm wide and 3,500 mm long was coated on a roller coater with a commercial, water-bearing, A-stage phenol-formaldehyde resin composition having a nominal viscosity of 800 to 1,000 cp, a pH of 8.0 to 9.0, and specific gravity of 1.25 at 20°C. The resin composition was capable of conversion to the B-stage in about 5 minutes at 120°C without significant loss of formaldehyde.

The coated base of textile sheet material was covered with 120-grit emery grains from an electrostatic spray gun, and the abrasive grains were bonded to the base by heating to 90°C until the resin layer solidified. The grains then were covered with a surface layer prepared by dispersing 10 parts micronized amide wax and 4 parts calcium carbonate powder in 20 parts of another commercial, A-stage phenol-formaldehyde composition and depositing the mixture by means of a roller coater, all parts herein being by weight unless stated otherwise.

The amide wax was Hoechst Wax C, a commercial product essentially consisting of di-octadecanoylethylenediamine, the stearoyl groups being bound to the nitrogen atoms of the diamine. The wax employed had a nominal dropping point of 139° to 144°C, an acid number of 3 to 7, a saponification number of 3 to 10, specific gravity of 0.99-1.01/20°C, and was free from measurable amounts of ash.

The A-stage phenolic resin used as a binder material in the covering or surface layer had nominal properties of 900 to 1,100 cp viscosity, a pH of 8.0 to 8.5, specific gravity of 1.26/20°C, and it converted to the B-stage in about 10 minutes at 120°C. The laminar belt structure so obtained was cured in an oven at a temperature gradually raised to 120°C, and held at 120°C for 2 hours.

Additional belts were prepared from otherwise the same materials employing 220, 320, and 400 grit emery. A second set of four abrasive belts was prepared for comparison purposes without the amide wax under otherwise identical conditions. The weight ratio of first binder layer, abrasive grains, and covering layer were as shown in the table below.

Grit	First Binder	Abrasive (g/m²)	Covering Layer
120	90	250	150
220	70	150	100
320	50	90	80
400	40	60	60

Test bodies of commercial aluminum (99% Al) were ground for 60 minutes under otherwise identical conditions on the eight belts prepared as described above. The weight loss of each test body by the grinding operation was determined as a measure of abrasive belt quality. The results achieved with and without amide wax in grams of removed aluminum are listed in the table below.

Grit	Weight Loss, g. With Wax	Weight Loss, g. Without Wax
120	257	100
220	92	14
320	32	4
400	10	2

Use of Halogenated Paraffins

An object of the process of *G. Bigorajski and J. Ellerkamp; U.S. Patent 3,888,640, June 10, 1975; assigned to Vereinigte Schmirgel- und Maschinenfabriken AG, Germany* is to improve the grinding efficiency of a grinding medium or abrasive on high duty alloyed steels.

This is achieved by the binder containing an addition of 1 to 50% by weight of organic halogen compounds, based on the weight of the complete mixture.

It has been found that by this means premature blunting of the grinding grains is substantially prevented or retarded. The grinding efficiency and the life of a grinding body are thus considerably improved.

Suitable as organic halogen compounds for this purpose are, for example, liquid and/or solid halogenated paraffins, more especially chlorinated paraffin hydrocarbons with a chlorine content of at least 30% by weight, and also brominated or fluorinated paraffin hydrocarbons with a halogen content which likewise is at least 30% by weight. Suitable are both halogenated paraffins of relatively high molecular weight, more especially chlorinated paraffins, with a viscosity at 20°C of at least 500 cp, such as those used as extender plasticizers for polymers, and low molecular weight types, e.g., hexachloroethane or mixed halogenated paraffins, and also halogen-containing organic acids, halogen-containing aldehydes, halogen-containing aromatic compounds, halogen-containing cycloaliphatic compounds, and also polymer dispersions, such as polyvinylidene chloride dispersions or solutions of halogen-containing polymers, such as chlorinated rubber in toluene.

If aqueous animal glues or resin systems, e.g., phenol-formaldehyde or urea-formaldehyde resins, are used in the binder, difficulties may arise in respect of the compatibility of the organic halogen compounds with such resins.

The aqueous phenol-formaldehyde or urea-formaldehyde resins are alkali-adjusted condensation products of phenol and urea, respectively, with aqueous formaldehyde solution (37% by weight) in a molar ratio from 1:1.3 to 1:1.8, preferably between 1:1.5 and 1:1.6. By vacuum distillation, the aqueous resin solution is increased in concentration to a solid content of about 80% by weight. As regards the animal glues, these are 30 to 40% by weight aqueous skin or hide glue solutions. Adopted as factors of quality are the viscosity according to Bloom, of at least 100 mp after 1 hour at 60°C, and the jelly strength of at least 300° Bloom. The aforesaid difficulties as regards compatibility are avoided by the substances being incorporated in pulverized or dispersed form into the mass of binder.

The incorporation of the resins is effected with use of solution-promoting additives, more especially low molecular weight alcohols or ketones, and particularly advantageously methanol and/or ethanol and/or acetone and/or methyl ethyl ketone. Depending on the required viscosity of the binder mass (between 500 and 2,000 cp), the addition of solvent varies between 5 and 30% by weight, related to the total weight of the mixture. The components of the binder mixture are homogenized by means of mechanical stirrer mechanisms or ultra-Turrax stirrers, without any particular additional measures. If aqueous, animal glue solutions are used in the binders, the incorporation can be effected in similar manner to the synthetic resins.

Furthermore, the halogen compounds can be reacted with ammonia and/or amines, in order to obtain a better compatibility with the aqueous resin systems. The reaction starts at temperatures above 30°C and can be carried on up to 130°C. The reaction is stopped before the free amino groups of the amines are completely neutralized by split-off hydrogen halide, due to salt formation or due to the hydrohalogen compounds, so that free hydrohalic acid is not formed.

The mixing of the components is again effected with normal mechanical stirrer mechanism or with an ultra-Turrax stirrer, it being possible to adjust the required viscosity with the alcohols or ketones.

The mixing ratio between the organic halogen compound and ammonia and/or amines is between 20:1 and 4:1 by weight. Amines to be considered for use in this connection are: ethylene diamine, diethylene triamine, triethylene tetramine, tetraethylene pentamine, ethanolamine, diethanolamine and other amines of similar structure.

The quantity of the additive according to this process of organic halogen compounds is between 1 and 50% by weight of the total binder mixture, but preferably between 10 and 30% by weight.

Example: For this experiment, two identical grinding bands on a flexible support were prepared, with corundum grinding grains of grain size 80 and a binder of the described phenol-formaldehyde resin (80% by weight solid substance) and calcium carbonate as filler in the ratio by weight 1:1.

One of these grinding bands was subsequently coated with a binder of the following composition: 30 parts by weight of the described phenol-formaldehyde resin (80% by weight solid substance); 10 parts by weight of chlorinated paraffin (60% by weight chlorine, density 1.41 g/ml at 60°C, viscosity 300 cp at 60°C); and 60 parts by weight of filler (calcium carbonate).

The viscosity of the total mixture was adjusted with methanol to 1,000 cp at 20°C. Both grindng bands were ground under identical conditions against alloyed steel of the material number 4301 (18% chromium, 10% nickel). Using the grinding band without the addition of organic halogen compounds, altogether 220 grams were ground off the steel during the life of this grinding band, which was 60 minutes.

With the second grinding band, subsequently coated with the binder mixture according to this process, the quantity ground off after 60 minutes was 405 grams. However, it was only after another 40 minutes that the grinding efficiency had fallen to the same value as the band without the subsequent coating. In this time, another 110 grams of the steel were ground off. The life of the subsequently coated band was consequently 100 minutes and the total quantity ground off was 515 grams. This corresponds to an improvement in the effective life of about 66% and of the quantity removed by grinding of about 134%.

Surface Deposit of Fluorocarbon Particles

G.C. Mullin, R.E. Wetherby and V.O. Chevalier; U.S. Patent 3,869,834; March 11, 1975; assigned to Avco Corporation provides a coated abrasive device having a surface deposit comprising an accumulation of fluorocarbon particles covering at least that portion of the surface not covered by the abrasive particles. The ratio of the fluorocarbon to abrasive particles on the surface is 1:8 to 1:2 by wt.

Example: A ⅝ by 90 inch continuous belt containing 60 grit aluminum oxide was provided with a deposit of a plurality of layers of submicron sized particles from an aqueous suspension of a fluorocarbon. Conventional spraying procedures were used. A precalculated weight of 3.7 grams of tetrafluoroethylene was deposited on the belt. This was accomplished by spraying continuously for 900 revolutions of the belt at 7,200 fpm.

The water was fully evaporated during the course of developing the deposit. Evaporation could be accelerated, however, by supplying gentle heat to the surface of the coating.

The belt requires no further processing before being used. The performance improvement of these coated belts is quite startling. The rate of stock removal of a coated belt is twice that of an uncoated belt. For a particular application grinding a nickel alloy under identical grinding parameters, at least a three-fold increase in life of the coated belt was observed. Whereas previously an untreated belt could be used for two to three workpieces the treated belt repeatedly performed well with nine to ten workpieces.

Water-Insoluble, Water-Resistant Presize

In accordance with the process of *N.E. Wisdom, Jr. and M.T. Gladstone; U.S. Patent 3,861,892; January 21, 1975; assigned to Norton Company* there is provided an open mesh cloth coated abrasive material of improved physical properties, in particular flexibility, comprising an open mesh fabric backing member coated with a water-insoluble and water-resistant presize material and abrasive grains adhered thereto by means of an adhesive binder. In the preferred aspect, the adhesive binder provides controlled friability. Such an abrasive material provides coated abrasive articles of unique physical characteristics.

The abrasive material of this process in its basic aspects incorporates a presize material which because of its resistance to wet grinding conditions, as well as its insolubility in water, results in abrasive material of less variation in use in its flexural characteristics. Thus, quite advantageously, an abrasive article manufactured from abrasive material of this process is subject to essentially no variation in fold resistance between its dry and water-saturated condition. The abrasive material finds rather wide application and can be used by operators under widely varying degrees of exposure to water.

In addition to the above desirable consistency in flexural characteristics and fold resistance, the presize composition used in this process, which has excellent adhesion to both nylon yarns and phenolic resins, provides coated abrasive material of excellent tear resistance.

The presize material which has been found most satisfactory is the reaction product of a composition comprising in admixture a terpolymer of ethylene, vinyl chloride, and acrylamide, and an aminoplast. Such a terpolymer and its manner of manufacture is disclosed in U.S. Patent 3,632,425. It is preferred that the interpolymer contains from about 5 to 70% ethylene, 30 to 95% vinyl chloride, and from about 1 to 5% acrylamide. Even more preferably, the interpolymer will contain from about 19 to 23% ethylene, about 74 to 78% vinyl chloride, and from about 2 to 4% acrylamide.

Example: An open mesh leno weave fabric (44 inches wide, 18 by 18 thread count, 210 denier high tenacity continuous filament nylon warp yarns and 420 denier high tenacity continuous filament nylon fill yarns, Warwick Mills, Style No. 1122) was coated with a presize composition having the following composition:

Components	Parts by Weight
Monoflex 4514, latex*	180
Aerotex M-3**	18
Foamkill 608, antifoaming agent	1
Witco LPA, antifoaming agent	1

*An aqueous dispersion (47% solids) of a terpolymer of ethylene (19.23%), vinyl chloride (74 to 78%) and acrylamide (2 to 4%).
**A clear, viscous, melamine-formaldehyde syrup (80% solids in water) having a weight of 10 lb/gal and a pH 8.5 to 9.

Coating was accomplished by means of conventional roll coating techniques, a sufficient amount of the presize composition being applied to the fabric backing member to result in, on drying, an add-on weight of from 2.0 to 3.0 pounds per sandpaper maker's ream. The wet coated backing member was then dried by passage through an oven at 225°F for about one minute. On visual examination, the presize composition is seen to have encapsulated the yarns of the fabric backing member. The yarns are also observed to be penetrated to some degree.

The dried coated backing member was then roll coated with an adhesive composition (maker adhesive) which, when dried and cured, results in unique friability characteristics. This composition comprises the following:

Components	Parts by Weight
Phenol-formaldehyde resol resin, caustic catalyzed, formaldehyde: phenol ratio, 0.94:1, solids content 78.4%	100
Rosin Amine D*	15

*High molecular weight stabilized abietylamine derived from pine resin acids, predominantly dihydroabietylamine.

On the adhesive coated backing member (gravity coat top side, electrostatic coat bottom side) aluminum oxide abrasive grain (grit 360) was then deposited in amount sufficient to give 20 to 30 pounds per sandpaper maker's ream. Afterwards, the adhesive layer was dried and cured (350°F, 1½ minutes), the amount adhesive composition applied giving an add-on (dry weight) of about 2.0 to 3.0 pounds per sandpaper maker's ream.

Next, a size composition, the composition being the same as that of the maker adhesive composition but of lesser viscosity was applied. Afterwards, it was dried and cured in the same manner as the maker adhesive layer. A sufficient amount of size composition was applied to result in, on drying, an add-on weight of about 10.0 pounds per ream (sandpaper maker's).

The thus formed abrasive material was then ready to be cut into appropriate abrasive articles, e.g., abrasive discs, for use in, in particular the finishing of auto bodies paint and sealer coats.

In use abrasive discs are found to possess improved tear resistance and much better overall flexibility characteristics.

BACKINGS

Particles Embedded by Gravity

The process of *F.J. Kronzer; U.S. Patent 4,240,807; December 23, 1980; assigned to Kimberly-Clark Corporation* is directed to an improved type of backing material for flexible abrasive products such as waterproof sandpapers, as well as dry finishing papers. A unique feature of the backing material is that it is coated with an adhesive for securing abrasive grit thereto which adhesive in its normal state is not tacky and does not flow readily at ambient temperatures, but becomes a viscous liquid at elevated temperatures. The adhesive is characterized by its ability to be softened sufficiently at a temperature which does not degrade the paper substrate, and which in its softened state permits grit particles deposited therein to sink into the softened adhesive by virtue of their own weight.

The coated backing material may be wound into rolls for storage and shipping without blocking. In the manufacture of abrasive paper using this coated substrate, the backing material is heated above the softening point of the adhesive to a point where the viscosity of the adhesive is reduced to a condition whereby when particles of abrasive grit are deposited on the backing and electrostatically aligned the particles will penetrate into the softened coating by their own weight and become bound therein and to the backing. No pressure other than the force of gravity on the particles is employed or desired.

Depending on the operating speed, it may be necessary to maintain heat on the backing material at a temperature above the softening point of the adhesive for about 30 seconds after deposition and electrostatic alignment to obtain suitable penetration of the grit particles. After deposition and embedment when the backing is cooled or the source of heat removed, the adhesive hardens to its original state, and the abrasive particles are thus bound firmly therein. A final step in the fabrication of the finished abrasive sheet product is to apply a conventional resin coating, or grain size over the abrasive particles as is done with conventional abrasive papers.

The substrate to which the adhesive is applied is a flexible sheet material, preferably a resin or latex-impregnated paper, which must not soften or degrade at the temperatures employed to soften the adhesive. The amount of adhesive applied to the backing is adjusted so that abrasive particles of the desired size will become only partially embedded in the adhesive but not be buried in it.

The process for manufacturing abrasive products using the coated backing material of this process is easily adapted to the continuous processes currently being used. For example, the adhesive coating equipment in the conventional process is replaced by a means for heating the moving web of coated backing material. An infrared heat source, banks of electric elements, gas panels or a heated metal cylinder may be employed as the heating means. Preferably, a second heating means for heating the web for several seconds after the grit has been deposited in the softened coating is also provided. Another option is to heat the grit particles before they are deposited. This speeds up the embedment process and requires a lesser amount of heat to maintain the viscous condition of the adhesive during and after application of the grit.

In carrying out the process, two types of base paper or substrate for the backing are preferred. The first, which is commonly employed for waterproof sandpapers, is a highly porous kraft paper containing from 15 to 40% nitrile rubber or an acrylic elastomer. Pigments may also be added. The preferred basis weight of the paper is from about 70 to 200 grams per square meter and the preferred Gurley porosity is less than 10 seconds per 100 cc of air for a single sheet.

The second type of paper which is commonly employed for dry finishing papers is a more highly refined kraft paper containing filler and dyes or pigments if desired and from 0 to 15% of a nitrile rubber or an acrylic elastomer. The basis weight of the latter paper is from 65 to 200 grams per square meter and the single sheet Gurley porosity is greater than 10 seconds per 100 cc of air. Functional additives which may be useful in this type of paper include sizing agents and starch. It is understood that the base papers used in wet finishing have the impregnant cured or thermoset to the extent that the impregnant will impart sufficient wet strength for such uses.

For the adhesive coating, a variety of waterbased, heat-sensitive adhesive systems have been tested and many of these have been found suitable. A number of others have not been found suitable and are specifically excluded. Examples of adhesives which are not suitable are: (1) Polyvinyl chloride latexes blended with various softer polymers such as acrylic elastomers or with conventional plasticizers. These do not soften sufficiently below 220°C to permit abrasive particles to sink therein by gravity alone and thus require too high a temperature to be practical. (2) SBR, ABS, polyvinyl acetate, and most acrylic latexes are not suitable for the same reason. (3) Wax, polyethylene, and low molecular weight ethylene-organic acid copolymer dispersions or latexes adhere poorly to the paper and/or the silicon carbide grits commonly employed in the preferred abrasive papers and are unsuitable for those reasons.

Among the useful adhesives, several of the ethylene-vinyl acetate copolymer latexes and ionomer resin dispersions such as Elvax D have been found suitable. These adhesives may also be blended with nitrile rubber or acrylic latexes to give coatings which have improved flexibility and provide improved adhesion to the paper substrates especially when the latter are impregnated with the same latexes. Blending these ethylene-vinyl acetate copolymer latexes or ionomer latexes with polyterpene emulsions provides such blends with very low softening temperatures which are desirable, and while these blends are useful the adhesion to the paper is not as good as with the blends with acrylic or nitrile rubber latexes.

Relatively high molecular weight ethylene-acrylic acid copolymer dispersions such as PCX 300, an alkali dispersible ethylene copolymer (Union Carbide), when blended with acrylic, nitrile rubber, or polyterpene latexes also are suitable. Blends of a novolak resin, such as Durez with Acrysol WS-32, also are suitable.

Other useful polymers include polyvinyl butyral, ethylene-ethyl acrylate copolymers and linear aliphatic polyamides and polyesters. Many other types of polymers or blends with properties similar to those given above are, of course, possible. The required characteristic is that the selected adhesive be a nontacky solid at ambient temperature, and when heated to a temperature low enough so as not to degrade cellulose substrates be softened sufficiently to permit abrasive particles to embed themselves therein by gravity alone.

Fabric Coated with Novolak Resin

The process of *M.A. Sandman and B.T. Loughlin; U.S. Patent 4,230,461; October 28, 1980; assigned to Eli Sandman Company* relates to improved fabrics used in the reinforcement of resin-bonded abrasive wheels. The fabrics are characterized by a coating of a novolak resin essentially free of added crosslinking agents.

The fabric, used in the form of a disc or any other convenient shape, may be any of those used in the prior art. However, cloth of a high strength material is preferred. Typical cloths comprise cotton, dacron, rayon, nylon, Kelvar and glass, glass being most preferred, especially open mesh glass fabric.

In the most preferred embodiment of this process, a novolak resin is used completely free of added crosslinking agent though amounts of crosslinking agent up to a maximum of 3% by weight can be tolerated with the understanding that the results obtained with this quantity of added crosslinking agent are less desirable than when the novolak is free of added crosslinking agent.

The resin is coated onto the fabric in conventional manner such as by immersing the cloth in a varnish of the resin where the varnish comprises a solvent such as an alcohol having a resin solids content varying within relatively broad limits dependent upon the desired percentage resin content of the reinforcement. The varnish may contain other additives as is conventional such as internal or added plasticizers. The solids content of the varnish can typically vary between about 25 and 80% by weight, but preferably ranges between about 60 and 75% by weight.

After the fabric is coated with varnish, it is dried preferably at elevated temperatures to more rapidly remove solvent. Temperatures of from 150° to 450°F are suitable for periods of time ranging between a fraction of a minute and 30 minutes. Following drying, the cloth may be cut to any desired shape, for the fabrication of an abrasive wheel reinforcement.

Example: A woven glass cloth identified as Style 500 of the Greenville Mills Division of Warwick Mills Corporation was selected. The cloth had a weight of 9.2 ounces per square yard of material. It was coated with a varnish consisting of 70% by weight of a novolak resin dissolved in methyl alcohol. The novolak resin used was identified as GP2173 of the Georgia Pacific Corporation. The varnish did not contain added crosslinking agent.

Following coating of the glass cloth with varnish, the cloth was dried by passing it through an oven at a speed of 20 feet per minute. The oven measured 20 feet in length and was maintained at a temperature of 240°F. The dwell time of the cloth in the oven was one minute. Following drying, the novolak content of the coated glass cloth was 30% of the total weight. The glass cloth was then cut into circular discs having a diameter of 9 inches. The discs are suitable for the reinforcement of abrasive wheels.

Calendering Woven Polyester

R.E. Pemrick; U.S. Patent 4,214,877; July 29, 1980; assigned to Norton Company has found that calendering of woven polyester fabric provides a smooth surface upon which fine grit abrasive may be applied to produce flexible coated abrasive articles for polishing applications.

Specifically, the use of from 600 to 2,000 pounds per lineal inch of cloth, and a temperature of 210° to 380°F, have been found satisfactory, when the web is passed between a set of conventional calender rolls at a speed of from 20 to 75 feet per minute. Preferably the pressure is from 800 to 1,600 pounds per lineal inch, and the temperature from 230° to 300°F.

After calendering the cloth is finished in a conventional manner, by application of adhesive and abrasive grain on one side, to produce a coated abrasive product.

Suitable cloth for this application is a dimensionally stable polyester woven fabric having a weight of up to 9.5 ounces per square yard (most cloth lighter than 3.5 ounces per square yard can be manufactured smooth enough). The weave can be square, twill, or sateen weave, but square weave is preferred. The yarn weight can vary from yarn numbers of 6 to 30 for staple yarns (cotton count), to 100 to 1,100 denier, for continuous multifilament yarns.

Resin Not Permeating Cloth

H.J. Swiatek; U.S. Patent 4,163,647; August 7, 1979; assigned to Norton Company describes a process for making cloth backing for coated abrasive articles wherein a backing material is coated on its front side, with a liquid thermosetting resin, in such a manner that the resin does not permeate the interstices of the cloth. As a result, the cloth retains most of its original pliability when incorporated in the finished coated abrasive article. A back sizing, of for example glue, may also be applied in the same manner to provide a coated abrasive backing material which is both front sized and back sized but which still possesses most of its original pliability.

Example: A standard cotton drill having a thread count of 76 x 48 (cotton count, 12½s warp thread; 18s fill thread) and of such weight that one yard of cloth 60 inches wide weighs 1.38 pounds was prepared for manufacture of a resin bond abrasive cloth backing by first dyeing and drying the same according to conventional techniques well known to those skilled in the coated abrasive industry.

The cloth was dyed brown and was "pulled down" during drying to a width of 54½ inches, such representing a 9.0% pull-down. The cloth backing was then ready for application of the back size.

The dyed cloth was forwarded at a linear speed of 150 feet per minute on the surface of a carrier roll through the nip formed by such roll and a rubber-surfaced applicator roll, rotating in the same direction of travel as the cloth, and at the same linear speed as the carrier roll in an aqueous glue solution. Glue solution was picked up on the surface of the applicator roll and was wiped from the surface by the cloth backing. A relatively open nip (⅛ inch) between the rolls was provided.

The aqueous glue solution 51%, 58 mp glue, was maintained in the glue tank at a temperature of about 140°F, and had a viscosity of approximately 2,200 cp (Brookfield Viscosimeter, No. 2 spindle, 30 rpm, 140°F).

The newly applied mass of glue solution was then spread and made relatively smooth and continuous and excess glue solution was removed by passing the coated backing in contact with a flexible knife. The knife, the end of which pressed directly against the solution-coated backing, was about 0.031 inch in thickness and extended across the width of the roll. It formed an angle of approximately 16° with respect to the horizontal.

The aqueous glue layer was allowed to cool to a temperature of about 100°F before being subjected to drying conditions. The glue-coated backing was passed in contact with (aqueous glue solution layer on outboard side) steam-heated rolls and was then passed alternately over hot and cold (room temperature) steel cans. With such manner of drying, the glue solution layer was heated for about 43 seconds at 230°F.

The "tack-free" glue-coated cloth was then wound up in a roll in conventional fashion in preparation for application of the front size. The glue solution, applied to the back side of the dyed cloth in excess, was sufficient to provide a weight (dry basis) of about 2.0 pounds glue per sandpaper maker's ream. The original cloth weighed (dyed) 15.3 pounds per ream and the glue-coated cloth weighed 17.3 pounds per ream.

The glue-coated cloth backing was unwound and forwarded at 150 feet per minute past a heat-hardenable resin solution applicator roll. The applicator roll, rotating at a linear speed of 150 feet per minute and in the same direction of travel as the cloth backing, was rotated in an aqueous phenol-formaldehyde resin solution having the composition shown below and a viscosity of 2,100 cp (Brookfield Viscosimeter, No. 2 spindle, 30 rpm, 85°F).

	Pounds
Varcum 2536*	550
$CaCO_3$	412
H_2O	58
Total	1,020

*A phenol-formaldehyde resin aqueous solution.

The resin solution was wiped from the surface of the applicator roll by the backing, after which it was then made relatively smooth and continuous, and excess solution was removed from the backing by passing the newly coated backing across and in contact with an inverted knife.

The knife, which was of stainless steel and about ⅝ inch in thickness, extended across the width of the backing at an angle of about 45° with respect to the leading end of the backing. The knife was tapered at the contact edge to form a bevel of 60°. The beveled edge faced the leading end of the backing.

The aqueous resin layer was then heated for about 4½ seconds by passing the backing under a plurality of hot air streams. The air, heated to a temperature of about 700°F, was directed onto the resin solution layer in 16 streams extending across the width of the backing from a conventional hot air heater located about 4 inches above the surface of the backing.

The backing was then passed over and under a plurality of cool (ambient temperature) cans whereby the temperature of the resin layer was reduced prior to winding. A "tack free," uncured resin layer was provided stabilized against premature curing. Resin solution, applied to the backing in excess, was sufficient to provide a weight (dry basis) of about 6 pounds per ream front size. The finished backing had a total weight of about 23.6 pounds per ream and a width of 53.5 inches, thus representing a total pull-down of about 10.9%.

Upon examination of a cross section of the finished coated cloth backing under a microscope, the interstices of the backing were observed to be unfilled, and the glue and resin layers were observed to be deposited substantially on the surfaces of the yarns on the respective sides of the backing. The glue and resin layers, while penetrating to a slight degree the surfaces of the yarns, did not contact one another in the interstices of the backing and permitted the yarns freedom of movement.

GRAINS

Fused Zirconia Grains to Abrade Iron

The process of *S.M. Leahy; U.S. Patent 3,996,702; December 14, 1976; assigned to Minnesota Mining and Manufacturing Company* provides a means of rapidly removing stock from iron castings containing free graphite, by a process wherein a rapidly moving supported coated abrasive product is forced against a workpiece at high pressures, on the order of at least 200 psi; at least a significant percentage of the abrasive grains consist essentially of fused zirconia. In this application, not only is stock removed at an extremely rapid rate, but the useful abrading life of the coated abrasive product substantially exceeds, sometimes by several fold, that of a product employing conventional coated abrasive minerals.

Grains of fused zirconia can be used as the sole abrading material in coated abrasive products employed to practice this process. To reduce the cost of the product without unduly reducing performance, however, it is considered desirable to include a substantial proportion of alumina or other diluent. Alumina grains can either be blended with zirconia grains or, alternatively, grains may be formed by crushing hardened fused blends of alumina and zirconia. If desired, softer grains (for example, flint), which function in a manner analogous to that of filler or diluent, may be blended with fused zirconia-containing grains, such softer grains being quickly rendered ineffective under the high pressures utilized in the practice of this process.

Fused zirconia grains may have a crystal lattice which is monoclinic, cubic, or a blend of the two. Zirconia which is made by merely fusing and crystallizing zircon ore is termed "unstabilized," and is mainly monoclinic. The addition of minor percentages of lime to the melt "stabilizes" the zirconia, causing it to be primarily cubic. Either stabilized or unstabilized fused zirconia grains may be employed in the practice of this process although unstabilized are generally preferred.

It is particularly interesting to note that zirconia grains actually detract from the effectiveness of coated abrasive products for abrading most metals and that the unexpected utility of zirconia grain is not attained until it is incorporated in coated abrasive products used at comparatively high pressures, on the order of at least about 200 psi. Below this pressure, both the coated abrasive product and the workpiece tend to burn. The pressure should generally not exceed 1,200 psi. To obtain an optimum combination of uniform cutting rate and coated abrasive life, it is generally preferred initially to apply the lowest effective pressure, gradually increasing it during use.

It is also interesting to note that this process is effective essentially only when the material being abraded is cast iron, or a modiifed form thereof, containing free carbon.

Blends of fused zirconia grains and fused alumina grains are effective when they contain as little as 10% zirconia by weight. As a compromise between cost and performance, however, the particularly preferred practice of this process includes the use of coated abrasive products in which the zirconia:alumina weight ratio in the abrasive grains is at least about 1:2 (i.e., 25 to 40% zirconia by volume). Because the primary utility of this product is in high pressure operations requiring heavy stock removal, it is also preferred that the abrasive grains themselves should be on the order of 100 mesh or coarser.

Example: Two lots of Grade 36 coated abrasive sheet material were prepared by electrostatically coating a standard volume of mineral. In control Lot A, 218 grains of conventional fused alumina (density 3.97) was applied, while in Lot B, 240 grains of a crushed cofused 40:60 zirconia:alumina mineral (Exolon AZ-40, density 4.50) was applied.

Preweighed bars of No. 32510 malleable iron (cast and fully annealed; minimum tensile and yield values 53,000 and 35,000 psi respectively) having cross-sectional dimensions of 1 by 1 inch, were forced at a pressure of 350 psi against the belt. Control Lot A removed 10.4 pounds of metal before failing after about 4½ minutes. In contrast, the experimental belt of this example cut 19.8 pounds of metal before failing after 9 minutes.

Spheroidal Granules

The process of *R.N. Howard and H.G. Sowman; U.S. Patent 3,916,584; November 4, 1975; assigned to Minnesota Mining and Manufacturing Company* provides wholly inorganic composite abrasive granules which are simple to manufacture and which can utilize such thermally unstable abrasive grains as diamonds and cubic boron nitride. The granules can be made substantially spheroidal and of extremely small diameter, e.g., 10 to 200 microns.

In accordance with this process, fine, hard abrasive grains are distributed uniformly throughout a matrix of softer microporous metal oxide (e.g., silica, alumina, titania, or zirconia) or mixtures such as alumina-boria-silica, zirconia-silica or others. In a preferred method of manufacture, the abrasive particles are mixed into an aqueous sol of a metal oxide (or oxide precursor) and the resultant slurry in turn added to an agitated dehydrating liquid such as 2-ethyl-1-hexanol.

Water is removed from the dispersed slurry and surface tension draws the slurry into spheroidal composites, which are thereafter filtered out, dried, and fired at temperatures sufficiently high to drive off the balance of the water, volatilize and decompose fugitive material and calcine the matrix material to a firm, continuous, microporous state, but insufficiently high to cause vitrification, or fusion. As a result, the matrix retains a degree of microporosity, as can be detected by the disappearance of the matrix when filled with an oil having the same refractive index as the matrix and viewed with an optical microscope. This oil-absorbing feature of the matrix permits the incorporation of lubricants, liquid grinding aids, etc., to enhance performance of the composite in actual abrading operations.

As previously indicated, the matrix of the composite granule is softer than the abrasive grains. Generally speaking, the hardness of the matrix will not exceed 1,000 Knoop and the hardness of the abrasive grain will be at least 1,500 Knoop.

The abrasive granule of this process overcomes the prior art deficiencies and provides a method for utilizing very fine abrasive grains, even thermally degradable grains, on the order of one-half micron or smaller, in a porous metal oxide matrix. The large abrasive granules do not become embedded in a carrier substrate, e.g., film or fiber, as do small abrasive grains.

The aqueous sols or solutions as prepared may be relatively dilute, generally containing the equivalent of about 10 to 40 weight percent equivalent metal oxide solids.

The fired abrasive compounds will normally contain about 6 to 65% by volume of abrasive grains.

Example: A mixture of 0.5 gram of 15-micron diamond powder, 3.3 grams of 30% colloidal silica dispersion in water (Ludox LS) and 3 grams of distilled water was stirred and sonically agitated to maintain a suspension. The mixture was carefully poured into 600 ml of agitated, stirred, 2-ethyl-1-hexanol drying solution. After about 15 seconds, 100 ml n-butyl alcohol was added to the drying solution. Particles containing SiO_2 and diamond grains were formed and the stirring was continued for about 5 minutes to further dehydratively gel the particles.

The resulting green granules were filtered from the drying solution using No. 54 Whatman filter paper and dried at 95° to 100°C for 2 hours to further extract water. The dried green granules were subsequently placed in an electric kiln in an air atmosphere, at room temperature, and the temperature raised to 500°C over a 1-hour period.

The resulting abrasive composite granules consisted essentially of spheroidal particles ranging in size from about 20 to 200 microns in diameter, the bulk of the spheres being between 75 and 100 microns. The silica matrix was transparent, porous and colorless; dispersed discrete diamond grains were clearly visible in the matrix. The spheres contained approximately 33% diamond and 67% SiO_2 based on the weight of the composite.

The silica-diamond abrasive composite spheres were bonded to a plain weave two-ply 65% Dacron cotton cloth. A make coat of base-catalyzed phenol-formaldehyde, 50% solids in Cellosolve and water, was applied to the cloth with a paint brush. The spheres were then dropped onto the make coat and the excess allowed to fall off. After the make coat was heated to 65° to 85°C to precure, a size coat of the resin was applied with a brush. The coated abrasive was then heated for 10 hours at 100°C to fully cure the resin. There were 1.5 grams of the abrasive composite granules per 150 cm^2. The resulting coated abrasive product was used as a disc in forming and polishing gem stones, e.g., sapphire and spinel. Compared to a conventional 15 micron resin-bonded diamond cloth, the product of this example showed a 30 to 40% higher rate of cut, and provided a better surface finish, upon visual examination.

Fused Alumina-Zirconia

J.R. Quinan and J.E. Patchett; U.S. Patent 3,893,826; July 8, 1975; assigned to Norton Company describe a coated abrasive article comprising a flexible backing having cofused alumina-zirconia abrasive grits adhesively bonded thereto. The abrasive is characterized as follows.

The alumina-zirconia abrasive has been solidified from the molten state at least as rapidly as is accomplished by casting molten alumina-zirconia abrasive into a mold filled with ¾ inch lumps of abrasive grains. The grits have an average maximum to minimum dimensional ratio of at least 1.2 to 1. The zirconia in the eutectic is in the form of rods or platelets with the average diameter or plate thickness being less than 3,000 Å as measured at the colony centers. The alumina-zirconia grits contain from 20 to 50% zirconia, and include eutectic alumina-zirconia colonies having an average size of less than 65 microns. The adhesive bond is sufficiently strong to hold the alumina-zirconia grits to permit fracture thereof during grinding with creation of new cutting edges on the individual grits as the result of the fracture.

MANUFACTURING PROCESSES

Spherical Bonded Abrasive

W. Löhmer and J. Schotten; U.S. Patent 4,132,533; January 2, 1979; assigned to The Carborundum Company provide a spherical bonded abrasive having a hollow core and consisting of abrasive grain fused on the periphery of a thermoplastic carrier.

The abrasive grains are preferably in size between 600 and 3 microns, especially the grain size from 300 to 3 microns being the preferred one. The abrasives recommended are known materials such as silicon carbide, beryllium oxide, boron carbide, and in particular aluminum oxide (corundum).

The individual abrasive grains are by preference fused into the carrier-material to an extent of from ⅕ to ½ of the volume of the individual grains, in particular from ¼ to ⅓. This leaves ½ to ⅘ of the volume, preferably ⅔ to ¾, exposed outside the carrier. The bonded abrasive consists of an internal thermoplastic encasement, which has on its outer surface a layer of abrasive grain, the volume of the abrasive grains being partly embedded in the thermoplastic. Such bonded abrasives are particularly suited for coated abrasives, wherein preeminently no conventional abrasive grain is present.

It was shown, that these abrasive grains offer a large supply of freely protruding abrasive grain per unit of area, while on the other hand a premature break-loose of a whole grain is prevented. If after some grinding time an individual abrasive grain detaches itself from the sphere, there is immediately a new whole individual abrasive grain available. During practical application it is shown that the individual bonded abrasives open themselves (or are broken open) uniformly during grinding. The hollow spheres, being affixed firmly to the grinding belt overwhelmingly by the second "size" coat adhesive bonding, are uniformly abraded down to the height of the base adhesive coat. The special manner of abrasive grain embedding is extensively utilized, and the grain after being worn out is replaced by the next in line.

Attention is directed to the fact that grinding performance decreases with the increasing proportion of carrier-material, i.e., the thermoplastic within the hollow sphere (degree of sphere filling). For larger amounts of thermoplastic being present within the sphere a smaller amount of stock removal is achieved. For this reason it is recommended that the void volume occupied by the individual, ready made spherical bonded abrasive amounts at least to 50%, preferably to 60%, and in particular to above 75% by volume.

As a carrier material a thermoplastic is therefore preferred, which after incipient melting is present as a hollow body and has, prior to fusion, an internal diameter from 0.1 to 3.0 millimeters. In particular, the inner diameter ought to be from 0.3 to 1.5 millimeters. Very good results are achieved if the carrier-material consists of polystyrene (especially Styropor or Vestipor).

It has to be taken into consideration that the size of the abrasive grain has to be purposely adapted to the thickness of the carrier material. Hence, if for example, the inner diameter of the thermoplastic sphere is 0.3 to 0.5 millimeters, abrasive grains from 63 to 150 microns in size are advantageously used.

A preferred process for the manufacture of the claimed bonded abrasive is as follows. The thermoplastic carrier, especially foamed plastic carrier with cellular structure, is surrounded by abrasive grain and heated within the thermoplastic range of the plastic and held at that temperature until part of the abrasive grain volume, at the most one half and preferably only up to one third, is fused into the carrier, whereupon it is cooled. As thermoplastic carriers foamed spheres with outer diameters (prior to fusion) of from 0.5 to 6 millimeters, particularly from 1 to 3 millimeters, are especially used. Such spheres are heated within an excess of abrasive grain and after fusing and cooling, the excess in grain is separated.

According to an especially preferred process, spheres of polystyrene foam are embedded or agitated tumbled or likewise in abrasive grain at a temperature within the thermoplastic range, preferably from 150° to 200°C, until the desired anchoring of the abrasive grain layer in the carrier material is achieved.

During the adjustment of temperature and time, attention is imperative, since the fusion process proceeds more quickly, the higher the foamy sphere is heated within the thermoplastic range. During this heating a distinct decrease in sphere volume occurs. During the manufacturing process a layer of abrasive grain clings or adheres first onto the foamy sphere being heated. On further temperature increase the marked volume reduction of the foamy spheres occurs, whereby their diameter drops below a half of their original diameter and simultaneously, the microcells present within the foamy spheres are contracted into a hollow spherical shell. During such a fusion of the foamy plastic a part of the volume of the adhering grains become anchored and firmly bonded.

The manufacture can be carried out especially simply, continuously, in a rotary kiln whereinto thermoplastic spheres and abrasive grains are continuously fed and rolled. For moving spheres of polystyrene foam and a melting zone temperature from 160° to 170°C, a melting zone residence time of at least 30 seconds up to a maximum of 3 minutes, preferably from 1 to 2 minutes, is recommended.

Cutting Teeth for Etched Abrasives

H.M. Bond and M.E. Rucinski; U.S. Patent 4,099,935; July 11, 1978; assigned to Buckbee-Mears Company describe improved cutting teeth for etched abrading material and a process for etching the abrading material. The improved teeth comprise substantially flat top posts having substantially straight or vertical sides. In addition, selected dimensional ratios are maintained between the etched teeth and the thickness of the base material to provide a flexible etching abrasive that has long life and will smoothly and efficiently abrade the surface of an article.

Referring to Figure 5.1a, reference numeral **10** designates a cross section elevation view of two of the improved cutting teeth of this process. Reference numeral **11** designates a single tooth having a flat top **12** and vertical sides **13**. It should be noted that the tooth has the appearance of a post projecting from the remaining base material.

In the preferred embodiment, tooth **11** has a circular top that ranges in size from about 0.0015 to 0.004 inch with the diameter of about 0.003 inch being the optimum. It has been found that if the diameter of the top is less than 0.0015 inch, the tops will penetrate into the article and produce a furrowed or roughened surface.

The total thickness of the teeth and the backing sheet is designated by $\mathbf{T_1}$ and the thickness of the remaining backing is designated by $\mathbf{T_2}$. The major dimension of the top of the tooth is designated by **D** and in the preferred embodiment the cross-sectional configuration is circular; however, an irregular shape such as ellipse, square or the like are also suitable for use.

Figure 5.1: Cutting Teeth for Etched Abrasives

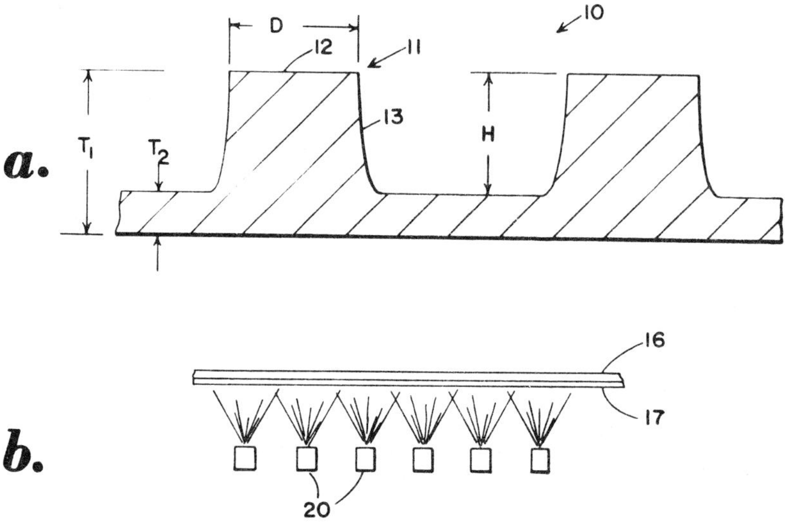

Source: U.S. Patent 4,099,935

The distance from the top of the tooth to the base is designated by **H** and corresponds to the amount of material that is etched away from the base sheet during the etching process.

The first dimension used as a guideline for designing an etched abrasive is the top diameter of the tooth **D**. Experimental results have shown that if **D** ranges from about 0.0015 to 0.004 inch, there is sufficient material and surface area so that the teeth do not penetrate the article being abraded but instead smoothly shear away the material on the top of the article being abraded.

The second dimension of significance is the etched depth **H**. It has been found that if the depth **H** ranges from about 0.0025 to 0.0065 inch, there is sufficient rigidity to shear the material properly yet there is also sufficient flexibility to enable one to smoothly follow the contour of the article being abraded.

If flexibility of the abrading device is desired, the dimension $\mathbf{T_2}$ should not exceed 0.0025 inch and preferably be in the range of about 0.001 to 0.0025 inch.

It has been found that the most efficient cutting is obtained if the density of the teeth is in the order of 700 to 750 teeth per square inch.

Referring to Figure 5.1b, reference numeral **16** indicates the metallic material to be etched and reference numeral **17** indicates the layer of photoresist having the desired tooth pattern thereon. Located beneath material **16** is a set of nozzles

20 which are spraying etchant vertically upward onto the protective coating on the abrading material sheet. The purpose of spray etching the abrading sheet while in a horizontal position is that the undercutting, etc., is minimized or substantially eliminated to thus give a post-like pattern appearance which is shown in Figure 5.1a. Thus, to obtain a post-like pattern of cutting teeth, it is preferred to etch the backing material from underneath.

Compressive Shearing Followed by Noncompressive Shearing of Backing

In accordance with the process of *G.L. Haywood; U.S. Patent 4,038,047; July 26, 1977; assigned to Norton Company* a thin sheet of foam, preferably after it has been laminated to a backing material, before the abrasive grain-binder slurry is applied, is subjected to a surface abrading step. Quite advantageously, this abrading step has been found to produce superior results in the action of the finished abrasive product although the physical appearance and feel of the abraded surface is not materially different from the surface prior to abrading.

More specifically, and referring to Figure 5.2a, the foam material **10** in the preferred embodiment of the abrasive material, comprises resilient particles or granules of foam. These may be formed by shredding waste or virgin foam to the desired particle size which in general will be in the order of $^1/_{16}$ to ¼ inch in diameter. Any foam material can be used but it is preferred to use polyester or polyether types of polyurethane foam.

Although not shown in the drawing, the particulate foam and a suitable resinous binder therefore are intermixed after which the binder coated particulate foam is fed into a compactor **11**. Various binders may be used depending somewhat upon the particular foam material utilized; however, those which will be found suitable include conventional polyurethane adhesive compositions.

In compactor **11**, the foam particles are bonded to each other by the resinous binder and a large cylinder **12** is reconstituted from the particulate foam. Such a cylinder of foam is extruded from the compactor. Cylinder **12** is usually cured in an oven **13** to set the resin binder after which the foam cylinder is then fed to a peeler or slitter **14** which utilizes a knife **15** to peel a continuous, long, thin, sheet **10** of foam from cylinder **12**. The density of the compacted foam granule sheet should be in the range of 6 to 30 pounds per cubic foot with a preferred density range of from 12 to 18 pounds per cubic foot.

The surface **16** of the sheet of foam material **10**, when it leaves peeler **14**, will be observed on close examination to possess a plurality of small waves therein such feature being more clearly shown by reference numeral **16'** in Figure 5.2b. This is a phenomenon found peculiar to the knife cutting of the foam material. The knife appears to form compression waves ahead of itself as it slices the sheet of foam from the cylinder. Where a similar layer of foam, using the same foam granules and binder is formed between smooth sheets this type of surface is not exhibited. Nevertheless, the abrading process, as hereinafter made clear, will find application therein as such last-mentioned manner of manufacture of foam material, whether of particulate foam or not, results in a foam layer having an outer skin of little or no porosity. Such an outer skin can be removed by abrading as disclosed herein thus resulting in a relatively smooth, porous surface.

Figure 5.2: Process for Flexible Resilient Abrasive

a.

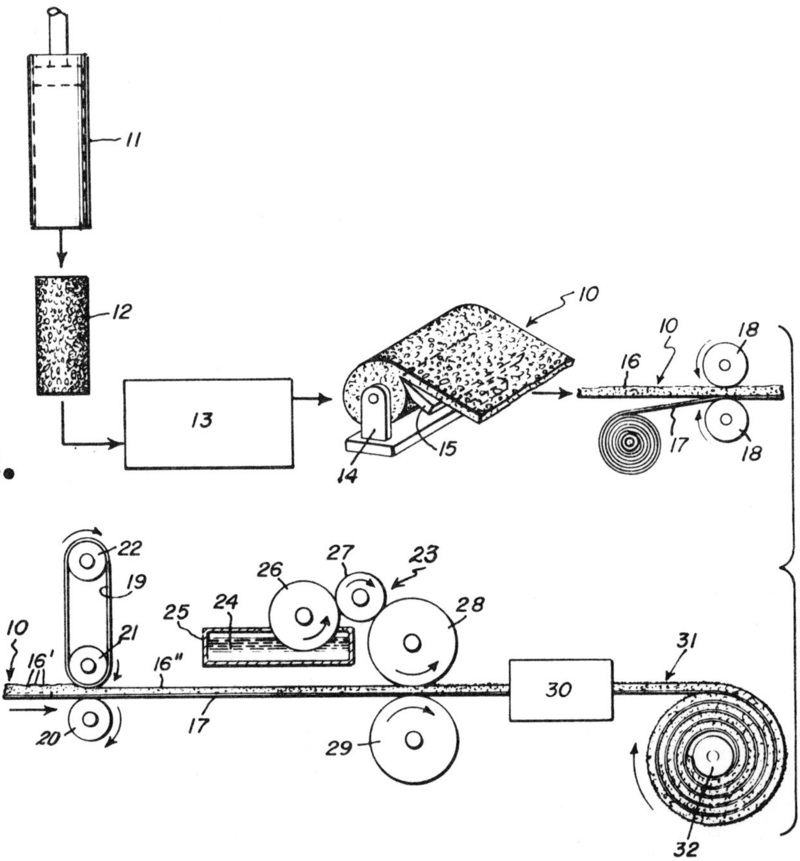

b.

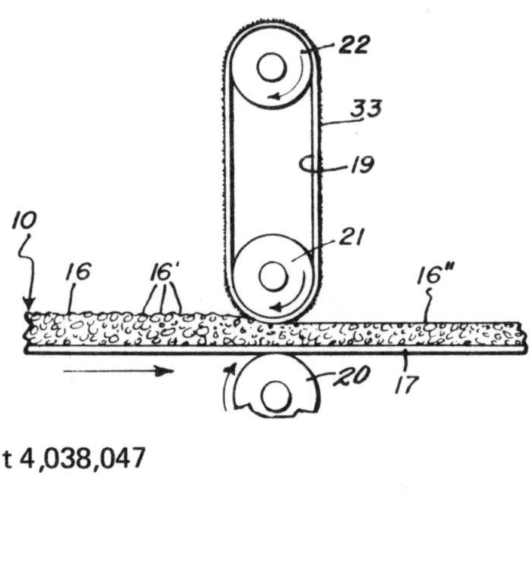

Source: U.S. Patent 4,038,047

Sheet **10** is laminated in conventional fashion to a supporting backing member **17**. This may be accomplished by passing foam layer **10** in superposed fashion, as shown, on backing member **17** through the nip formed by laminating rolls **18**.

Backing member **17** may be any flexible reinforcing and supporting member such as paper, woven cloth, polymeric films, fiber or any of the backing member materials conventionally used for coated abrasive manufacture.

The thus formed lamination is then passed, in contact with a suitable abrasive medium, this being, in the preferred aspects, an endless abrasive belt **19**. A backup roll **20** is provided to give support to the foam layer **10** as it passes under abrasive belt **19** and the associated driver contact and idler rolls **21** and **22**. The clearance between abrasive belt **19** and roll **20** is such as to permit the belt **19** to engage with and cut the upper surface **16** of sheet **10**. It will be noted that surface **16** is one of two surfaces generated by knife **15** of peeler **14**.

Sheet **10** and its associated backing **17** is then passed to roll coating station **23** where the abrasive grain and adhesive binder are applied to the abraded surface **16"** by means of a slurry **24** thereof contained in a suitable vessel **25**. Slurry **24** may be transferred in conventional fashion to the abraded surface **16"** by means of rolls **26, 27, 28,** and **29**; however, obviously, other means of coating the abrasive grain, resin binder slurry may be used to advantage.

The slurry coated substrate is then passed through oven **30** after which the abrasive material **31** is wound up in roller form **32** in preparation for subsequent conversion into belts, sheets, discs, etc.

The abrading medium, as before disclosed, is preferably a coated abrasive belt and is driven so that it "up-cuts" the surface of the sheet of foam material. This feature is more clearly shown in Figure 5.2b, wherein the abrasive belt **19** is driven in the direction shown by the arrows and the foam material **10** moves into the belt from left to right. This view is exaggerated to show the manner in which abrasion takes place. Because of the multitude of sharp cutting points **33** on the abrasive belt, cutting takes place without generation of the compression wave found in knife cutting and the surface of the foam layer is sheared or abraded in a relatively noncompressive manner by the teeth of the abrasive media and while in a substantially uncompressed state. This results in a surface **16"** which is much more uniform in its compressibility than the somewhat rough, undulating surface **16'** which was formed previously by knife cutting.

Preferably, the grit size of the abrasive media on belt **19** lies in the range of grits 50 to 240 with the optimum for most foam sheets being from grit 120 to grit 180. While any abrasive grain material can be used, silicon carbide grain is preferred due to its extremely sharp cutting edges.

In practice, the feed rate of the foam layer can be generally about 65 to 95 feet per minute. The abrasive belt speed or speed of the abrasive particles relative to the foam surface being abraded is from 2,000 to 6,500 surface feet per minute with a preferred range of from 5,000 to 5,500 surface feet per minute. The back-up roll used to support the foam lamination during the abrasion step may vary from 1½ to 14 inches.

Graphite Outer Layer to Reduce Electrostatic Change

Flexible coated abrasive articles generally comprise a flexible backing material, usually made of woven or nonwoven fibers or paper or a combination of these materials, one side of which is provided with adhesively anchored abrasive grits. When these articles are used in an endless belt machine to sand wood and plastic, and often when they are used to grind some metals, such as light metal alloys, voltages in the order of magnitude from 50 to 100 kilovolts may be generated. The voltage is generated where the endless belt travels between a back support and the workpiece.

The high electrostatic voltage is of considerable inconvenience. A sudden discharge may take place and cause a serious accident by shocking the workman to a sudden, uncontrolled movement. The charge also causes the working space to become dusty and dirty due to electrostatic accumulation of dust on machines and workpieces, which, due to their electrical charge, are not removed by the usual suction means. If the dust is very fine, it may also be held in suspension in the air for a very long time, contaminating the entire working area and presenting an obvious health hazard as well.

It is an object of the process of *E.L. Markoo, H.S. Ekblom and T.W. Sandell; U.S. Patent 3,992,178; November 16, 1976; assigned to Fabrika AB EKA, Sweden* to provide an abrasive article that has antistatic properties and can be manufactured using conventional production equipment.

The abrasive article may be prepared by coating the outer surface of the abrasive side of a conventional abrasive article with the particles of the electroconductive material so as to form a layer of particles of electroconductive material over the abrasive grits. The resulting antielectrostatic layer provides a strong, electroconductive cover over the abrasive article. After use, when the points of the abrasive material have been laid bare, a network of electroconductive material remains between the grits. This network is as effective for the elimination of electrostatic charges as is a continuous layer.

The antistatic layer comprises an adhesive containing graphite particles having a size from 0.1 to 100 microns, the amount of graphite being from 1 to 100 grams per square meter of the abrasive article and sufficient to provide the layer with an electrical resistance of less than 5 megohm as measured between points 1 centimeter apart.

Alternatively, graphite may be coated over the abrasive article without any glue or other binder, and heated to 100° to 130°C for 1 to 3 minutes to secure the graphite to the abrasive article.

Electroconductive and Insulating Layers

It has been found by *E.L. Markóö, T.G.H. Strand and T.W. Sandell; U.S. Patent 3,942,959; March 9, 1976; assigned to Fabriksaktiebolaget EKA, Sweden* that in a conventional multilayered abrasive article of the kind normally used for grinding or polishing wood, in particular of the endless, flexible belt type, the electrostatic surface charge may be reduced from a typical voltage of 10 kV or higher to a residue voltage of about 50 V or lower.

This can be accomplished by including in the abrasive article, at any level thereof except an outer layer, a layer of a material or compound having good electrical conduction properties. The layer of conducting material or compound does not have to be "conductive" in the usual sense of the word; i.e., it does not have to be a metallic conductor but may be, for example, metallic pigment or an electrolytically ionizable salt. On both sides of this layer will be an insulating layer of significantly lower electroconductivity.

Referring to Figure 5.3, there is shown in cross section a flexible abrasive article according to the process. A backing member **1** is covered on one side with a layer of adhesive **2** in which abrasive grains **4** are embedded or anchored with a second layer **3** of glue partly filling the space between the grains, leaving the abrading points free. So far the abrasive is of conventional design.

Over the layer **3** of glue, however, there is shown a further layer **6** of a material having comparatively high electroconductivity. This layer can comprise metallic particles, metal pigments or metal salts or a combination of these.

One skilled in the art need only select a material or compound for this layer meeting the following criteria: (1) The material or compound should have essentially similar mechanical properties as the glue or adhesive normally used for anchoring the abrasive grains; i.e., when dried or cured it can be somewhat elastic but should preferably crack freely and have no tendency of peeling from its bond to layer **3**.

(2) The material or compound should have a comparatively high electric conductance but need not be anything like a metallic conductor. As a matter of fact, metallic and ionic conductors seem to work equally well in meeting the object of this process.

(3) The material or compound should have no adverse chemical effects on the glue and other materials of the abrasive article with which it may come in contact.

(4) The material or compound should preferably be cheap and easy to apply in the same manner with which glue is applied to the abrasive article in its manufacture.

(5) The material or compound should preferably require a short curing time.

Figure 5.3: Electroconductive and Insulating Layers

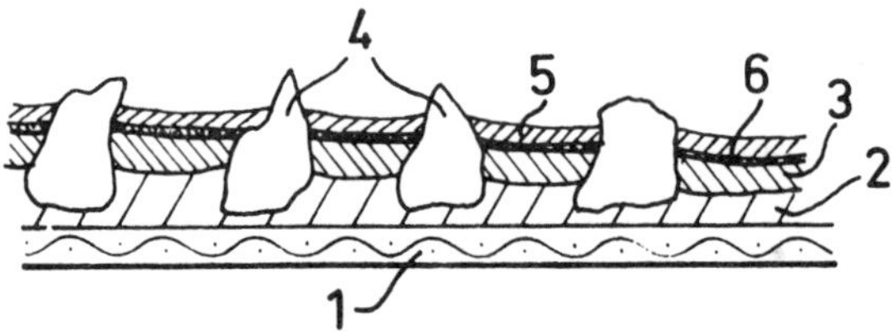

Source: U.S. Patent 3,942,959

The conductive layer **6** is covered with a protective layer **5** of a material or compound having comparatively less electric conductivity. It is usually sufficient to use for layer **5** the same glue as used for anchoring the abrasive grains **4**, in the layers **2** and **3**, without, of course, any additives increasing conductivity.

Abrasive Grains in Walls of Hollow Bodies

According to the process of *E. Wagner; U.S. Patent 3,928,949; December 30, 1975; assigned to Norddeutsche Schleifmittel-Industrie Christiansen & Co., Germany* there is provided a grinding material comprising a multiplicity of hollow bodies whose walls are formed of abrasive grains and a bonding means and are arranged to be stable in resistance to grinding forces.

Preferably all the abrasive grains forming the grinding material are disposed in the hollow body walls.

Thus there is provided a grinding material having a large pore volume without the mechanical strength of a grinding tool made from the material being endangered by a diminution of the grain coherency. The pore volume is not restricted by the size of the abrasive grains, since even grinding materials with fine grains can be provided with pores of large individual volume. The size of the individual pores and their distance from the individual abrasive grains can be made so as to vary statistically only within narrow limits, and the individual pores can lie in a predetermined size range, their distances from the individual grains being equal within narrow limits.

In this way, a grinding material can be obtained whose abrasive properties, insofar as they are determined by the grain size and the pore volume, are predetermined over wide ranges, as is the case with known grinding materials.

The thickness of the abrasive-grain-holding wall of the hollow body may lie in particular in the range up to 10, but preferably up to 3, abrasive grain diameters. As dense an abrasive grain arrangement as possible is sought. In most cases a substantially single-layer arrangement of the abrasive grain in the walls of the hollow body is sufficient.

It has therefore been sought to produce as stable and dense as possible a hollow body sheath consisting of abrasive grains and a suitable bonding material. The arrangement is all the denser or more tightly-packed, the less is the distance apart of the abrasive grains, and the more completely the wall surface appears coated with the abrasive grains. A coating of more than 50% of the surface available in each abrasive grain layer is aimed at. The maximum, where all the grains are in contact with one another, lies at 80%.

A hollow-spherical construction has proved particularly favorable with those grinding materials which have to stand up to high grinding pressure or resist powerful centrifugal forces, and also with those grinding materials where high requirements are set regarding uniform particle size, such as for instance abrasive or polishing belts.

The hollow bodies may consist exclusively of a closed wall formed from abrasive grain and bonding means. Such hollow bodies are obtained for example by blowing up granulates consisting of abrasive grain and bonding material or by ensheathing carrier particles which disappear as the bonding material binding the grain hardens out, as they evaporate or decompose. For example, ceramically-bonded abrasive grain sheaths can be built up on organic carrier particles which volatilize on combustion. Again when using synthetic resin as bonding material, which is subjected for hardening out to a high temperature, the carrier body can disappear or lose its original form if it is fluid or gaseous at the final hardening temperature. No harm is done however if the finished hollow body still contains completely or partially, a carrier particle, provided the carrier particles do not act unfavorably in the grinding process.

The size of the hollow bodies should be as uniform as possible. It should further lie in a given proportion to the mean diameter of the abrasive grain. It has proved advantageous if the mean hollow body overall diameter is 3 to 50 times, and in particular 6 to 20 times, as great as the mean abrasive grain diameter.

The mean overall diameter of the hollow body, measured parallel to the grinding surface, preferably lies between 0.1 to 8 mm, in particular between 0.2 and 2 mm.

In the production of the hollow bodies the preferable procedure is for carrier particles to be mixed in a tacky condition with the abrasive grain. The carrier particles can be brought into the tacky condition by themselves, i.e., without the addition of an adhesive, if for example, they are heated to their softening temperature before mixing with the abrasive grain, when for instance they consist of a thermoplastic material. Whether the carrier particles consist of a thermoplastic or a nonthermoplastic synthetic resin, they can also be mixed with the abrasive grain before their material is completely polymerized and they are therefore still tacky. In those cases in which the carrier particles themselves are tacky and soft, they are preferably mixed with the abrasive grain in a fluidized bed process.

Obviously this process is also applicable when the particles have been rendered tacky by wetting with a bonding material. However, in such a case it is usually simpler and therefore more advantageous to effect the mixing mechanically during the filling, for example, in a moving drum.

When using the hollow balls for producing a grinding body, the hollow balls can be processed in the same way as has long been done with the abrasive grains, i.e., during the production of a grinding body they are mixed with a bonding material and brought into the shape of the grinding body, after which the bonding material is allowed to harden out, and during the production of an abrasive belt the hollow bodies are bonded in the usual manner to a base material.

Example: Hollow beads which have been produced by expanding vinyl chloride/ethylene copolymers in accordance with Belgian Patent 702,673, with diameters from 0.05 to 1.00 mm, were wetted in a mixer with a bonding material consisting of a 40% solution of an epoxy resin of the bisphenol-A-epichlorohydrin type. The resin had an epoxy value of 0.200; the hardener was the polyamino-amide Versamid 115. The solvent mixture consisted of 74% xylene, 13% butanol and 30% diacetone alcohol.

The hollow beads wetted with bonding material were mixed with electrocorundum of grain size 400 (mean grain diameter 35 μm) so that the abrasive grains are fixed substantially in one layer and tightly packed on the ball shell. The ratio by weight of hollow beads to solid bonding material to corundum is 1:10:100. The abrasive grain balls were carefully dried and screened, only balls with a mean diameter of 0.425 mm being used.

A cotton twill fabric conventionally provided for producing abrasive belts was uniformly strewn, after the application of a layer of a phenol-formaldehyde resin, with the ready-prepared abrasive grain balls, which were given an intermediate drying in the usual manner, and, after being coated over with a further thin layer of the same resin, but thinned this time, hardened until final definitive anchoring of the hollow balls was obtained. After the completion of the hardening which was carried out at temperatures of more than 120°C, the thin skin of the bead was melted, leaving only the bead sheath consisting of abrasive grain and bonding material.

Two Binders with Differing Heat Resistance

C.W. Marshall, L.A. Henderson and W.F. Zimmer, Jr.; U.S. Patent 3,906,684; September 23, 1975; assigned to Norton Company have discovered an abrasive mixture and a process for using it to manufacture an improved abrasive article which not only remains free flowing, thus making it particularly suitable for storage and the continuous coating onto a flexible backing member, but which also allows the manufacture of an abrasive material in the nature of a coated abrasive material having an abrasive layer of improved cutting and wearing characteristics.

The abrasive mixture, on formation of a layer thereof and after curing, comprises basically abrasive grain coated with a binder in a matrix binder of greater thermal resistance. Thus, the binder forming the abrasive grain socket or coating, on being subjected to heat and pressure, deforms and permits the abrasive grain on meeting sufficient resistance to penetration by the workpiece to be picked out of and shed from the abrasive layer.

Referring to Figure 5.4a, there is disclosed a portion of abrasive material **10** having a physical appearance somewhat similar to that of conventional coated abrasive material. Abrasive material **10** has a flexible backing member **11** on the front side of which is provided an adhesive layer **12** which anchors a porous, compacted abrasive layer **13** to the backing member.

The backing member **11** may be any of various flexible materials conventionally used in the manufacture of coated abrasives. Particularly good results have been obtained in backing members of woven natural fibrous material.

Adhesive layer **12** is a thin layer of suitable adhesive material preferably a rigid, high modulus thermosetting resin, the more preferred one being a heat-hardened phenol-formaldehyde resinous composition.

Abrasive layer **13**, contrary to that conventionally found in coated abrasive articles, has abrasive grains **14** unoriented and distributed throughout the abrasive layer. The abrasive grains are surrounded by or encapsulated in a socket **15** of adhesive binder. These binder coated abrasive grains are in turn bonded together in abrasive layer **13** in a matrix **16** of binder. The amount of binder matrix is insufficient, however, to fill the interstitial space between the abrasive grains thus leaving pores **17** in the abrasive layer. The structural configuration of the abrasive layer is better seen in the greatly enlarged view of a section thereof in Figure 5.4b.

Figure 5.4: Abrasive with Combination of Binders

a.

b.

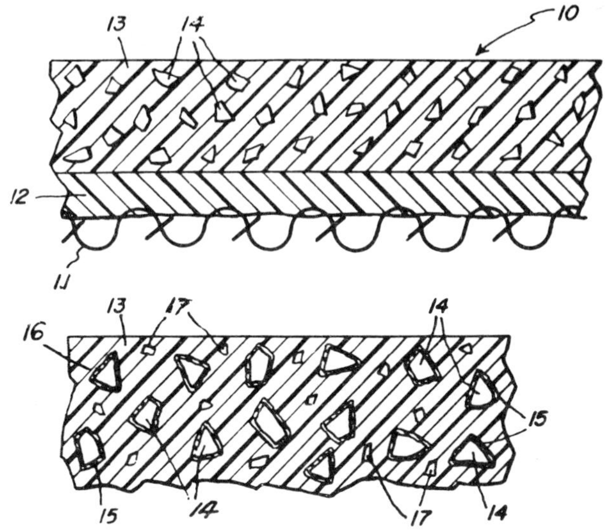

Source: U.S. Patent 3,906,684

Binder material **15** surrounding the abrasive grains, although desirably somewhat heat resistant, is, at the temperatures usually attained during coated abrasive usage, distortable, particularly under pressure. This feature, i.e., the thermal distortable nature of the binder material, results in breakdown of the abrasive layer. In application of pressure during grinding, abrasive grains **14**, as they become excessively dull, become hotter. This in turn imparts more heat to binder material **15** and permits abrasive grains **14** to be picked out, so to speak, on distortion of the binder **15** from the socket formed thereby.

The picked out abrasive grains are then shed thus providing new cutting surfaces. Binder matrix **16**, meanwhile, remains heat resistant at the grinding temperatures attained and, moreover, tends to resist fracture. Nevertheless, with this combination of binder materials, i.e., one more heat resistant than the other, an abrasive layer is obtained having the desired combination of performance characteristics.

Various binder material combinations may be used so long as they possess the necessary differences in thermal resistance. A particularly useful combination has been found in using an epoxy resin in conjunction with a phenol-formaldehyde resinous composition. A suitable, and one of the preferred epoxy resins for use in this process is Epon 812. Such an epoxy resin is preferred because of its relatively low viscosity (9 to 15 poises) and its rather high level of solubility of powdered phenol-formaldehyde resin. Other epoxy resins of higher viscosity may be used, for example, Epon 1009. Regardless of what binder material is used, however, to coat the abrasive grain, it must be less heat resistant than the matrix binder and, in general, will be deformable at temperature above about 240°F.

For the matrix binder, a two stage, heat reactive, powdered resinous phenolic composition, as mentioned before, is preferred. One such material, which can be used with good results in combination with Epon 812 is BRP5417. This material, once cured, is heat resistant and will char, i.e., degrade and pyrolyze, at about 500°F.

Although the amount of and ratio of different binder compositions will depend upon the particular components used in combination, where the combination is the abovementioned epoxy-phenolic combination, the epoxy component should be at least about 0.6% by weight of the abrasive grain. In any event, the amount of binder should be sufficient to provide a coating around, or socket for, the abrasive grains.

Carrier Partially Embedded in Backing

The process of *D.H.G. Prowse and S.D. Prowse; U.S. Patent 3,860,400; January 14, 1975; assigned to D.H. Prowse and Company Limited, England* provides an abrasive article comprising abrasive carrying material combined with non-abrasive carrying material in a manner whereby spaced portions only of the carrying material are exposed from the noncarrying material on one surface of the article and abrasive particles secured only to the exposed portions of the carrying material.

By securing the abrasive particles only to the exposed portions of the carrying material, control of the distribution of abrasive particles over an area can be obtained. Further, where the carrying material is a flexible mesh or perforated sheet, and the abrasive is in the form of diamond particles to be plated onto the mesh, by exposing only portions of the mesh on one surface of the backing member and plating the diamond particles to these portions only, the plating process does not substantially affect the flexibility of the mesh intersections, so that if a flexible backing member is used, a flexible sheet of diamond plated mesh can be obtained.

Numerous techniques may be employed for combining the materials so that only portions of the carrying material are exposed on one surface of the noncarrying material. For example, a sheet of carrying material may be embedded in a backing member by extrusion or by pressing the sheet into the backing member, i.e., by vulcanization. The sheet may be coated or sprayed with backing member material or the sheet can be dipped into molten backing member material or simply adhered to the backing member with solvent or adhesive.

The abrasive particles may comprise diamond particles either natural or synthetic, cubic boron nitride, tungsten carbide, glass, sand or any other abrasive or combination thereof or the abrasive particles may comprise conglomerates of abrasive particles impregnated into a carrying medium. To attach the abrasive particles to the exposed portions only of the carrying material it is preferred, particularly in the case of a diamond abrasive which it is envisaged will be the most useful form of abrasive for articles according to this process, to plate the particles to the carrying material, by conventional electroplating techniques where a metal carrying sheet is used with a nonconductive backing member or for a nonmetallic carrying sheet first by chemically depositing a metal onto the exposed sheet portions followed by an electroplating process for the abrasive particles.

Figure 5.5 shows carrying material in the form of a sheet of wire mesh **1** embedded in noncarrying material in the form of a backing member **2** of plastic material in such a manner that only regularly spaced peak portions **3** of the mesh at its intersections are exposed on the upper surface of the backing member. These exposed portions of the mesh are plated with abrasive particles **4**. This form of the process is useful for grinding wheels or as an abrasive covering or belt, if the backing member is flexible.

Figure 5.5: Carrier Partially Embedded in Backing

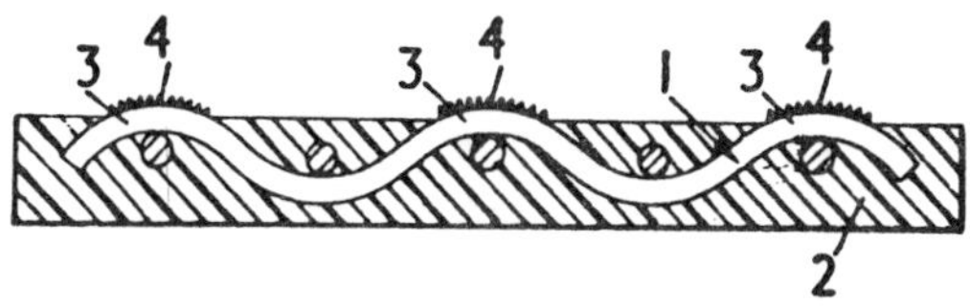

Source: U.S. Patent 3,860,400

Etched Abrasive with Intermixed Cutting Points and Planing Edges

J.J. Frantzen; U.S. Patent 4,240,806; December 23, 1980; assigned to Minnesota Micro Metal, Inc. provides a metal abrading material formed from acid etching with a resist pattern comprising a plurality of polygonal shapes with accentuated corners dispersed in a pattern chosen to intermix cutting points with planing edges.

In Figure 5.6a, one embodiment of the resist pattern is shown. The pattern includes a number of elements of resist **10** which are formed in the shape of three or four sided polygons having accentuated corners. In Figures 5.6a and 5.6b, a triangular shape is shown with accentuated corners **12**. In Figure 5.6c, a four sided polygon or square **14** is shown for the element of resist with accentuated corners **16**. Typically the chosen pattern of Figure 5.6a is multiplied many times over and applied to a suitable metallic base **18** as shown in Figure 5.6d. The base **18** with the resist elements **10** thereon is then subjected to an etching spray from a suitable nozzle in a manner well known to those skilled in the art. If a triangle type resist element is employed the tooth **14** shown in Figure 5.6e results after the residual resist is removed. As shown in Figure 5.6f tooth **24** has sharp cutting points **26** which operate to quickly remove material during the abrasion process.

Figure 5.6: Etched Abrasive with Intermixed Cutting Points and Planing Edges

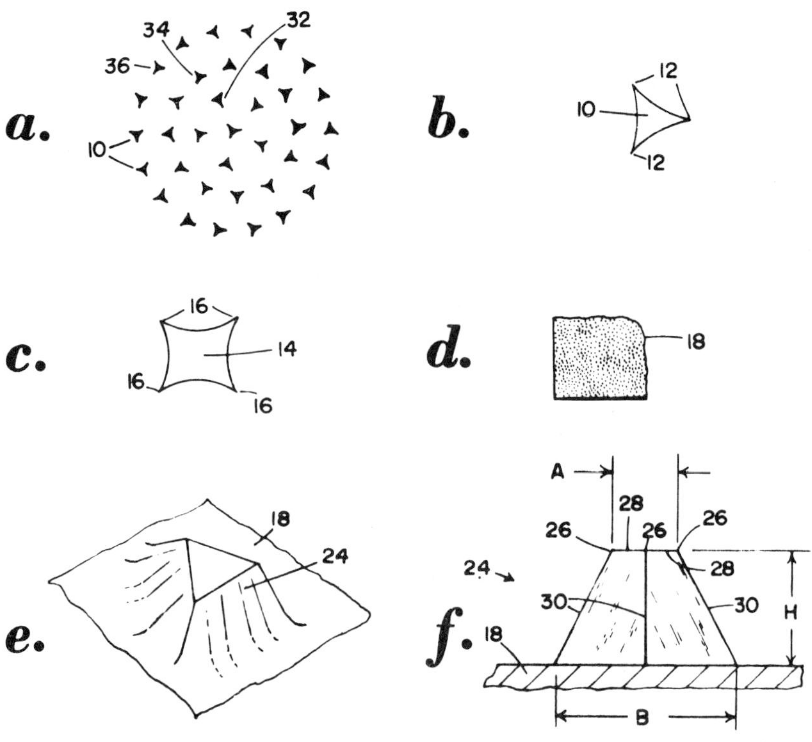

Source: U.S. Patent 4,240,806

However, tooth **24** also has sharp planing edges **28** which simultaneously smooth the roughened surface and remove material between the gouged tracks made by the points **26**. To properly intermix the fast cutting points **26** and the smooth planing edges **28**, the pattern of Figure 5.6a is carefully arranged so that the elements of resist are all positioned in slightly different directions.

For example, in Figure 5.6a, it will be noted that the first concentric ring of resist elements **10** designated by the numeral **32** point inward while the second concentric ring of elements **34** all point outward. The third concentric ring of elements **36** are tilted relative to the first two rings. This randomizing process may be continued outward and then repeated numerous times to form the pattern shown in Figure 5.6d. In this way all of the elements of resist and all of the resulting teeth are randomly oriented so that an even intermixture of deep cutting points and smooth planing edges is assured.

To assure that the teeth remain strong and tough they are hardened by heat treating in a conventional manner to produce a hardened layer. The shape of the resist pattern and the etching process are carefully controlled to leave a substantial amount of nonhardened metal inside the tooth so as to properly support the cutting points and the planing edges.

Typically, dimension **A** in Figure 5.6f is about 3 mils with the width of the base **B** being about twice **A**. The height of the tooth **24**, dimension **H**, is also about twice **A**. Because of the accentuated corners in the resist pattern and with proper etching control, the edges of the tooth are kept relatively straight, extending from the corners outward to the base at an angle sufficient to provide good mechanical support to the cutting corners **26** and the planing edges **28** extending therebetween. This mechanical support resists the forces of abrasion which is especially important when the tooth is hardened and becomes more brittle.

Polishing and Finishing

METAL FINISHING AND POLISHING

Plate for Polishing Samples for Microscopic Examination

The process of *S. Molnar and W. Rieger; U.S. Patent 4,162,899; July 31, 1979; assigned to Swiss Aluminum Ltd., Switzerland* concerns a polishing foil or plate which is used together with a polishing substance for preparing the surfaces of materials, in particular for polishing metallographic, petrographic and ceramographic samples.

The object of the process is to produce a polishing foil or plate which is able to ensure optimum polishing results, at the same time consuming the minimum of polishing substance and with the minimum of edge rounding, and which can also polish hard materials made up of a plurality of layers.

This object is achieved by way of the process in which a metallic base is provided with a nonmetallic layer which acts as a polishing substrate, covers the polishing surface and is intimately bonded to the metallic base.

By metallic base is to be understood the metal layer which carries the nonmetallic layer which serves as the polishing substrate. The metallic layer is, depending on the type of polishing plate desired, from several hundred angstroms to some centimeters thick.

Any metal, the surface of which can be provided with a nonmetallic layer by chemical, electrochemical or physical means, can be employed for the metallic base. It is made preferably from one of the metals of the group aluminum, zinc and iron, whereby both the pure metal and the alloys of these metals are included in this choice.

The metallic base is preferably in the form of foil or sheet (plate). To achieve specific properties, however, the base can be produced by suitable chemical or physical processes, e.g., by evaporating, electrolytic or nonelectrical deposition etc. on to another form. This other base could, for example, be metallic foil, sheet or plate, or foil and plate of a suitable plastic.

In a preferred version the metallic base consists of a foil which is bonded to a plastic foil in order to increase the mechanical stability, at the same time maintaining the flexibility. In a further version the foil is made self-adhesive for bonding to the polishing wheel.

Preferred nonmetallic layers are anodically or chemically produced oxide, phosphate, chromate, molybdate or oxalate layers. The polishing plate of the process exhibits the following significant advantages over the classical polishing cloths and plates.

The amount of time required to achieve comparable results is shortened.

In the case of hard materials (e.g., ceramics) the number of steps required is reduced and, correspondingly the number of pieces of equipment required is also reduced.

The outlay of consumable materials, in particular for polishing substances is considerably reduced.

Because no deformation occurs in the polishing plate of the process, the samples can be polished without problem with perfectly flat surfaces right up to their edges.

Just as conventional polishing cloths, discs and plates require matching of the polishing substance and the sample, with the polishing plate of the process the type of substrate, the metallic base, the nonmetallic layer and the thicknesses of these are chosen in keeping with the requirements of the sample to be polished.

Example: A sample of a surface treated 50 CrV 4 spring steel, which had a Vickers hardness of 620 kp/mm^2 was mounted in acrylic resin and prepared by wet grinding for subsequent polishing.

The polishing plate comprised a 0.1 mm thick aluminum foil which had been provided with a 3 μm thick molybdate layer and had been adhesively bonded to a 0.5 mm thick plastic foil.

The molybdate layer was produced by a conventional two-step process. The first step was activation of the aluminum foil in a solution of the following composition in grams per liter and at room temperature.

Zinc oxide	20
Sodium hydroxide	120
Potassium sodium tartrate	50
Iron(III) chloride	2
Sodium nitrate	1

The second step consists of deposition of the molybdate layer in 3.5 min in a solution of the following composition in grams per liter and at room temperature.

Ammonium heptamolybdate	50
Ammonium borofluoride	15
Triethanolamine	60

The polishing plate was stuck down onto the polishing wheel of a conventional polishing device and provided with 0.25 μm diamond paste before starting to polish. A commercially available lubricant was dripped continuously onto the polishing surface during the polishing operation.

After 1.5 hr the surface to be examined had been polished perfectly flat right up to the edges and with almost no signs of scratches.

Flowable Abrasive Composition

K.E. Perry; U.S. Patents 4,087,943; May 9, 1978; and 3,909,217; September 30, 1975; both assigned to Winfield Brooks Company, Inc. describes a flowable abrasive composition comprising a polyboronorgano silicon compound and dispersed abrasive granules. The composition is modified with a gel formed from a metallic soap (8 to 22 carbon atoms) and a hydrocarbon oil. The composition is especially useful for honing, abrading, deburring or polishing surfaces, especially metallic surfaces.

The gel used to modify the polymer is known in the art and formed by reaction of a metal soap with a hydrocarbon oil.

The metallic soap component of the gel is a salt, preferably substantially water insoluble, of a fatty carboxylic acid usually having from 8 to 22 carbon atoms. Typical examples include aluminum, beryllium, cadmium, calcium, lithium, magnesium and zinc salts of acids such as lauric acid, stearic acid, oleic acid, linoleic acid, palmatic acid, behenic acid and mixtures thereof. Frequently, these salts or fatty acids or metallic soaps are by common usage referred to as metallic stearates which term includes not only the insoluble metallic salts of stearic acid but also a wide variety of other high molecular weight organic acids as well. The aluminum salts are preferred as they form the most stable gels, and those materials known in the art as aluminum stearates are most preferred.

The hydrocarbon oil, preferably an aliphatic oil, used as the second component in the formation of the gel is preferably one of the paraffinic oils, but may be any of the other hydrocarbon oils known to form gels with metallic soaps such as one of the naphthenic oils.

The gel is formed by mixing from 1 to 35% by wt, preferably from 3 to 15% of the metallic soap with the hydrocarbon oil together and heating the mixture to a temperature in excess of the gelation temperature, preferably to between 10° and 20°F in excess of the gelation temperature and holding at this temperature for a period of at least 15 min.

The polymer is made by reacting a silicone oil, e.g., polydimethylsiloxane with a boron compound, e.g., pyroboric acid, boric acid anhydride and others in the presence of a catalyst, i.e., a Lewis acid catalyst, with heat and agitation followed by treatment with ammonium carbonate or ammonia to neutralize any residual Lewis acid catalyst.

The abrasive component of the composition or grit is of any conventional abrasive material used in such honing formulations, i.e., silicon carbide, alumina, boron carbide and such. The size of the grit may vary over a wide range depending on the job to be done and the physical properties desired in the composition. Thus, the size may vary from 20 to 600 mesh, and in some cases, it may be as low as 2 to 5 μ. It has been found that often larger abrasive sizes are employed for heavier work. Substantially better results are achieved by using finer abrasive sizes in combination with the larger particles.

The amount of abrasive relative to the remainder of the composition (gel and the polymer) is at least 25% by wt of the total composition, preferably varies between 50 and 300% by wt of the composition and most preferably varies between 100 and 190% by wt of the composition.

To provide the flowable abrasive composition in accordance with the process, the polymer is thoroughly kneaded together with the gel modifier until it is homogenous and has the desired ductility, softness, flowability and viscosity. The grit or abrasive is then thoroughly and intimately mixed with the composition with stirring to provide a composition ready for use.

Example: A polymer was formed from 50 lb of a dimethyl silicone oil, one-half pound of $FeCl_3 \cdot 6H_2O$ (Lewis acid catalyst) and 3 lb of pyroboric acid by reaction at a temperature of 200° to 250°C with agitation to form a highly viscous boron organosilicon heteropolymer. This was reacted with 1.5 lb of $(NH_4)_2CO_3$ to neutralize any acid residue. The resulting material was somewhat brittle, stiff and crumbly.

A gel was formed by adding 3.6 lb of aluminum stearate (No. 801, S.B. Penick and Co.) to 60 lb of hydrocarbon oil (Drakol 9) at room temperature with stirring. The mixture was then heated to the gelation temperature of 194°F, heated further to 240°F and held at that temperature for 15 min to form a relatively thick gel.

20 lb of the polymer were mixed with 30 lb of the gel at a temperature of 215°F. After thorough mixing, the mixture was cooled and kneaded until it was homogenous.

Thereafter, 75 lb of silicon carbide abrasive particles having an average particle size of 220 mesh were mixed with the polymer composition. At this point, the mixture had a pastelike consistency.

The resulting abrasive mixture was then pumped at a pressure of 450 psi back and forth 10 times through the one-eighth inch holes of test washers made of steel; the pressure drop through these holes causing the mixture to squirt through the holes at an extremely high velocity.

The resultant action was complete removal of burrs from the edges of the hole and excellent surface finish on the wall of the hole.

Chemically Impregnating Refractory Oxide Base Materials

The process of *P.K. Church and O.J. Knutson; U.S. Patent 4,007,020; Feb. 8, 1977; assigned to Kaman Sciences Corporation* involves materials that are formed by multiple chemical impregnations of a relatively soft porous body of finely divided refractory oxide base materials, each followed by a low temperature cure to convert the impregnant to an oxide. The resulting ceramic structure formed in this manner has been shown to exhibit extreme hardness, a high compressive strength and a dimensionally stable material over a wide temperature range. In addition, a number of these ceramic materials show an inherently small coefficient of friction coupled with a very low wear rate characteristic.

These ceramic materials will withstand repeated water quenching from 1000°F as well as prolonged exposure to temperature extremes of 2000° to -300°F. Mohs scale hardness is in excess of 9, normally being about equal to that of silicon carbide. Rockwell hardness can be as high as A-85 to A-90, with associated compressive strengths in excess of 125,000 psi.

A variation in the chemical hardening and salt impregnation method of the process results in an excellent abrasive material for the honing and finishing of metals. When treated correctly, such an abrasive stone will remove metal very rapidly in addition to having the feature of not loading the stone with metal particles.

The basic material found to be most satisfactory for this purpose is Coors AHP-99 aluminum oxide, although others have been found to be satisfactory. After machining to the desired shape, the ceramic material is first treated with a reduced strength or starved phosphoric acid solution of less than 85% and then cured at at least 600°F. Following this, a salt impregnation may be added and the piece recured at a temperature sufficient to convert the salt to an oxide. (For standard ceramic pieces, the salt impregnation is made prior to the chemical hardening treatment.) This procedure has been found to retain a high degree of porosity, in addition to providing the necessary low loading of the pores and rapid cutting properties. 85% phosphoric acid saturated with aluminum phosphate will also provide the necessary starved acid solution.

The most significant advantages of this new abrasive material should be in the fact that the desired shapes and close tolerances can be generated prior to the hardening of the abrasive. When using sintering techniques in the production of abrasives, a high proportion of the expense is in the grinding and finishing operations. A further advantage of the process may be in the low temperature-short time cycle involved in the curing and hardening process, as compared to the presently employed high temperature sintering method.

Accelerator Solution for Barrel Finishing

The process of *D.E. Semones and W.H. Safranek; U.S. Patent 3,979,858; September 14, 1976; assigned to International Lead Zinc Research Organization, Inc.* relates to the barrel finishing or vibratory finishing of metal objects. In the process, metal workpieces are placed in a rotating barrel or vibratory machine which contains abrasive media and an aqueous solution of a lower aliphatic acid having at least 2 and not more than 6 carbon atoms. The accelerator solution may also contain sodium bisulfate.

In accordance with the process, metal workpieces, such as zinc die castings, are subjected to vibratory action in a machine containing abrasive media and a chemical accelerator solution, or are agitated in a rotating barrel containing abrasive media and a chemical accelerator solution. Typically, zinc die castings are alloys containing predominantly zinc and minor amounts of other metals such as aluminum, copper or magnesium.

Included among the preferred accelerator acids are the dicarboxylic and hydroxy acids, such as tartaric, citric, glycolic, lactic, malic and maleic. These acids may be characterized as having at least 2 carboxyl groups or as having at least 1 carboxyl and 1 hydroxyl group. The accelerator solution contains 5 to 25 g of aliphatic acid per liter of solution. The pH of the solution is maintained in the range of 1.1 to 1.9. Sodium bisulfate may be added to the solution at a concentration in the range of 1 to 40 g/ℓ of solution as a pH buffer to maintain the pH range above. This has the effect of maximizing the rate at which metal is removed.

Abrasive media which may be used in the process include ceramic, quartz, aluminum oxide, carborundum and resin-bonded abrasive particles. These particles may range in size from 5 to 100 μ in diameter, preferably 5 to 10 μ in diameter, embedded in preformed cubes or cones of plastic material, such as polyester.

The vibratory finishing machines used may vibrate in the range of 1,200 to 3,600 cpm with an amplitude of 0.0625" to 0.25". Using the chemical accelerator described, the surfaces of the zinc metal workpieces can be polished to a surface finish of 3 to 5 microinches (0.075 to 0.125 μ) after 10 to 60 min of controlled vibration. The process results in metal removal rates of approximately 0.10 mil (2.5 μ) to 3.0 mils (75.0 μ) per hour. These rates are 2 to 40 times faster than the metal removal rate for vibratory finishing in the absence of a chemical accelerator.

In the processs described, barrel finishing machines with diameters of 6" to 18" typically are rotated in the range of 20 to 60 cpm; barrels with diameters of 36" to 48" generally are rotated at 5 to 15 cpm. Using a sodium bisulfate-tartaric acid accelerator, the metal removal rate in barrel finishing is increased from less than 0.01 to about 0.3 mil per hour.

Example: An electromagnetic tub with a capacity of 2 cubic feet designed to vibrate a mixture of abrasive media and metal workpieces in a liquid accelerator solution at 3,600 cpm with an amplitude of 0.25" was used. Plastic cones with a height of ⅜" and a diameter of ¼" containing abrasive particles having diameters of 5 to 10 μ were added to the tub to provide a media to workpiece volume ratio of approximately 5 to 1.

The accelerator solution was prepared by dissolving 10 g of tartaric acid and 10 g of sodium bisulfate per liter of water. With concentrations of each compound at approximately 10 g/ℓ, a pH of 1.5 to 1.6 is obtained. At a pH above 1.9, which is obtained with lower concentrations of tartaric acid and sodium bisulfate, the metal finishing rate decreases considerably; below 1.1, which corresponds to higher concentrations of tartaric acid and sodium bisulfate, the highly acidic solution causes excess roughness on the metal surface.

The accelerator solution was added to the contents of the two cubic foot electromagnetic tub at the constant rate of 3.0 gph, equivalent to 1.5 gph/ft^3 and allowed to drain from the bottom of the tub at the same rate. Using the above-prepared accelerator solution with plastic media in the electromagnetic tub, a surface finish of 0.075 to 0.125 μ was obtained after subjecting zinc die castings to treatment for 30 min.

The metal removal rate was calculated at 0.5 mil (12.5 μ) per hour producing zinc workpieces which had smooth, smut-free surfaces suitable for subsequent plating. The metal removal rate was 7.6 times faster when the accelerator was flushed through the tub than the rate measured for the same load in the same machine flushed with a detergent solution instead of the accelerator solution. The accelerator solution contains the biodegradable substance tartaric acid; therefore, chemical treatment of the waste solution prior to disposal is not necessary.

Finish Grinding Steel Balls for Bearings

The ability to formulate compositions described below, that may be cold pressed to the densities indicated that fall within the enclosed area shown in Figure 6.1a enables *C.M. Jones; U.S. Patent 3,913,281; October 21, 1975; assigned to Norton Company* to produce a wheel with a specified modulus of elasticity after firing that has the proper durability and relative hardness or softness of cut to satisfy any of a customer's steel ball bearing finish grinding requirements. Further, this ability to design variations in durability and speed of cut can be realized by making use of abrasive grains of different grit sizes.

As shown in Figure 6.1c, for example, with 100 U.S. Standard screen fused crushed alumina grits, the modulus decreases somewhat more rapidly with increasing percentages of abrasive content relative to the weight of bond in the wheel as compared with the gentler slope of the curve when increasing percentage additions of 240 grit (corresponding to U.S. Department of Commerce Commercial Standard CS-371-65) fused crushed alumina grits are added to the given weight of bond in the wheel mix. The same kind of relation between percent abrasive addition and weight of bond exists when silicon carbide in any given grit size is used except that the slope of the curve showing a reduction in the modulus of elasticity is somewhat steeper than the curves shown for a correspondingly sized alumina grain. Abrasive grains of alumina oxide are preferred, however, because they are much more compatible with the disclosed ceramic bond composition than are silicon carbide grains.

Figure 6.1: Modulus of Elasticity vs Parameters of Process

a.

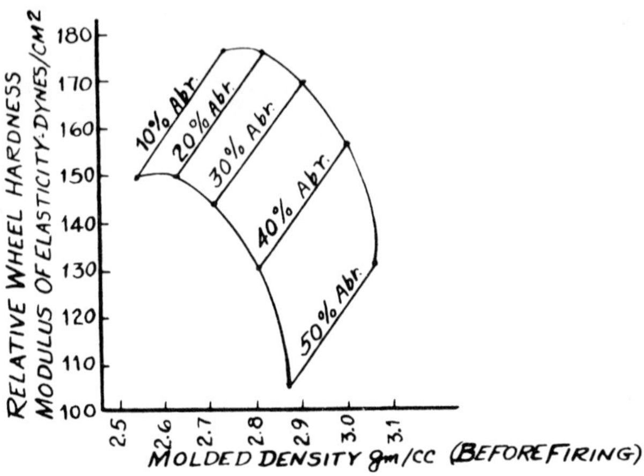

b.

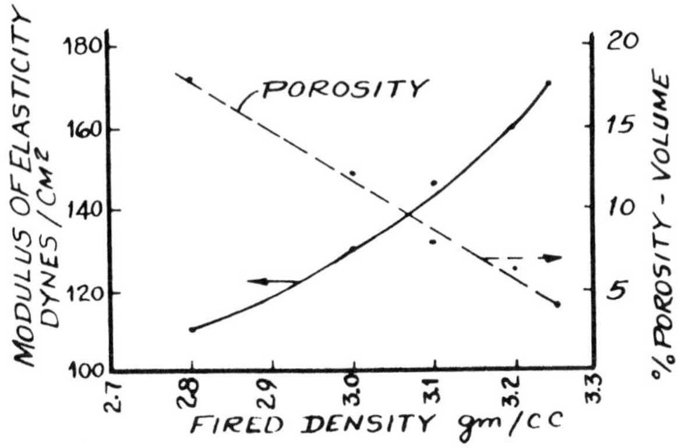

(continued)

Figure 6.1: (continued)

c.

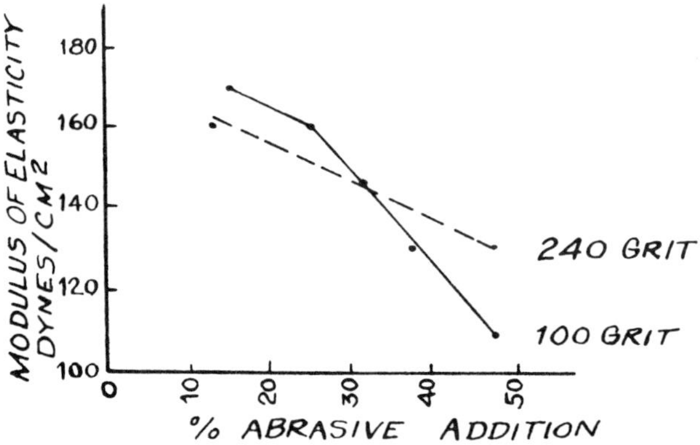

(a) Graph showing the relationship of the modulus of elasticity (times 10^{-10}) that will result upon firing "green" wheels that have been cold pressed to have the molded density shown

(b) Graph showing the relationship of fired density of the grinding wheel, modulus of elasticity and the volume percent of pores in the fired wheels of the process

(c) Graph showing variation of the modulus of elasticity in relation to variations in the percent of abrasive content relative to the weight of bond and the effect of different grit sizes of abrasive therein

Source: U.S. Patent 3,913,281

In Figure 6.1b the modulus of elasticity of the wheels is shown to be variable with respect to both the fired density as well as the porosity. These three characteristics of modulus of elasticity, density and porosity of the wheel vary in a regular manner. For any particular composition within the range of the process, the modulus and/or the volume percent porosity can be predicted from the density of the cold pressed green wheel, as shown in Figure 6.1a. The percent addition of the abrasive grit relative to the weight of bond within the conventional range of cold pressing pressures, determines the fired density so that a faster cutting softer wheel or slower harder cutting wheel can be produced as needed.

In order to control the properties of these steel ball bearing finishing wheels for any given molding pressure, there are four ingredients that can be varied in the wheel mix to control the properties in the fired product, namely: the clay of the bond, with a bentonite additive; a flux of manganese dioxide and soda or soda ash; a fine grit alumina crystalline additive; and the abrasive grit.

In formulating the bond for the wheels, one preferably starts with a conventional vitrifiable bonding clay that not only supplies alumina, silica and various low melting oxides in the fired bond that assist in the bonding of the abrasive grits but one that also gives some moldability during green wheel processing; that is, the clay compacts well and gives good green strength during the pressing operation. The type of clay is not critical but a conventional clay for bonding, a so-called ball clay, should be selected. For convenience and economics, Mississippi ball clay has been found to be entirely satisfactory.

To formulate a wheel, one starts with an amount of clay equal to about 20% by wt of the final mixture and preferably adds about 1% bentonite in order to give better compaction properties to the final mix that may be either cold or, in some cases, hot pressed, to form a wheel. The bentonite incidentally functions to lower the melting point of the clay bond mix, but the bond composition can be made without bentonite. When bentonite is used, it is added in percentages no greater than 4 or 5% of the weight of the final mixture simply because it is a clay that shrinks excessively upon firing. The ball clay content of the completed mix can be varied within the range of 10 to 30% by wt of the final mixture.

The second basic ingredient included in the mix is a manganese compound, preferably added as manganese dioxide, which upon firing, acts as a suitable fluxing agent for the clay. This flux material controls the densification of the product upon firing and the degree of density produced is related to the amount of flux put into the mix and the firing temperature as compared to the firing of the same mix with no manganese. For example, when no manganese is present, a hard dense wheel would be produced only if fired at a temperature of 1600°C or 1650°C. With a 5% manganese dioxide addition, the firing temperature can be lowered significantly. It is preferred to use a 7% manganese dioxide addition because that is the amount that fires most satisfactorily in a standard Orton cone 12 kiln.

Also soda is separately added as an ingredient of the filler material described below. The soda ash in an amount of about 1.5% of the dry weight of the final wheel mix serves as an additional fluxing component to make the bond more fluid. When soda is present in a selected filler ingredient of the bond, a separate addition may not be required provided the ultimate dry mix for the wheel includes approximately 1.5 to 1.6% soda content.

This glassy ceramic bond developed under the influence of the MnO_2 and soda forms a strong adhesion to the Al_2O_3 abrasive particles. It is known that a glue or adhesive must wet a surface in order to stick to the surface. This viscous glass forms at the lower temperatures under the influence of the flux and has such a degree of surface tension that it draws the abrasive particles together although it does not wet the abrasive surfaces in the normal manner of a glue. The viscous glassy bond is in fact also a powerful solvent for the Al_2O_3 abrasive so that the surface of the abrasive is dissolved by the fluid glass, thus insuring an intimate contact or wetting of the abrasive by the glass.

The third basic ingredient of the wheel of the process is crystalline alumina in the form of dust collector fines and/or Bayer process alumina. Such crystalline alumina will have some abrasive characteristics when the fired wheel is in use and it is added to the mix of the process primarily as part of the bond material. Also, the component of the mix gives the essentially crystalline bond its hard, durable, long-wearing properties as well as supplying some of the abrasive action of aluminum oxide crystals for grinding the balls.

Any crystalline form of fine grit aluminum oxide having particles sometimes as large as 100 μ, but mostly in a size range of under 50 μ, can be used for this purpose in the mix with the average particle size in the range of from 7 to 12 μ. The alumina filler is present in an amount of from about 15 to 60% of the dry weight of the final mix.

The fourth ingredient to be added to the mix is the primary abrasive grits, which material serves a two-fold function. First, the larger abrasive grits do increase the rate of stock removal in the grinding of steel balls for bearings. Secondly, the larger grits of fused crushed aluminum oxide serve in conjunction with the bond mix to control the modulus of elasticity by controlling the degree of porosity resulting from firing the wheel composition of the process. A percentage range of larger sized abrasive grit additions is shown clearly in Figures 6.1a and 6.1c.

The compositions are mixed, and then are preferably cold pressed in a mold within a range of from ½ to 1 ton pressure per square inch, are then shaved and dried, utilizing conventional grinding wheel manufacturing procedures.

Pressures within a range of from $\frac{2}{10}$ ton/in^2 to 2 tons/in^2 can be used. If hot pressing facilities are available pressures within a range of from 500 psi and lower should be used to accommodate the strength of the mold. The firing procedure for these compositions is similar to those used for firing conventional ball wheel compositions. In the preferred processing, tunnel kilns are used, the wheels to be fired being set in sand on a suitable batt to be fed through the kiln in a normal procedure at about Orton Cone 12.

While the wheels described herein do shrink from 5 to 9% lineally, the geometric integrity of the wheel is maintained and there is no obvious melting or slumping, bubbling or frothing or any evidence whatever that the bond melts or liquefies during the firing process. It should be noted, in fact, that all blemishes, chips, marks, scratches and such that are on the surface of these wheels in the unfired state will be clearly evident in the wheels after they are fired.

A steel ball bearing grinding wheel of the composition disclosed herein with 10% weight addition of fused crushed 100 mesh alumina abrasive grits has been found to last 4 times as long as a conventional wheel would be expected to last with the same abrasive grit, size and content.

POLISHING SEMICONDUCTOR AND PHOTORECEPTOR SURFACES

Ferromagnetic Pellets

N.S. Khomich, F.J. Sakulevich, B.N. Putimtsev and V.A. Silaev; U.S. Patent 4,187,084; February 5, 1980 prepare a ferromagnetic abrasive material containing a ferromagnetic component and an abrasive component. The ferromagnetic component includes an iron-aluminum-silicon-phosphorus-carbon alloy, and the abrasive component includes aluminum and silicon oxides, carbides, nitrides and phosphides with the following proportions of the ingredients in percent by weight:

Iron	75-97
Aluminum	0.2-16.0
Silicon	0.2-18.0
Phosphorus	0.005-1.5
Carbon	0.005-4.0
Oxygen	0.05-3.0
Nitrogen	0.01-0.5
Oxides, carbides, nitrides and phosphides	0.01-5.0

The ferromagnetic abrasive material according to the process exhibits high magnetic properties which are 1.1 to 2.5 times greater than those of prior art ferromagnetic abrasive materials.

The abrasive component exhibits high hardness (average microhardness 2,200 kg/mm^2).

The ferromagnetic abrasive material according to the process exhibits chemical neutrality, low diffusion and adhesive capacity towards the surface being treated. No carbon saturation of the surface of a workpiece occurs during an abrasive treatment in a magnetic field as carbon is in the bound form as silicon and aluminum carbides.

The method for preparing the material, comprises melting an iron-aluminum-silicon-phosphorus alloy and overheating it above the temperature corresponding to the beginning of solidification by 50° to 200°C, atomizing the melt jet to obtain pellets of a size smaller than 1.5 mm with a gas jet with subsequent cooling of the resultant pellets. Carbon is added to the iron-aluminum-silicon-phosphorus melt. The gas jet has a pressure of from 2 to 20 atm and a velocity from 150 to 500 m/sec and contains from 99.999 to 75% by wt of nitrogen (N_2) and from 0.001 to 25% by wt of oxygen (O_2) which form aluminum and silicon oxides and nitrides on the surface of pellets during the atomization of melt and cooling of pellets.

The method insures the preparation of pellets in which the ferromagnetic component is disposed centrally thereof, and the abrasive component is disposed on the periphery of the ferromagnetic component. Moreover, by modifying the melt composition and atomization conditions, a preset shape of pellets may be varied over a large range: from spherical one to an elongate splinterlike shape. Thus, an increase in content of aluminum in the melt and reduction of gas velocity in the atomizing jet and nitrogen content thereof contributes to the formation of elongate splinter-shaped pellets, whereas an increase in silicon content of melt and a greater gas velocity in the atomizing jet and increased nitrogen content thereof contributes to the formation of nearly spherical pellets.

By varying the composition of the gas jet, the quantitative composition of nitrides and oxides in the abrasive component may be changed. By increasing the oxygen content in the gas jet, the resultant pellets will have a thicker layer of abrasive component and may be advantageously used for an abrasive treatment of hard-to-machine materials in a magnetic field. With an increase in the nitrogen content of the gas jet, the resultant pellets have a thin layer of the abrasive component, better magnetic properties and higher cutting capacity in an abrasive treatment of aluminum and its alloys in a magnetic field.

Example: A ferromagnetic abrasive material according to the process had the following composition in percent by weight:

Iron	83.3
Aluminum	8
Silicon	5
Phosphorus	0.1
Carbon	0.1
Aluminum and silicon carbides, nitrides and phosphides	3.5

The above material was used for an abrasive treatment of the surface of printed circuit blank in a magnetic field prior to the printed circuit application. The abrasive treatment resided in the removal of copper oxides film and in providing a preset microrelief of the surface for application of a coating, that is, the printed circuit. Thus, the abrasive treatment in a magnetic field was conducted in the microcutting mode.

The pellets of the material were of an elongated splinterlike shape with acute-angle edges and small radii. As a result of the abrasive treatment in a magnetic field, a relatively thick layer of copper and its oxides was removed from the blank surface (2 to 20 μ). The microprofile of the resultant surface was characterized by comparatively small spacing of irregularities vs their height, large angle of inclination of the sides of irregularities and small radii of indents and projections. This treatment insured high quality of preparation of blanks for further production steps.

Colloidal Silica Sol plus Water-Soluble Amine

U.S. Patent 3,170,273 shows that colloidal silica and silica gels are useful as polishing agents for the surfaces of semiconductor crystals which are most often used in the manufacture of semiconductor devices. The siliceous polishing agents of that patent are particularly suited to producing highly polished surfaces on silicon crystal wafers, thus rendering these surfaces suitable for the disposition of an epitaxially deposited crystal layer.

The silica sols and silica gels used have an ultimate particle size of 4 to 200 mμ. From a commercial standpoint, it is preferred to use as the polishing agents in the process of the patent aqueous colloidal silica sols having particle sizes within the range of 4 to 100 mμ.

In accordance with the process of *C.C. Payne; U.S. Patent 4,169,337; October 2, 1979; assigned to Nalco Chemical Company* it has been found that semiconductor surfaces of the type previously described and, most particularly, silicon, can be efficiently polished by using as the polishing agent a combination of either a colloidal silica sol or silica gel having an ultimate particle size of 4 to 200 mμ and preferably 4 to 100 mμ with a water-soluble amine. The amount of amine in relation to the silica present in the silica sol or gel may range between 0.1 to 5% based on the SiO_2 content of the sol or gel. The preferred amount of the amine is 1.0 to 5% and, most preferably, 2.0 to 4.0%.

The amines should contain between 2 to 8 carbon atoms and are preferably aliphatic in character. Diamines containing 2 to 8 carbon atoms and, preferably, aliphatic in character will also give excellent polishing results. Most preferably, the amines containing one primary amino group or containing several amino groups should be used. Preferred amines contain at least one hydroxy alkyl group and one primary amino group.

Metal-Ion-Free Amorphous Silicon Dioxide

In the manufacture of semiconductor devices, thin monocrystalline silicon wafers are made by cutting a crystal of monocrystalline silicon. For further processing these wafers are fixed with wax on a polishing block. The exposed surface of each wafer is then polished in order to remove the surface irregularities which are caused by the cutting of the crystal of monocrystalline silicon.

Generally, the silicon wafers are processed in successive steps with grinding or polishing agents of different particle size. These process steps comprise the lapping and/or the so-called Blanchard grinding by which rough irregularities are removed from the surface, and finally one or several mechanical or chemical precision polishing processes which produce an extremely smooth and defect-free surface which subsequently is subjected to the known processing methods for making semiconductor devices.

As the polished surface of a silicon wafer is later supplied with a mask and etched, it is necessary for it to show a minimum of scratches and irregularities.

For polishing silicon wafers, polishing agents containing a silicon dioxide powder are highly desirable because silicon dioxide shows a Mohs hardness of 7 which is very close to the hardness value of silicon and, therefore, shows a relatively high material abrasion rate. During the polishing process it is, therefore, not embedded in the surface of the polished silicon wafers. However, a disadvantage of the silicon dioxide-containing polishing agents is their sodium ion content which is due to the presence of $Na_y(SiO_x)$ compounds in the polishing suspension. Low quantities of the sodium ions adhere to the polished silicon surface and diffuse into the substrate during the subsequent high temperature process. Due to the high integration density of integrated circuits, however, even these low quantities of metal ions cause high leakage currents and thus the failure of the highly integrated circuits.

It is the object of the process of *F.C. Brunner, F.W. Schwerdt and H.-H. Steinbeck; U.S. Patent 4,117,093; September 26, 1978; assigned to International Business Machines Corporation* to provide a method for making amorphous silicon dioxide free of metal ions, with a specifically uniform grain size, which is suitable for the mechanical polishing of semiconductor surfaces. A polishing agent is also provided.

The object is achieved by a process for making silicon dioxide which is characterized in that a mixture of an organic amine in surplus, a hydroxybenzene, water and high-grade silicon are boiled under reflux with bubbling of purified air, and that the precipitate of silicon dioxide is rinsed with water and alcohol and dried in a known manner.

The polishing agent comprises amorphous silicon dioxide free of metal ions; the dioxide being made in accordance with the process, in aqueous suspension.

As the amine, 1,2-ethylenediamine, which has a pK_1 value of 9.93 may be used; other amines, e.g. 1,2-propylenediamine or diaminobutanes or also monoamines can be used in the process. It is advantageous to use the amine in a large surplus, for instance in an approximate ten-fold molar surplus, referring to the quantity of the silicon used and to the quantity of the hydroxybenzene used.

The use of pyrocatechol (1,2-dihydroxybenzene) in the process is particularly advantageous. Other useful dihydroxybenzenes are resorcin (1,3-dihydroxybenzene) and hydroquinone (1,4-dihydroxybenzene). The use of monohydroxybenzenes is advantageous, too.

The silicon, which is advantageously added to the aqueous alkaline solution in the form of thin disks, should be made oxide-free with diluted hydrofluoric acid and rinsed with deionized water to the conductance (absence of ions) of the deionized water prior to its use.

Example 1: The production of the silicon dioxide polishing agent is described by means of a laboratory scale reaction.

160 ml deionized water are added to a 1 ℓ beaker and mixed with 340 ml ethylenediamine. The mixture heats up in that process to approximately 60°C. 60 g pyrocatechol are added to the mixture with stirring. The solution is then placed in a heated reflux apparatus, and 20 g purified silicon is added in the form of

thin disks. Prior to the admixture, the silicon was made oxide-free by means of diluted hydrofluoric acid, and subsequently rinsed to the conductance of deionized water.

With continuous bubbling of purified air, the mixture with the silicon is boiled under reflux for approximately 8 hr. After a short time silicon dioxide precipitates in the form of a fine white deposit. After the reaction period the unreacted silicon is removed, and the precipitated white deposit of silicon dioxide is filtered off and thoroughly rinsed with deionized water. For better drying, the isolated silicon dioxide is finally suspended in isopropanol, filtered off once more, and dried in the oven at approximately 50°C. The reaction product is an amorphous silicon dioxide of high purity and uniform grain structure. It has a grain size of less than 2 μm.

Example 2: A crystal of monocrystalline silicon was cut into wafers with a thickness of approximately 600 μm. The surfaces of these wafers were lapped as a preparation for the subsequent polishing process, a lapping agent with a grain size of approximately 10 μm being used. In order to remove the surface defects on the silicon wafers, they were etched in a known manner which removed 30 μm per side of silicon. After having been cleaned by ultrasonics in water, the silicon wafers to be polished were fixed on a plate with wax.

A polyurethane foam-covered linen cloth was placed on the polishing head of a standard polishing device. This cloth was moistened with an aqueous suspension of the amorphous fine-grain silicon dioxide made in accordance with the process described above. During the polishing process, a sufficient amount of suspension was added to keep the polishing cloth moist. The polishing head with the polishing cloth was firmly pressed against the surfaces of the silicon wafers. During the polishing process the polishing head and plate were rotated at a rate of about 40 cpm.

After polishing for approximately 20 min, silicon wafers with an extremely smooth surface free of any scratches and defects were obtained.

Two-Step Polishing Process

It is the object of *R. Griesshammer, H. Kirschner and G. Lechner; U.S. Patent 4,070,797; January 31, 1978; assigned to Wacker-Chemitronic Gesellshaft fur Elektronic Grundstoffe mbH, Germany* to provide a process for making polished "haze-free" surfaces on semiconductors with the use of odorless nontoxic agents. This object is accomplished by polishing the surface of a semiconductor body in a first step by means of a polishing agent containing at least one of the following: quartz, silica, a silicate, and a fluosilicate, and polishing the surface in a second step by means of a polishing agent containing in addition to the above, from 0.1 to 10% by wt, calculated on the polishing agent of an anionic and/or nonionic surfactant that is free of nitrogen.

Surprisingly, the additives used in the second polishing stage eliminate the microroughness of the surface and hence the semiconductor surfaces are substantially haze-free (optically smooth), and the yields of electronic components manufactured from semiconductor material polished in this way are not appreciably diminished through defective surface quality.

Known polishing suspensions may be used for the first polishing step of the process. The polishing agent used for the second polishing step fundamentally contains the same constituents as employed in the first step, but it additionally contains from 0.1 to 10%, preferably from 0.5 to 5% by wt of an anionic or nonionic non-nitrogen-containing surfactant. The percentages relate to the polishing agent without additive.

Suitable anionic surfactants are, for example, salts of long-chain carboxylic acids, sulfuric acid esters, such as alkyl sulfonates and alkylaryl sulfonates, and corresponding fluorinated compounds. Suitable nonionic surfactants are, for example, various ethylene oxide adducts, for example nonylphenol polyglycol ether or oxethylated polypropylene glycols, fatty acid monoglycerides, carbohydrate fatty acid esters, for example, saccharose monofatty acid, and corresponding fluorohydrocarbons. Nonionic surfactants are preferably used.

As in the known polishing process, the duration of the first polishing step may be from 10 min to 2 hr. In the second step, polishing need be carried out, in most cases, only from 2 to 10 min.

Example 1: Silicon discs of 50 mm diameter and 350 μ thickness, which had been manufactured by sawing a monocrystalline silicon rod, were attached by means of wax onto a plane carrier plate of stainless steel and placed on the rotating plate of a polishing machine, over which a polishing cloth was stretched, and were subjected to a pressure of 0.2 kp/cm^2. The rotating plate was rotated at 90 rpm.

A polishing suspension that had been prepared from 50 ℓ of water glass (30% SiO_2) and 25 kg of calcium chloride hexahydrate in 600 ℓ of water, was dripped onto the rotating plate at a rate of about 25 ml/min. After polishing had been continued in this manner for about 1 hr, the surface of the silicon disc had been polished in the conventional sense: no flaws could be detected under a microscope with a thousand-fold magnification. However, on examining the surface under a focused light beam, a milky-grey haze was observed, caused by light scattered by the slightly rough surface.

The discs were again placed on the polishing machine and 2% by wt of nonylphenol ether was added to the polishing suspension. Polishing was then carried out for 5 min. Re-examination of the disc after this second polishing step showed that the surface was free of all scattered light: the surface appeared jet black.

Example 2: 50 kg of quartz powder between 10 and 40 mμ particle size were suspended in 500 ℓ of water and sodium hydroxide solution was added to the suspension until its pH was between 9.5 to 10.5. The suspension thus prepared was used to polish gallium arsenide discs of 3 mm diameter, in the manner described in Example 1. The polishing agent was added at a rate of 30 ml/min. After polishing had been continued for about 1 hr, 1% by wt of isododecyl benzene sulfonate was added to the polishing suspension. Polishing was then continued for another 7 min. Examination of the discs under a focused light beam showed that the discs had flawless haze-free surfaces.

Polishing (111)-Oriented Gallium Arsenide Crystals

I. Lampert; U.S. Patent 4,064,660; December 27, 1977; assigned to Wacker-Chemitronic Gesellschaft fur Elektronik-Grundstoffe mbH, Germany provides a process for polishing semiconductor surfaces, especially (111)-oriented gallium arsenide discs having haze-free surfaces, in a single operation.

The process comprises polishing the surface by means of a polishing agent containing a mixture of an aqueous suspension of a compound selected from a member of the group consisting of quartz, silica, silicate and/or a fluosilicate, and having a pH from 6 to 8 and from 2 to 15% by wt of an aqueous solution of hydrogen peroxide having a pH from 6 to 8.

By means of this process, it is possible to polish semiconductor surfaces, especially (111)-oriented gallium arsenide crystals to give a haze-free surface and at a higher polishing rate than was hitherto possible.

Known aqueous polishing suspensions may be used in the polishing agents used according to the process. They advantageously contain precipitated silicates or fluosilicates.

The size of the solid particles in the polishing suspension is advantageously from 5 to 200 μ.

The aqueous suspension of the quartz, silica, silicate or fluosilicate may suitably be prepared by mixing 1 pbw of the solid component with from 1 to 10 pbw, preferably from 3 to 5 pbw, of water. The pH of the suspension may then be adjusted to a value preferably from 6.5 to 7.5 with the use of an acid. Suitable acids for this purpose are, for example, phosphoric acid, sulfuric acid, acetic acid, oxalic acid and, especially, hydrochloric acid.

The pH of the solution of hydrogen peroxide may be adjusted to a value preferably from 6.5 to 7.5, by means of an alkali. Suitable alkalies for this purpose are aqueous solutions of sodium hydroxide, potassium hydroxide and ammonia.

The aqueous suspension and the hydrogen peroxide solution may then be thoroughly mixed in order to give a polishing agent used according to the process. These two components may be mixed in widely varying proportions, but it is preferred to mix them in such proportions that the polishing agent contains from 2 to 15%, especially from 4 to 8%, by weight of hydrogen peroxide. Both components of the polishing agent may be used for at least 24 hr at the given pH range. Advantageously, the two components namely the aqueous suspension and the hydrogen peroxide solution, are separately conveyed to the polishing machine by means of suitable pumps and are mixed only immediately prior to use, that is immediately prior to flowing out onto the polishing disc: this has the advantage of insuring a constant composition of the polishing agent throughout the polishing process.

Polishing may be effected using the type of machine conventionally used for polishing semiconductor discs, and also using conventional polishing cloths. The rate of abrasion is, however, generally on the order of 25 μ/hr which is relatively high.

Example: 5 ℓ of water glass (30% by wt SiO_2) and 2.5 kg of calcium chloride hexahydrate were suspended in 40 ℓ of water and adjusted to pH 7 with concentrated hydrochloric acid. 30 ℓ of a 30% by wt aqueous hydrogen peroxide solution were then adjusted to pH 7 with an aqueous ammonia solution. Both solutions were conveyed separately to a polishing machine by means of pumps and were mixed together only immediately before flowing onto the polishing disc. Various semiconductor discs were polished with this polishing agent using a speed of 90 rpm and a contact pressure of 0.2 kp/cm^2.

(100)-orientated gallium arsenide discs doped with tellurium were polished. The rate of abrasion was 15 μ/hr using a soft polishing cloth and 10 μ/hr using a hard polishing cloth.

(100)-orientated gallium arsenide discs doped with chromium were polished. The rate of abrasion using a soft cloth was 17 μ/hr and using a hard cloth was 10 μ/hr.

(111)-orientated gallium arsenide discs doped with selenium were polished. The rate of abrasion with a soft polishing cloth was 25 μ/hr and with a hard polishing cloth was 20 μ/hr.

In general, 80 to 100 μ/hr were abraded from the discs. The polished gallium arsenide discs showed surfaces which had been perfectly polished so that they were haze free when examined under a focused beam of light.

Silica plus Dihaloisocyanuric Acid Salt

J.S. Basi; U.S. Patent 4,057,939; November 15, 1977; assigned to International Business Machines Corporation provides a polishing composition comprised of an aqueous dispersion of colloidal sized polishing abrasive particles (e.g., 5 to 200 mμ) admixed with an alkali metal base, preferably the carbonates, and an oxidizing agent selected from the group consisting of sodium and potassium salts of halo derivatives of isocyanuric acid, alternatively known as the corresponding salts of halotriazinetriones, preferably the corresponding salts of dihaloisocyanuric acid.

Typical abrasives are cerium oxide, aluminum oxide, zirconium oxide, and such, among which the preferred abrasive is silica (e.g., silicon dioxide).

Preferably, the silicon substrates or wafers are polished with the compositions of this process by the free polishing technique such as illustrated in U.S. Patent 3,691,694 including modifications thereof in accordance with the planetary configuration of Goetz et al shown in the IBM Technical Disclosure Bulletin, pp. 1760-61, Vol. 15, No. 6, November 1972.

By this technique, the wafers are placed in openings with circular carriers which are placed between two counter rotating polishing pads. In turn, the periphery of the carrier is geared to mesh between teeth of an independently rotating spindle and gear teeth in a circumscribing periphery of the polisher so as to impart planetary motion to the wafer carriers resulting in relative motion of the carrier and wafers between the rotating polishing pads. In operation, the polishing slurry of the process is fed continuously to the surface of the polishing pads. The completion of the polishing operation is followed by washing and removal of the wafers or substrates from the polishing mechanism.

It is to be understood that although the polishing compositions of this process have been developed for and are described for the polishing of silicon substrates, such compositions can be advantageously employed for the polishing of other substrates and materials, particularly quartz and the various glasses.

A typical composition in accordance with this process will comprise a silica (e.g., SiO_2) sol, sodium carbonate and sodium dichloroisocyanurate. With ratios of the carbonate to the dichloroisocyanurate of about 0.5:1-3.2:1, polishing rates for silicon can be obtained of about 2 mils/hr to as high as 5 mils/hr.

The components of the polishing composition can be readily prepared simply by mixing and normally in lots as low as 20 ℓ for convenience in commercial application. Preferably, the freshly prepared compositions are utilized for polishing, and if previously prepared compositions are to be used, they should be stored in the absence of light which tends to decompose or coagulate the polishing slurry, via the oxidizing agent.

The aqueous dispersion of silicon dioxide can be conveniently employed in the form of silica sols and silica gels having a silicon dioxide particle size of 1 to 500 mμ, in weight percent concentrations of 10 to 50 wt %. A conventional source of the aqueous dispersion of the silicon dioxide is the composition Syton HT-50. This Syton HT-50 has a silica content of about 50 wt % with maximum particle size of 40 to 45 mμ.

Calcium-Titanium-Zirconium-Oxygen Compounds

It has been found by *W.A. Shelton; U.S. Patent 4,022,625; May 10, 1977; assigned to NL Industries, Inc.* that compounds having the empirical formula $CaTiZr_3O_9$ are, when in finely divided form, very good polishing agents for semiconductors. Such calcium-titanium-zirconium-oxygen compounds are obtainable by calcination of suitable oxygen-containing compounds of the three metals.

Such products and the production thereof are described in U.S. Patent 3,870,789. For convenience, the abovementioned products will hereinafter be referred to as "caltizir products."

It has been found that for satisfactory results the polishing composition should have a pH of at least 10, preferably between 12 and 13. If necessary, the pH can be raised to the desired point by addition of an alkali such as KOH.

The proportions of the ingredients in polishing compositions according to the process may vary. The caltizir product may comprise from about 10% to 70% of the composition. However, it must be realized that large percentages of the solid polishing agent will increase the viscosity of the composition to such an extent that modification of the polishing process may be necessary. Usually from about 10 to 60% caltizir product will be preferable. The glycerine, when used, preferably constitutes from about 2 to 10% of the composition.

The potassium hydroxide (KOH) content is variable since it is used only in sufficient amount to establish the pH of the composition in the desired range. Normally no more than about 0.55% is sufficient. The sodium hypochlorite (NaOCl) is desirable as an ingredient since it increases the stock removal rate of the composition. It may be used for such purpose in amounts up to about 5% of the composition. The presence of sodium metasilicate is desirable when sodium hypochlorite is employed since it tends to prevent deterioration of such compositions. The percentage of sodium metasilicate in the composition may vary from zero upwards, though amounts above about 3 to 4% appear to produce the desired viscosity in the slurry.

Example 1: A polishing composition consisting of 561.6 g finely divided caltizir product (-5 μ particle size), 203.0 g of 18.3% aqueous solution of NaOCl, 32.1 g sodium metasilicate, 9.0 g KOH and 850.0 g H_2O was prepared by adding the caltizir product to a solution containing the other ingredients and stirring to form a slurry with a pH of 12.5.

The composition was used in polishing silicon wafers. After polishing for 1 hr, it was found that the average stock removal per wafer was 0.0127 mm. The wafer surfaces were mirror finished and showed no pits.

Example 2: A polishing composition consisting of 700.0 g finely divided caltizir product (-5 μ particle size), 267.0 g of 15% aqueous solution of NaOCl, 11.2 g KOH, 100.0 g glycerine and 922.0 g H_2O was prepared by mixing the last four ingredients and forming a slurry by stirring the caltizir product into the mixture. The composition had a pH of 12.5.

When tested on silicon wafers the polishing composition was found to produce an average stock removal per wafer during a 1 hr period of 0.0203 mm. The wafer surfaces after polishing were free from pitting and had mirror finishes.

The silicon wafers used in the polishing test described above had been sawn from silicon crystals and lapped with Al_2O_3 abrasive to obtain substantially parallel faces and to reduce the surface roughness to approximately 5×10^{-5} to 25×10^{-5} mm. The wafers or slices were approximately 25.4 mm in diameter and about 0.25 mm thick.

Five slices, after careful determination of thickness, were mounted on the pressure plate of a planetary type polishing machine adjusted to provide a pressure of about 1.40 g/mm^2 on the slice surface. Polishing was then carried out for 1 hr at a speed of about 60 rpm using a Geoscience Politex polishing pad and with an average slurry flow rate to the pad of 10.8 ml/min. After polishing, the wafer thickness was measured with a micrometer at three points, and the results were averaged to obtain an average loss per wafer. The accuracy of measurement is believed to be within $\pm 2.54 \times 10^{-3}$ mm. Inspection of the wafer surfaces was made with a binocular microscope.

Polishing Alpha-Alumina Wafers

It has been found by *H.W. Gutsche; U.S. Patent 4,011,099; March 8, 1977; assigned to Monsanto Company* that a chemical reaction between silica and alpha-alumina proceeds to render possible a chemical polishing method for alpha-alumina at relatively low temperatures.

The alpha-alumina surface to be polished is preferably in the form of a thin wafer cut from a sapphire boule by means of a diamond-impregnated saw blade. In order that no diamond particles can be carried over into the chemical polishing sequence, all "raw" alpha-alumina wafers are heated to about 1000°C for about 10 min in oxygen to oxidize any diamond present.

The polishing fluid is a colloidal silica fluid, preferably, a hydrated silica colloid in a solvent such as water. Fluids with 10 to 50% solids content are preferred. The polishing fluid of the process has two primary purposes. The first is to carry the SiO_2 reactant to the surface of the alpha-alumina, and the second is to transport the product of the reaction away from the thus-polished alpha-alumina face.

Preferred silica polishing fluids are the Syton colloidal suspensions of hydrated silica in water. Since Al_2O_3 displays an amphoteric nature, it is not surprising that the reaction proceeds most rapidly at a pH of about 7. However, since Syton has a pH of about 10.5, an acid is preferably added to the colloidal silica fluid before use. While acids, such as sulfuric acid, have been used, hydrochloric acid is preferred. The pH of the polishing fluid used is preferably 7 to 8.

Thus, while silica, having a hardness of about 7 on the Mohs scale, has been known to be an effective mechanical polishing agent for materials having a Mohs scale hardness of 2 to 7, it has been found that a silica colloid can provide a superior polished surface on alpha-alumina by chemical action. Further, since the hardness of the alpha-alumina greatly exceeds that of the silica, no fragmentation can occur.

The above reaction is endothermic to the extent of about 14.6 kcal/mol at 400°K, and thus the speed depends on the heat supplied. Since all protruding portions of the surface of the alpha-alumina wafer receive more local frictional heat and are subject to more fluid removal action than the nonprotruding portions, damaged areas react faster and are removed faster than nondamaged areas.

Related to the problem of damage is the appearance of contamination on the surface. As the protrusions and depressions are eliminated, the tendency of the surface to retain contamination is also lessened and easier cleaning results. Thus, the three factors which most interfere with the performance of the final product, damage, contour and contamination, are significantly affected by the process.

After removal of the alpha-alumina wafer from the oxidizing treatment for removal of diamond particles described above, the cooled wafer is polished with the colloidal silica fluid and a polishing cloth or other material under pressure to create frictional heat sufficient to maintain the desired rate of reaction; thus a desired rate of surface removal. The pressure used can vary according to the thickness of the wafer. For a wafer 2" in diameter and 350 μ thick, pressures of 20 to 30 kg will suffice. Pressures should preferably be adjusted so that the temperature on the surface of the polishing turntable used is about 80°C, although temperatures of 60° to 100°C can be maintained.

After polishing with the colloidal silica fluid, the fluid residues can be removed by an immersion in hydrofluoric acid followed by a rinse with deionized water.

The finished product can then find application as an insulating substrate for high performance, silicon-on-sapphire semiconductor devices, transistors and microcircuits in particular. The semiconductor applied to the finished wafer can be a thin, e.g., less than 1 μ, (100)-oriented film which is heteroepitaxially deposited on the alpha-alumina surface.

Repairing Selenium Photoreceptors

J.F. Byrne; U.S. Patent 3,971,169; July 27, 1976; assigned to Xerox Corporation provides a method for the repair of an electrostatographic photoreceptor comprised of a conductive substrate with a uniform dispersion of selenium or a selenium alloy on its surface as the photoconductive layer in which the photoreceptor has been damaged by the formation of a scratch, fissure, aggravated surface conductivity, or depression partially through the photoconductive layer.

The composition includes a mixture of magnesium hydroxide as a primary suspending agent, a secondary suspending agent, and an abrasive dispersed in a liquid medium. The liquid medium generally comprises water, isopropyl alcohol, and propylene glycol. The method involves providing a composition comprising a mixture of magnesium hydroxide as a primary suspending agent, a secondary suspending agent and an abrasive dispersed in a solvent; applying the composition to a physically damaged area of a selenium or selenium alloy photoreceptor;

and rubbing the composition about the damaged area so as to effect a smoothing of the physically damaged area of the photoreceptor.

Generally, the composition should comprise from about 30 to 50% magnesium hydroxide as it occurs in milk of magnesia, 25 to 200% abrasive based on the weight of the milk of magnesia, and a small amount of a secondary suspending agent. The solvent for the composition generally can go as high as up to $\frac{4}{5}$ water and $\frac{1}{5}$ isopropyl alcohol by weight. Additionally, about $\frac{1}{10}$ as much propylene glycol, based on the amount of alcohol, can be added to prevent premature drying of the composition during use.

Any abrasive may be used in combination with the magnesium hydroxide primary suspending agent. Suitable abrasives include feldspar [aluminum sodium (or potassium) silicate] products containing feldspar and aluminum oxide. A preferred abrasive is feldspar C-6, or feldspar G-200 supplied by the Feldspar Corporation.

Suitable secondary suspending agents for the composition include Santocel Z (silica aerogels). Other suitable secondary suspending agents include Cab-O-Sil (colloidal silica particles).

While water and isopropyl alcohol are used as a preferred solvent mixture for the composition, alcohol is not necessary for the composition and may be omitted if the composition of the process is never permitted to become frozen. If the alcohol is omitted, the addition of a fungicide, such as Dowacide A may be added in quantities of approximately 0.01 to 0.1% by wt of the liquid present.

Example: An electrostatographic photoreceptor consisting of an aluminum cylinder 8" in diameter and 12" long with a uniform 60 μ layer on its surface of a photoconductive selenium alloy containing 0.33% As and 20 ppm chlorine, is scratched to a depth of about 20 μ. The photoreceptor is used in the normal xerographic mode with unsatisfactory results due to toner buildup in the depressions created by the scratches with consequent failure to discharge in these areas causing them to appear as black marks on the copies produced. The toner is first removed from the scratches by wiping with a cloth or by use of a cotton swab wetted with isopropyl alcohol.

A polish concentration of the process utilizes the following ingredients: 700 g of milk of magnesia or an equivalent milk of magnesia made by hydrating the light grade of magnesium oxide; 45 g of secondary dispersant (Santocel Z); and 1,050 g of abrasive (feldspar C-6). The above composition is mixed with a solvent of 2,400 g of water, 1,200 g isopropyl alcohol and 120 g of propylene glycol. The mixture is stirred to form a thick paste.

The specific preparation of the composition is as follows:

(1) Prepare a solvent medium by mixing 600 g water, 300 g reagent

grade isopropanol, and 30 g propylene glycol, Union Carbide technical grade.

(2) Mix 700 g solvent (1), 700 g milk of magnesia (McKesson), and 45 g Santocel Z for 2 min using an Eppenbach Homo-Mixer.

(3) Stir slurry of Step (2) manually using a spatula, while gradually adding 1,050 g feldspar C-6, which forms the final paste.

The paste is then applied to the 20 μ scratch on the photoreceptor referred to above by means of a gauze cloth, a soft sponge, or alternatively, a cotton pad. The damaged area is rubbed vigorously, with further liberal application of paste, with continual scrutiny of the damaged area. When the crack appears relatively smoothed, the buffing is terminated, and the residual polish contaminated with particles of the selenium alloy removed. The damaged area of the photoreceptor demonstrates a smooth scar in place of the original scratch.

The repaired photoreceptor is employed to produce copies in the normal xerographic mode. Inspection of the copies produced discloses that the scratched area which appeared black before repair now is undetectable. The image areas on the copies are uninterrupted since the repaired area has discharge properties substantially equivalent to the undamaged areas of the photoreceptor.

In an earlier work *J.F. Byrne; U.S. Patent 3,959,934; June 1, 1976; assigned to Xerox Corporation* describes a similar composition. In that composition zinc oxide is the primary suspending agent and the liquid medium is an organic one.

The organic liquid medium generally includes a mixture of isopropyl alcohol and a kerosene based solvent such as Sohio 3440 solvent.

While kerosene based solvents and isopropyl alcohol are used as a preferred solvent mixture, any hydrocarbon liquid may be used as the suspension medium.

Silica Coated with Chemically Combined Aluminum

It has been discovered by *G.W. Sears, Jr.; U.S. Patent 3,922,393; November 25, 1975; assigned to E.I. DuPont de Nemours & Company* that silicon and germanium semiconductor materials can be polished to a high degree of surface perfection by a process comprising polishing the material with a modified colloidal silica sol having a pH between about 11 to 12.5 and having its silica particles coated with chemically combined atoms of aluminum to give a surface coverage of about 1 to 50 aluminum atoms on the surface per 100 silicon atoms on the surface of uncoated particles. The modified colloidal silica sol used in the process has a silica concentration of about 2 to 50% by wt with the silica particles having a specific surface area of about 25 to 600 m^2/g.

Example 1: To a 580 ℓ stainless steel tank were charged 404 kg of a silica sol containing 52.4% SiO_2 by wt, the silica particles of which had a specific surface area of 132 m^2/g. The sol was stabilized with sodium hydroxide until the weight ratio of SiO_2:Na_2O was 230. The sol was then diluted with 171 kg of water. A turbine mixer with a rotor having a diameter of 5.1 cm and capable of operating at 10,000 rpm was installed with the rotor immersed several centimeters below the surface of the liquid and adjusted so as to force liquid upward through the rotor housing and against a baffle plate also positioned below the surface of the liquid.

A delivery tube for the introduction of a solution of sodium aluminate was installed with its exit end immediately below the turbine housing so as to provide almost instantaneous mixing of the sodium aluminate with the colloidal silica. A sodium aluminate solution was prepared by dissolving 8.3 kg technical sodium aluminate (Reynolds Aluminum Co.) which contains 72% $NaAlO_2$ by wt in 33.2 kg water. The solution was then pumped through the delivery tube at about 0.125 ℓ/sec while operating the turbine mixer at full speed. The resulting product was a stable sol having a pH of 11.7, containing 35% solids, the silica particles of which were coated to an extent of 19.5% with aluminate ions.

Silicon wafers 7.5 cm in diameter were initially polished with alumina having a particle size of 10 to 15 μ and thereafter with diamond paste and a lubricant using a soft felt polishing pad and standard polishing equipment having 4 platens for mounting the wafers. The wafers were cleaned and then polished with the abovedescribed modified sol which was previously diluted with a NaOH solution to a pH of 12 and a solids content of 10 to 12% by wt.

The polishing platen was operated at a rotational speed of 150 rpm and the wafer-mounted platens were run at a speed of 50 rpm with a pressure of 450 g/cm^2 being exerted on the surface of the wafers. The polishing was conducted at a surface temperature of 60°C with the modified silica sol being applied to the wafers by allowing the sol to drop onto the polishing pad at a flow rate of 95 ml/min. On the average a polishing or removal rate per wafer of 150 μ/hr was achieved whereas removal rate per wafer of only 50 μ/hr was achieved with an unmodified colloidal silica sol using the same technique. The wafers polished with the modified sol displayed a high degree of surface perfection.

Example 2: To a 75 ℓ plastic tank were charged 40 kg of a silica sol containing 50.8% SiO_2 by wt, the silica particles of which had a specific surface area of 130 m^2/g. The sol was diluted with 21.3 kg of water. Using a turbine mixer and delivery tube similar to those described in Example 1, a sodium aluminate solution which was prepared by dissolving 400 g of technical sodium aluminate in 1,600 ml of water was introduced while operating the mixer at full speed.

The resulting product was a stable sol containing 30% solids and having its silica particles coated to an extent of 10% with aluminate. The pH of this modified sol as prepared was 12.0 but after storage for 24 hr had drifted to 11.4 and after 48 hr had drifted to 11.2. In comparison, an unmodified sol of the type from which the starting material was selected underwent a pH drift to a pH value of 11.3 in 2 hr after alkalization to a pH of 12 with dilute NaOH.

The modified sol was diluted with an equal weight of water and used to polish silicon wafers with equipment similar to that described in Example 1. The rate of removal of silicon was 75 μ/hr and polished surfaces of the wafers displayed a high degree of perfection.

Precipitated Silicates and Silicofluorides

H. Deckert and H. Jacob; U.S. Patent 3,877,183; April 15, 1975; assigned to Wacker-Chemie GmbH, Germany have discovered a method of polishing semiconductor surfaces by means of efficient mechanical polishing agents. The method is characterized by the use of precipitated silicates and/or silicofluorides as polishing means. With this method it is possible to obtain uniform particle sizes of the maximum possible hardness.

The performance of this method has proved the advisability of using silicates that are difficultly soluble, having been precipitated by adding silicate-forming metallic salts or metallic salt solutions to silicic acid solutions, their salts or derivatives.

Salts of metals of the second and third group and subgroup of the Periodic System, as well as silicate-forming salts of heavy metals, for instance, zirconium, iron, lead, nickel, cobalt, are suitable as silicate-forming metallic salts. Besides, soluble salts of magnesium, calcium, strontium, barium, zinc or aluminum may also be used with good results.

In contrast to crystalline polishing powders, polishing suspensions produced in accordance with the process contain no oversize grains to cause scraping.

In view of the fact that polishing suspensions are manufactured by the simple process of pouring together solutions of raw materials previously filtered or decanted, any impurities caused by foreign bodies are easily eliminated.

Example 1: The polishing suspension is manufactured in the following manner. 120 ml of water glass solution (33% SiO_2) are diluted by the addition of 1,800 ml of deionized water. Then a solution of 25 g of $AlCl_3 \cdot 6H_2O$ in 100 ml of water is added slowly. The precipitate obtained is very finely divided; the grain size ranges about 20 mμ.

Round silicon disks of 32 mm diameter and 300 μ thick, obtained by sawing of a monocrystalline silicon rod, are fine-polished with corrundum powder (grain size of 5 μ) abrading or wearing each side by 50 μ. Every 12 disks are cemented on with beeswax to a circular refined steel alloy supporting plate of 180 mm diameter, then applied to a rotary plate or dish (450 mm diameter) of a polishing machine covered with a polishing cloth and finally loaded with 1.5 kg. The rotary dish is made to revolve at a speed of 90 rpm. The polishing suspension drips out onto the rotary dish at the rate of 1 to 2 drops per minute.

After a polishing period of 20 min the surface of the silicon disk is polished smooth and free of scratches and ready for epitaxial coating.

Example 2: Silicon disks, previously pretreated as described in Example 1, are polished with a polishing suspension which has been manufactured in the following manner.

12 ml of sodium silicate solution (33% SiO_2) are diluted with 1,800 ml of deionized water and heated to 80°C. While stirring a solution of 150 g $CaCl_2 \cdot 6H_2O$ in 200 ml of water is added slowly. A white crystalline precipitate having a grain size of about 60 mμ precipitates. The composition of the precipitate corresponds to about the formula $CaO \cdot 2SiO_2$. The polishing suspension can be used immediately after it has cooled. After a polishing period of 15 min the silicon disks are so well polished that they are ready for the epitaxial charging.

Example 3: Disks of a thickness of about 300 μ are sawed off a monocrystalline rod of gallium arsenide and fine-polished with corundum powder of a grain size of about 2 to 3 μ. In this way about 30 μ were worn off each side. The polishing is performed as in Example 1 and an aqueous suspension of barium hexafluosilicate, manufactured as indicated below, is used as the polishing means.

560 ml of an aqueous hydrofluoric acid (48% HF) are diluted with 500 ml of deionized water and mixed with 120 g of SiO_2. The dissolution of the silicon dioxide results in hexafluosilicic acid. As a solution of 75 g of $BaCl_2 \cdot 2H_2O$ in the form of a saturated aqueous solution is slowly added, an x-ray crystalline precipitation of $BaSiF_6$ precipitates. Next, the strongly acid polishing suspension is neutralized. After a polishing period of 15 min, the gallium arsenide disks display a completely undisturbed, smooth and shining surface.

Polishing Agent Modified with Alcohol and Polyvinyl Alcohol

H. Deckert, W. Heinke, H. Jacob and H. Kirschner; U.S. Patent 3,874,129; April 1, 1975; assigned to Wacker-Chemitronic, Germany produce semiconductors with exceptionally smooth haze-free surfaces in a multistage polishing process. The process comprises polishing the semiconductor with a first and conventional polishing agent containing a silicon compound suspended in water, followed by polishing with a modified polishing agent comprising the first polishing agent in major proportion containing minor but effective additions of polyvinyl alcohol and a 3 to 5 carbon monohydroxy aliphatic alcohol.

The time for the first polishing step, as is known in the industry, amounts to about 10 min to 2 hr. The polishing in the second step usually takes only about 2 to 10 min.

Most of the suitable monohydroxy alcohols are saturated alcohols, e.g., propyl-, isopropyl-, n-butyl-, isobutyl-, tert-butyl-, amyl- and isoamyl alcohol. Mixtures of any of these alcohols may also be used.

A polyvinyl alcohol additive containing a major proportion of polyvinyl alcohol, e.g., about 88 to 100 mol % saponified polyvinyl ester is typically employed.

Copolymers of vinyl esters, e.g., with ethylene, which subsequently are saponified, may also be used.

Example 1: Round silicon discs of 50 mm diameter and 350 μ thickness, which have been produced by sawing off sections of a monocrystalline silicon rod, are fastened by means of wax on a round, flat carrier plate of stainless steel. This plate is placed on a polishing machine turntable which is covered with a polishing cloth with the silicon discs in contact with that cloth and subjected to a pressure of 0.2 kp/cm^2 with the turntable rotating at 90 rpm.

A polishing suspension consisting of 50 ℓ water glass (30% SiO_2) and 25 kg calcium chloride ($CaCl_2 \cdot 6H_2O$) in 600 ℓ water, is dripped onto the turntable at the rate of about 25 cc/min. After polishing for about 1 hr, the surface of the silicon disc is polished to a normal extent. However, upon looking at the silicon surface under a spotlight, a milky haze can be seen, which is caused by diffusion of the light. The discs are again placed on the polishing machine, and the polishing process continued for 6 min with a new polishing composition, namely a mixture of 100 pbv of the first suspension to which have been added 5 pbv of n-butyl alcohol and 1 pbv of a solution of 10% by wt solution of polyvinyl alcohol in water.

When the disc is inspected again after this process step it is observed that the surface is free from all diffused light, i.e., the location of the spotlight beam reflected by the surface of the disc cannot be identified.

Example 2: Round gallium arsenide discs of 30 mm diameter are subjected to the same two-stage polishing process as that described in Example 1. These discs also display a milky haze under the beam of a spotlight after the initial polishing step but not after the second polishing step with additives, according to the process.

Example 3: A quantity of 50 kg of powdered quartz with a grain size between 0.01 and 0.04 μ is stirred into 500 ℓ of water and caustic soda is added until the pH value amounts to between 9.5 and 10.5. The resulting suspension is used to polish round silicon discs of 60 mm diameter on the device described in Example 1. The quantity dripped on per minute is 30 cc. After 50 min, the polishing is interrupted; then to 100 pbv of the polishing suspension are added 4 pbv of isobutanol and 3 pbv of a 10% by wt polyvinyl alcohol solution in water. Polishing is then continued for 7 min. A test of the disc under the light beam shows that the discs have been polished until they are haze-free.

BEARING SURFACE FINISHING

Metal Fluoride in Plasma Jet Applied Coating

The process of *H.E. McCormick; U.S. Patent 4,115,959; September 26, 1978; assigned to Ramsey Corporation* relates to a method of manufacturing an article with a bearing surface. The surface is formed by plasma spray coating a metallic substrate with a ceramic oxide composition containing at least one alkaline earth metal fluoride, cooling the coated article, and then grinding the coating with a silicon carbide grinding wheel to smooth and finish the bearing surface.

The process provides for increasing the life of silicon carbide grinding wheels when operating against a plasma applied coating of mixed oxides of titanium and aluminum derived from a powder mixture of the oxides and applied on a cast iron substrate to a thickness up to about 0.030". The process comprises the steps of incorporating from 10 to 15% by wt of a metal fluoride into the mixed oxide powder prior to plasma application to the substrate. The metal fluoride is selected from the group consisting of alkaline earth metal fluorides and mixtures of alkaline earth metal fluorides.

More particularly, the process then contemplates grinding off up to about one-half the thickness of the oxide coating with a silicon carbide grinding wheel to smooth and true the bearing surface thereby greatly reducing or eliminating entirely in-cycle dressing of the wheel. The process is particularly useful in the manufacture of piston rings.

By utilizing this method, substantial savings in the cost of silicon carbide grinding wheels is experienced with only very minor sacrifice in the wear characteristics of piston rings coated with mixed oxides of titanium and aluminum and containing from 10 to 15% by wt of other materials such as calcium fluoride. With coatings consisting only of the oxides of titanium and aluminum, the grinding operation utilizing a silicon carbide grinding wheel has required up to five in-cycle dressings of the wheel in the course of removing the plasma applied refractory metal oxide coating to the desired depth and finish for piston rings. The incorporation of the metal fluoride has resulted in the virtual elimination of the in-cycle dressing of the wheel when operating against plasma applied coating of such mixed oxides.

As indicated above, the metal fluorides of the process are the alkaline earth metal fluorides per se, or mixtures of such alkaline earth metal fluorides. Thus, alkaline earth metal fluorides which may be used in accordance with the process include calcium fluoride (the preferred material), magnesium fluoride, barium fluoride and strontium fluoride. Also, mixtures of such fluorides may be used; for example, a 50/50 mixture of calcium fluoride and magnesium fluoride; calcium fluoride-barium fluoride mixtures, e.g., a 38% calcium fluoride-62% barium fluoride eutectic mixture.

The fluorides are conveniently powdered so as to have a particle size such that 98% will pass through a 100 mesh screen. For example, with calcium fluoride, a desired particle size distribution is such that a minimum of 98% of the calcium fluoride shall pass through a 120 mesh screen. The titanium dioxide-aluminum oxide composite desirably has a particle size distribution such that 98% of the oxide composite will pass through a 200 mesh screen and preferably a 270 mesh screen. With the mixed fluorides, it is desirable to fuse the fluorides together and then pulverize the solidified composite to a fineness whereby at least about 98% of the fluoride passes through a 120 mesh screen. The screen sizes herein are U.S. standard sieve series.

The blending of the metal fluoride powder with the titanium dioxide/aluminum oxide may be simply a physical mixing of the components to obtain as uniform a distribution of the metal fluoride in the refractory metal oxide composite as possible prior to the plasma application thereof.

Alternately, the TiO_2/Al_2O_3 may be formed into a composite powder by a technique well known in the art and utilizing an organic binder, e.g., a phenolic varnish binder (10% solids).

The powders are applied to the rings on a mandrel or arbor as shown in U.S. Patent 3,697,091 and then allowed to cool to room temperature. While still on the mandrel, they are ground with an on-center plunge feed silicon carbide grinding wheel.

Typical examples of powders containing from 8.5 to 22.5% TiO_2, 76.5 to 67.5% Al_2O_3 and 10 to 15% by wt of alkaline earth metal fluoride, and which may be applied by the plasma spray technique using the foregoing spray parameters are as follows:

(1) Refractory metal oxide	90% by total weight
Titanium dioxide	12%
Aluminum oxide	88%
Metal fluoride	10% by total weight
Calcium fluoride	100%

(2) Refractory metal oxide	87.5% by total weight
Titanium dioxide	13.5%
Organic solids	3.0% maximum
Aluminum oxide	78.0% minimum
Other oxides total	5.5% maximum
Metal fluoride	12.5% total weight
Calcium fluoride	100%

Particles Embedded in Grooves

According to the process of *J.E. Tanner; U.S. Patent 3,961,104; June 1, 1976* there is provided a method of producing a cylindrical or other swept bearing surface on a component. The method consists in utilizing a slurry or paste of abrasive or hard powder or grit screened accurately to maximum size. The slurry is pressed into engagement with the surface on a helical path from one end of the surface to the other repetitionally with a pressure such that grooves on crisscross helical paths are cut in the surface and particles of the powder or grit embedded in the surface.

Thus in carrying out the process, a number of microgrooves are cut into the surface on opposite hands by the particles and the particles will be embedded or welded in the surface. The grooves provide oil retention grooves and the particles a hard wearing bearing surface. It will be understood that the grooves are of very small dimensions and their spacing close, the depth and spacing depending on the particle size and speed of rotation and reciprocation of the tool relative to the workpiece.

For example, the particle size and speed can be such that grooves in the region of 0.001 to 0.003 mm in depth width and spacing are formed. The grooves will not be in general continuous, but may extend only a short distance along the surface. It will be seen that a very large number of particles are embedded to form the bearing surface. It is possible, first to produce substantially continuous grooves with substantially no particle embedment and then follow up by using greater pressure, to cause particles to be embedded in the formed grooves. Pressure to cause embedment is in the region of 10 to 20 psi or more.

It has been found that in use, embedded particles remain in the surface and are not dislodged during operation of the engine or other machine. The grooves also serve for oil retention.

It has been found that by subjecting the surface to a second operation with a powder grit of smaller size, there is produced a low friction smooth finish and a hard wearing surface with a great number of particles embedded. This second run effectively smooths off the sharp peaks between the grooves without disturbing to any appreciable extent, the grooves cut by the first run. It can result in the embedding of further particles in grooves on a helical path. It also will force in any particles further into the surface which project excessively and clean off sharp peaks on particles. Further, if any particles are loosely embedded they will either be forced in or removed.

The particles virtually form the bearing surface and provide a large proportion of such surface area, e.g., 50%.

Referring to Figure 6.2, it will in the first place be understood that this view is purely diagrammatic, bearing in mind that in practice the grooves are close-spaced and of a depth and width of between 0.001 and 0.003 mm, and the particles can constitute a large proportion of the final surface area.

In Figure 6.2, the cylinder **10** will be set up with a conventional lapping tool with spring-loaded lapping sticks and arranged so that the tool is rotated and reciprocated relative to the cylinder axis and within the cylinder, while a slurry of silicon carbide in liquid medium is poured in at the top of the cylinder and collected at the bottom and recycled. The grit is of a screen size in the region of 220 and should be as free from sharp peaks as possible. The action results in the grit particles cutting grooves **11** on helical paths and on opposite hands to produce a diamond pattern.

Figure 6.2: Particles Embedded in Grooves

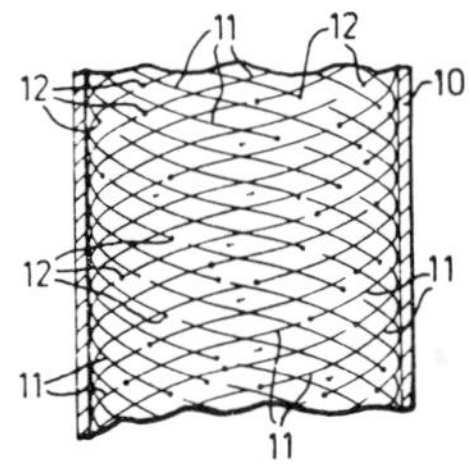

Source: U.S. Patent 3,961,104

An initial run with a light pressure, e.g., 10 psi or less may be done to start grooving without substantial particle embedment, and subsequently the pressure increased so that particles increase the groove depths and widths and finally become embedded or welded in as at **12**. Generally a particle is embedded at the end of a noncontinuous groove and a following particle will then commence to cut a following groove. The initial pressure can be great enough to result in embedment without prior grooving, i.e., the one operation both grooves and embeds the particles, it being understood that the pressure used and length of time the process continues will be dependent mainly on the material of the cylinder.

The operation results in the production of a great number of helical grooves and the embedment of a great number of particles. The grooves may be continuous or some may extend only around a part of the surface, the length of time the operation is carried out and the pressure being the controlling factors.

The method according to the process is not intended to size the cylinder bore or other component as there is no appreciable stock or surface removal, the powder or grit grooving the surface. Thus, the method can be applied to engine cylinder blocks, or wet and dry type cylinder liners of metal or cylinders of ceramic or such material.

LENS GRINDING AND FINISHING

Xanthan Gum and Dispersing Agent

A solution for use in suspending 0.05 to 15 μ ceria or alumina particles to produce a slurry for polishing lenses and semiconductor wafers is disclosed by *G.B. Regan; U.S. Patent 4,161,394; July 17, 1979.* The solution consists of water containing from 0.06 to 0.75% by wt of xanthan gum and from 0.5 to 5.0% by wt of a dispersing agent selected from the group consisting of the condensate of formaldehyde with naphthalene monosulfonic acid, polymethacrylic acid and their ammonium and alkali metal salts. The slurry is prepared by mixing one volume of the solution with about two volumes of water and then dispersing alumina, ceria, or zirconia particles in the resultant mixture. The quantity of ceria or alumina dispersed is in the range 10 to 20% by wt based on the combined weight of solution and added water.

The abrasive particles are small, being in the range 0.05 to 15 μ in diameter. The particle size of the abrasive should be selected with a view to the character of the material to be polished. For polishing plastic lenses it is preferred to use abrasive particles ranging in size from 0.05 to 1 μ. For polishing glass lenses it is preferred to use abrasive particles ranging from 1 to 2 μ in size and for polishing semiconductor wafers it is preferred to use abrasive particles about 7 μ in diameter.

Two practical considerations relating to the use of the polishing slurries make it desirable in many instances to supply the user with a concentrated solution of the xanthan gum and the dispersing agent in water and permit him to disperse abrasive particles of the size appropriate for the intended use in the concentrate diluted with water all within several days of the time of intended use

While the slurries are very stable, the particles do settle out after about 7 days standing, and it must be redispersed before the slurry is used. Also, since the particle size of the abrasive material is varied as indicated above depending on the character of the material to be polished supplying the user with a concentrated solution of the xanthan gum and dispersing agent in water permits the user to dilute the concentrate and disperse abrasive particles of the appropriate size in the diluted concentrate just prior to use.

A suitable concentrate consists of water containing from about 0.06 to 0.75% by wt of xanthan gum and from 0.5 to 5.0% by wt of the dispersing agent. One part of this concentrate is mixed with about 2 parts by volume of water and then the abrasive particles of appropriate size are added in amounts ranging from 10 to 20% by wt of the diluted concentrate and dispersed in it.

Longer Time to Retruing of Wheel

The process of *R.C. Wiand; U.S. Patent 4,131,436; December 26, 1978* relates to an ophthalmic flat roughing wheel particularly adapted for long use in grinding the edges of glass or plastic lenses. The wheel of the process has a circumferential grinding layer made of a diamond-containing metallic matrix having a relatively greater concentration of diamond in the center than toward the edges. This provides a wheel which tends to remain flat or true longer during use and which can be retrued easier than conventional wheels.

An abrasive wheel is made by bonding to the circumferential surface of a wheel blank, a diamond-containing metallic matrix of relatively high diamond concentration sandwiched between diamond-containing metallic matrixes of relatively low diamond concentrations thereby forming a circumferential grinding layer having a greater concentration of diamond at the center than toward the edges. The circumferential grinding layer can comprise 3, 5, 7, 9 or a multiplicity of metallic matrixes having different diamond concentrations increasing from the edge to the center.

Example 1: 4 carats of synthetic diamond dust and 84 g of a mixture of 80% iron and 20% bronze metal powder are mixed to form a first mixture. 16 carats of synthetic diamond dust and 121 g of a mixture of 80% iron and 20% bronze metal powder are mixed to form a second mixture. One-half of the first mixture is layered onto a round metal mold having a 6" inside diameter and an inner core having a 5$^{13}/_{16}$" outer diameter. Next, all of the second mixture is layered in the mold on top of the first mixture and then the remaining half of the first mixture is layered on top of the second mixture. The layered mixtures are cold pressed with about 85 tons and then the inner core is removed and replaced with a metal blank of the same dimensions.

The mold, metal blank and metal matrix mixtures are heated at about 1600°F for about 1½ hr and then hot pressed at temperatures with 30 tons. After the mold has cooled, the blank which now has bonded to it the diamond-containing metallic matrix is removed and excess metal on the sides is removed to yield a roughing wheel about ⅝" wide having two edge sections of relatively lower diamond concentration ⅛" in width and a center section having relatively higher diamond concentration ⅜" in width. The wheel has a 6" outer diameter and a $^{3}/_{32}$" thick diamond-containing grinding layer.

A roughing wheel made in accordance with this example has about twice the useful life of a conventional roughing wheel having the same dimensions and diamond content, but with uniform diamond concentration in its grinding layer.

Example 2: An ophthalmic flat roughing wheel is made as in Example 1 except that a third mixture of synthetic diamond dust and metal powder is layered in the mold intermediate each layer of first and second mixtures. The third mixture contains 6 carats of synthetic diamond dust and 84 g of 80% iron, 20% bronze metal powder. The resulting roughing wheel is about ⅞" wide and has a section of intermediate diamond concentration between each edge section and the center section.

Layer of Rare Earth Oxide on Aluminum Silicate Particles

The process of *L.T. Barfield; U.S. Patent 3,934,988; January 27, 1976; assigned to London & Scandanavian Metallurgical Co., Ltd., England* relates to polishing powders and particularly to such products which include or are based upon rare earth compounds. The polishing powders of the process are especially useful for polishing glass objects, such as optical lenses and mirrors.

The process can be carried out by precipitating any one rare earth compound on to Molochite or other mullite-type alumina silicate and converting it into the corresponding rare earth oxide, but in practice the products usually contain mixtures of rare earth oxides.

The alumina silicate used is preferably the product known as Molochite which has been found to be particularly effective. This product is formed by the prolonged calcination of china clay at elevated temperatures, for instance in the range from 1400° to 1600°C. It has a crystalline structure corresponding to mullite, although its ratio of silica to alumina is substantially higher than that required for mullite, the excess silica being in an amorphous form. Other silicate materials which can be used in carrying out the process include calcined and uncalcined fireclay and fly ash.

The degree of effectiveness of the composition depends upon the amount of rare earth compound precipitated on the Molochite or other alumina silicate material. While it has been found that the use of quite small amounts (5%) of rare earth produces a dramatic effect in improving the polishing performance compared with that of Molochite alone, the polishing performance is improved further though less dramatically by higher levels of rare earth compound. Therefore, a polishing powder according to the process preferably contains at least 5% by wt of rare earth elements.

Example: A rare earth sulfate solution containing excess sulfuric acid was run into a suspension of -325 mesh Molochite in a heel of water, simultaneously with a caustic soda solution, in such a way as to maintain mildly acid conditions.

The precipitated mixture was decanted free of sodium sulfate and sufficient extra caustic soda solution was added to convert the resultant basic rare earth sulfate to a rare earth hydroxide, but while avoiding adding so much as to hydrolyze the Molochite. The mixture of Molochite and rare earth hydroxide (which is in effect an oxide) was then decanted and washed free from sodium with water. The slurry was then dried, and the product calcined to rare earth oxide at a temperature from 800° to 1000°C and, in this instance, at approximately 900°C. After sieving to break down agglomerates, the product was then ready for use. The percentage of rare earth oxide in the product was adjusted by altering the proportions of the Molochite to the rare earth solution.

Grinding Aids and Lapping Compounds

GRINDING AIDS

Sodium Nitrite

J.C.J. Wirth; U.S. Patents 4,239,501; December 16, 1980; and 4,095,961; June 20, 1978 provides a grinding and cutting process and articles employing an effective amount of a grinding aid, typical of which is sodium nitrite which, when applied to the abradant or cutting edge of a tool, accelerates the modification of a metal workpiece and prolongs the useful life of the tool. The grinding aids of this process are compounds free of sulfur and/or halogen and having a melting point in the range of 70° to 1000°F, a decomposition temperature at least 100°F above the melting temperature and a latent heat of melting greater than 10 calories per gram.

It was discovered that sodium nitrite undergoes phase transitions when subjected to elevated temperatures. In such a phase transformation, the energy necessary to accomplish the transition is in the form of heat absorbed and is derived from surrounding objects.

Sodium nitrite, because of its excellent thermal properties in heat transfer, is used as the principal component in the formation of a grinding or cutting accelerator according to this process. The quantity of sodium nitrite used is optimized to absorb, as much as possible, the frictional heat generated during the abrasive machining and/or cutting of metals and it was found that when sodium nitrite was applied, e.g., to a coated abrasive belt via a solid vehicle in the form of a cerate, the sensible surface temperature of the abraded metal can be reduced by 500°F. The reason for this is that a relatively large amount of heat generated by the abrading process is absorbed as the latent heat of melting of sodium nitrite.

It has been found desirable in fabricating the product of this process, that the wax or grease cerate be first heated 20°F above its melting point and, while in such melted state, heated grinding aid, e.g., sodium nitrite in crystalline, granular or micropulverized form added thereto and uniformly dispersed therein. By preheating the aid to the same temperature as the wax prior to introduction, the temperature of the melt will not be prematurely lowered, thereby assuring proper uniform distribution of the aid throughout the wax matrix. The use of thickening or suspension agents to control the viscosity prevents the salt particles from settling while the mass is cooling.

The principles of this process can also be employed in ordinary machining processes such as drilling, milling and lathing by conducting the machining process in a manner such that the interface of the cutting edge and the workpiece is immersed in a liquid or waxy vehicle containing at least 10% of the grinding aid. The vehicle may be a hard or soft wax as noted above, or a liquid such as water or preferably a natural or synthetic oily material such as a liquid hydrocarbon, or a Carbowax containing wetting and suspending agents to aid in the formation of a stable suspension of the aid.

The grinding aid may be incorporated into and/or applied to the cutting or grinding edges in other vehicles and forms. It may be applied as a coating with or without a top coating to act as a vapor barrier to prevent pickup of water. It may be impregnated into porous grinding tools simply by soaking or pressure impregnating them in a nitrite molten or solvent solution which may contain a wetting agent to provide proper wetting and penetration.

In addition, the aid can be incorporated directly into known grinding and/or cutting tools, as for example, during fabrication, e.g., in the size and/or make coat of a coated abrasive, and in the bonding resin of a bonded grinding wheel or cut off wheel.

When applied as a coating, it can be admixed with a suspending agent and an inert liquid vehicle and can be applied by brush, doctor or roll coating, or even through an aerosol spray. It can also be applied as a 100% solid by using pressed or melted salts in a bar form or molten salt may itself be sprayed on the workpiece or the cutting tool or abrasive. For example, molten salts can be directed to the grinding interface when grinding or scarfing stainless steel billets with coarse grit resinoid wheels.

Wax Treated Grindstone

O. Kunimasa; U.S. Patent 4,190,986; March 4, 1980; assigned to Daichiku Co., Ltd., Japan provides a grindstone impregnated with a solution containing a higher aliphatic acid having 5 to 30 carbon atoms per molecule and a higher alcohol having 5 to 30 carbon atoms per molecule mixed together at a range of ratios by weight of 10:90 to 90:10.

A significant increase of grinding efficiency and the prevention of substantial elimination of burning during the grinding operation are observed in a resinous grindstone which has been thus treated with the mixture prepared according to this process.

Example 1: A commercially obtainable offset resinous grindstone was obtained. The grindstone standard was A-36-P while the size of the grindstone was 100 by 6 by 15 mm.

Stearic acid, which is solid at room temperature, and isostearic acid, which is liquid at room temperature, were used as the higher aliphatic acids. A mixture containing 60% cetyl alcohol and 40% stearyl alcohol by weight, which is also solid at room temperature, was used as the higher alcohols. The above aliphatic acids-higher alcohols were mixed in equivalent ratios by weight and heated to 120°C to obtain a homogeneous liquid.

A grindstone as described above was then totally immersed in the specified treatment solution for three to five minutes, removed from the solution, drained, and then dried at a temperature of 100° to 120°C. By this treatment, it was found that approximately 2 to 4 grams of treatment solution was impregnated and retained.

Example 2: The ability to maintain the grinding efficiency and the amount of grinding as measured by ground weight as against the predetermined amount of wear of the grinder were examined. In other words, the amount of stainless steel removed in a time which is necessary for an offset grinder such as described above having a diameter of 100 mm to wear down to a diameter of 70 mm was measured. The results are tabulated in the table below. The test was conducted in the manner as follows. The grinding was continued for one minute followed by one minute of nonoperation repeated five times. During the one minute of nonoperation or pause, the amount of ground material removed and the amount of wear on the grinder were measured.

	Example 2 (A)	Conventional (B)	A/B
Treatment	Same as in Ex. 1	No treatment	–
Time required to reduce the diameter from 100 mm to 70 mm (min)	20.5	10.3	2.0
Amount of wear of the grinder (g)	35.4	33.2	–
Amount of stainless steel removed (g)	103.8	36.1	2.9
Removal rate of stainless steel (g/min)	5.1	3.5	1.5
Grinding efficiency	2.94	1.08	2.7
Luster of the grinding surface	Metallic luster-brown	Blue-bluish purple	–
Estimated temperature of the surface (°C)	500-600	1000-1200	–

The results of the table indicate that when compared to an untreated conventional grinder, the rate of wear on a treated grinder is half as much, the amount of removal of stainless steel is three times greater, the removal rate about 1.5 times greater, and the grinding efficiency approximately three times greater than the conventional grinder.

For Grinding Coal or Ores in Liquid Medium

W. Manfroy and R.R. Klimpel; U.S. Patent 4,162,044; July 24, 1979; assigned to The Dow Chemical Company provide a process for grinding coal or ores containing metal values which comprises carrying out the grinding operation in the presence of a liquid medium and a grinding aid comprising polyacrylic or polymethacrylic acid or an anionic derivative thereof. The grinding aid is dispersible in the medium and is employed in an amount effective to provide increased grinding efficiency. The use of such grinding aids results in a substantial increase in the rate of particle breakage and permits higher density (solids) slurries of coal or ore to be ground, thereby achieving a greater volume throughput of solid ground per unit of time with a corresponding increase in the recovery rates of the desired metal value where ores are ground.

The resulting improved efficiency in the overall grinding operations, i.e., in the use of mill capacity and particularly in the consumption of energy per unit of product, is achieved with the present grinding aids without encountering a decrease in grinding kinetics normally observed when higher density slurries are ground.

It has also been found that the polyelectrolyte grinding aids of this process usually do not detrimentally affect downstream processing operations which are performed, particularly on mineral ores, after the mineral leaves the grinding mill. Thus, for example, the polyelectrolyte grinding aids generally do not detrimentally affect processes such as, for example, froth flotation processes in which select metal values such as copper, lead, zinc or gold are recovered from the ore with the aid of flocculating and deflocculating agents.

Neither do the polyelectrolyte grinding aids of this process have any counterproductive effects in subsequent operations such as, for example, in pelletizing iron ore. Since the polyelectrolyte is adsorbed on the solids, the system does not contribute to downstream pollution problems upon discharge of the aqueous medium such as might be the case with phosphatic materials, for example.

The method is preferably carried out in the presence of a polar liquid medium in which the grinding aid is sufficiently dispersible to produce an improvement in grinding efficiency, although the use of a liquid medium comprising a liquid which itself is not a solvent for the grinding aid may be feasible provided that a solvent or dispersant for the grinding aid is also present. Accordingly, the term dispersible means the aids are soluble or dispersible in the medium employed to an extent sufficient to provide adsorption thereof on the solid particles and increased grinding efficiency. Water is ordinarily employed and is the preferred medium. The concentration of the solids, e.g., ore or coal, in the liquid medium may vary within wide limits and it is usual to operate with a slurry solids content in the range of from about 40 to 95, most preferably from 70 to 88 percent by weight of the slurry.

The polyelectrolyte grinding aid used in this process is suitably any polyelectrolyte derived from acrylic acid which is inherently dispersible in the liquid medium employed. Preferably, the polyelectrolyte is dispersible in the liquid medium without the aid of surfactants. Exemplary polyelectrolytes include the water-soluble or dispersible polymers or copolymers of acrylic or methacrylic acid or the salts thereof. Preferably, the homopolymers are employed. The salt of the acid may be that of an alkali metal, for example, sodium, potassium, lithium or the like, or may be an ammonium salt. The alkali metal salts, especially the sodium salts, are preferred. Preferably, the grinding aids are employed in the form of water-soluble salts. Preferred grinding aids are the sodium salts of polyacrylic acid or polymethacrylic acid.

Copolymers of acrylic or methacrylic acid with other ethylenically unsaturated monomers such as, for example, acrylamide, vinyl acetate, acrylonitrile, methyl acrylate, ethyl acrylate, 2-methylpropanoic acid and the like are also within the scope of this process. Such copolymers usually contain about 25 weight percent or more of the acrylic or methacrylic acid monomer.

Generally, the anionic group of the grinding aid has a pKa of about 8 or less, wherein pKa is a negative logarithm of the acidity constant for the acidic (anionic) group. The average molecular weight (as determined by the Mark Houwink equation) of the water-soluble or dispersible polymers and salts thereof usually ranges from about two or three thousand up to about fifty thousand. Preferably, grinding aids having an average molecular weight from about 5,000 to 20,000 and more preferably from about 5,000 to 10,000 are employed.

Generally, the effective amount of grinding aid employed to increase the rate of ore grinding can be as low as about 0.002% by weight (of actual polymer) based on the dry weight of the ore present. The maximum amount of grinding aid employed is usually limited by economic constraints, i.e., the high cost of the grinding aid. Preferably, the grinding aids of this process are employed in the range of from about 0.003 to 0.08% by weight of actual polymer (i.e., from about 0.03 to 0.8 mg/g) preferably, from about 0.01 to 0.04% by weight.

Example: To illustrate the effectiveness of the chemical grinding aids of this process on an industrial scale, reference is made to a trial conducted on taconite ore in an industrial scale overflow discharge rod grinding mill in which the ore was ground on a continuous circuit basis. In control grinding runs without the addition of a chemical grinding aid, an optimum range of grinding ore in an aqueous slurry without chemical addition to obtain a particle size less than 35 mesh (U.S. Standard) was obtained between 80 to 82% solids in the slurry. Any further increase in the percentage of solids in the slurry quite expectedly necessitated a reduction in the throughput of the ore ground in order to match the constant percent less than 35 mesh of the product. At an 86% solids density, the throughput falls off to about 400 tons per hour from a previous high of about 410 tons at a solids concentration of about 80 to 82%.

In grinding runs, a chemical grinding aid, sodium polyacrylate, was continuously added to the slurry in an amount of 0.13 lb/ton of dry ore (0.065 mg/g). The results were initially quite expected in that the addition of the grinding aid, at a concentration of 80 to 82% solids in the slurry, resulted in only a slight increase in throughput as compared to the control run in which no grinding aid was added to the slurry. However, an unexpected and surprisingly sharp increase in the throughput was observed when the percent solids concentration of the slurry was raised to above 82%. At an 84% solids density, the throughput had increased to about 420 tons per hour while still maintaining the same percent less than 35 mesh in the product. At an 86% solids density, the throughput reached about 450 tons per hour, which represents an increase of about 10% over the 82% solids level of the control run.

An increase in the amount of the sodium polyacrylate grinding aid to about 0.20 lb/ton was found to further increase the throughput, particularly at a percent solids concentration in the slurry of about 82 to 86%.

In related work *W. Manfroy and R.R. Klimpel; U.S. Patent 4,136,830; January 30, 1979; assigned to The Dow Chemical Company* describe the polyelectrolyte grinding aid to be any copolymer of styrene with maleic anhydride, itaconic or citraconic acid which is inherently dispersible in the liquid medium employed.

Preferably, the polyelectrolyte is dispersible as colloidal size particles in the liquid medium without the aid of surfactants. The water-soluble salts of the copolymers are preferably employed and may be that of an alkali metal, for example, sodium, potassium, lithium or the like, or may be an ammonium salt. The alkali metal salts, especially the sodium salts, are preferred. Such copolymers preferably contain approximately one to one mol ratio between the styrene and the maleic anhydride, itaconic, or citraconic acid salts for use as grinding aids herein. Usually, styrene-maleic anhydride copolymers must be posthydrolyzed and contain about 17 or more mol percent of maleic anhydride.

A further modification of the process entails the partial sulfonation of the styrene fraction of the copolymer. The sulfonation of the styrene can be done either before or after the polymerization. A preferred grinding aid is styrene-maleic anhydride. In another embodiment, copolymers of styrene with itaconic and citraconic acids are preferred. Styrene-maleic anhydride copolymers wherein the styrene is partially sulfonated constitute a further preferred embodiment.

Generally, the anionic group of the grinding aid has a pKa of about 6 or less. The average molecular weight of the water-soluble or dispersible copolymers and salts thereof usually ranges from about one thousand up to about fifty thousand. Preferably, grinding aids having an average molecular weight (as determined by the Mark Houwink equation) of from about 1,000 to about 20,000, and preferably from about 1,000 to about 10,000, are employed.

In earlier work the grinding aid used by *W. Manfroy and R.R. Klimpel; U.S. Patent 4,126,278; November 21, 1978; assigned to The Dow Chemical Company* is suitably any polymer of sulfoethylmethacrylic acid (hereinafter "SEM") or any polyelectrolyte derived therefrom which is inherently dispersible in the liquid medium employed. Preferably, the polyelectrolyte is dispersible in the liquid medium without the aid of surfactants. Preferably, an SEM homopolymer is employed. The salt of the acid may be that of an alkali metal, for example, sodium, potassium, lithium or the like, or may be an ammonium salt. The alkali metal salts, especially the sodium salts, are preferred. Preferably, the grinding aids are employed in the form of water-soluble salts. A preferred grinding aid is the sodium salt of a sulfoethylmethacrylic acid polymer.

Copolymers of SEM acid with other ethylenically unsaturated monomers, such as vinylidene chloride, acrylonitrile, acrylamide, methylol acrylamide, N-hydroxyethyl methacrylamide, styrene sulfonate, styrene, vinyl acetate, acrylic acid or methacrylic acid can also be used.

Generally, the anionic group of the grinding aid has a pKa of about 6 or less. The average molecular weight of the water-soluble or dispersible polymers and salts thereof usually ranges from about 2,000 or 3,000 up to about 50,000. Preferably, grinding aids having an average molecular weight (as determined by the Mark Houwink equation) of from about 1,000 to 20,000 and more preferably from about 1,000 to 15,000 are employed.

In a similar process *W. Manfroy and R.R. Klimpel; U.S. Patent 4,126,277; November 21, 1978; assigned to The Dow Chemical Company* use as the grinding aid any sulfonated monovinylidene aromatic polymer such as sulfonated polystyrene or sulfonated polyvinyl toluene or polyelectrolyte derived therefrom which is inherently dispersible in the liquid medium employed. Preferably, the polyelectrolyte is dispersible in the liquid medium without the aid of surfactants. Preferably, homopolymers of sulfonated polystyrene or sulfonated polyvinyl toluene are employed. The polymers are preferably used in salt form and the salt may be that of an alkali metal, for example, sodium, potassium, lithium or the like, or may be an ammonium salt. The alkali metal salts, especially the sodium salts, are preferred. A preferred grinding aid is the sodium salt of sulfonated polystyrene.

Copolymers of sulfonated polystyrene or sulfonated polyvinyl toluene with ethylenically unsaturated compounds such as acrylic or methacrylic acid, acrylamide, methacrylamide, acrylonitrile, methacrylonitrile, maleic anhydride, fumaric acid or other ethylenically unsaturated monomers such as ethylene, propylene, butadiene can also be utilized as grinding aids. The copolymers preferably contain about 25 weight percent or more of the sulfonated polystyrene or sulfonated polyvinyl toluene.

Generally, the anionic group of the grinding aid has a pKa of about 8 or less. The average molecular weight (as determined by the Mark Houwink equation) of the water-soluble or dispersible polymers and salts thereof usually ranges from two hundred up to about fifty thousand. Preferably, grinding aids having an average molecular weight of from about 200 to 20,000 and more preferably from about 5,000 to about 20,000 are employed.

In still another process *W. Manfroy and R.R. Klimpel; U.S. Patent 4,126,276; November 21, 1978; assigned to The Dow Chemical Company* describe as the grinding aid a polyelectrolyte polymer comprising any poly N-(sulfoalkyl)acrylamide or poly N-(sulfoalkyl)methacrylamide or copolymers containing the same which is dispersible in the liquid medium employed.

Preferably, the polyelectrolyte grinding aid is dispersible in the liquid medium without the aid of surfactants. Preferably, polyelectrolyte homopolymers are employed. Additionally, the grinding aids are preferably employed in the form of water-soluble salts. The salt of the polymer may be that of an alkali metal, for example, sodium, potassium, lithium or the like, or may be an ammonium salt. The alkali metal salts, especially the sodium salts, are preferred. A preferred grinding aid is sodium poly N-(sulfomethyl)acrylamide.

Copolymers of poly N-(sulfoalkyl)-acrylamide or -methacrylamide monomers together or with other ethylenically unsaturated monomers, such as styrene, acrylic acid or methacrylic acid and the like can also be used.

Generally, the anionic group of the grinding aid has a pKa of about 6 or less, wherein pKa is a negative logarithm of the acidity constant for the acidic (anionic) group. The average molecular weight of the water-soluble or dispersible polymers and salts thereof usually ranges from about five thousand up to about fifty thousand. Preferably, grinding aids having an average molecular weight (as determined by the Mark Houwink equation) of from about 5,000 to 20,000 and more preferably from about 5,000 to 15,000 are employed.

Grinding Agent Carrier

G. Joschko and P. Albrecht; U.S. Patent 4,133,145; January 9, 1979; assigned to TED Bildplatten AG AEG-Telefunken-Teldec, Switzerland provide a process for producing a grinding agent carrier comprised of a carrier body and a layer of grinding agent on the surface of the carrier body, the grinding agent layer being initially produced on a planar substrate by sedimentation from a suspension comprised of a suspension medium and grinding agent and then transferred from the substrate and bonded to the carrier body.

The process comprises maintaining the substrate in a horizontal position, and effecting the sedimentation on the horizontal substrate from a liquid column of suspension in which the liquid column has a height in millimeters which is at most around 0.4 times the square root of the width in millimeters of the liquid column. The height of the liquid column can be less than one-tenth of its width.

It has been found that the last phase of drying of the liquid column is decisive for the quality of the grinding agent layer on the planar substrate. It is therefore not necessary nor advisable to sediment from a high liquid column but it is better to effect the sedimentation onto the planar and horizontally disposed substrate in but a thin layer, for example, in the form of a puddle. The sedimentation is then effected exclusively by evaporation so that no other measures are required to remove the suspension medium. Malfunctions produced by sucking away the excess of suspension medium or similar methods are thus eliminated.

The uniformity of the resulting grinding agent layer can additionally be improved by effecting the evaporation in a quiescent atmosphere.

Since the evaporation speed also influences the uniformity of the grinding agent layer, a suspension medium is advisably selected which is volatile at room temperature in the same order of magnitude as ethanol. The sedimentation is then effected approximately at room temperature.

In the practice of this process, the layer of grinding agent preferably comprises a layer of diamond powder as the grinding agent, but any other similarly fine-grained grinding agent, for example, one based on emery or the like can also be used.

For producing the suspension, easily evaporatable solvents have been found to be particularly advantageous as the suspension medium. A preferred medium for producing a suspension of fine-grained diamond powder has been found to be ethanol which can be easily removed from the sediment at room temperature.

If ethanol is used as the suspension medium, for example, the evaporation speed at 22°C is just great enough that a uniform layer is produced from a mixture of ethanol and diamond powder having a diamond grain size of 1.5 to 3 μ.

In producing a grinding agent carrier with a substantially single-layer grinding agent coating, it is necessary to select the concentration of the suspension so that with the selected height of the liquid column, a single-layer grinding agent coating is produced. If the sedimentation is effected, for example, from a suspension puddle of 3.3 cm^2 with diamond grain sizes between 1.5 and 3 μ, 3 to 5 mg, especially 5.7 mg diamond grains should be contained in one milliliter of ethanol in order to obtain a single-layer grinding agent coating.

The auxiliary carrier which is employed as the horizontal, planar substrate advisably has a roughness which is adapted to the grain size of the grinding agent. For example, for diamond grains having a grain size of between 1.5 and 3 μ, a roughness depth of approximately 0.25 μ is applicable which corresponds in the order of magnitude to one tenth of the size of the diamond grains.

The upper limit for the ratio of width to height is given by the fact that the height must be essentially greater (especially about 100 times greater) than the grain size of the diamond powder and that the width must not be so great that an essential evaporation begins before the suspension is completely applied to the planar substrate.

The preferred range of the height in millimeters in relation to the width d in millimeters of the column is $0.25\sqrt{d}$ to $0.35\sqrt{d}$.

Suitable heights of the puddle are 0.3 to 3 mm, using widths of 10 to 100 mm. It is preferred to use a ring-like substrate with an outer diameter of 32 mm and an inner diameter of 15 mm; then the preferred height of the column is 0.9 mm.

Referring to Figure 7.1a, there is shown a planet gear assembly, generally **20**, which has a stationary outer ring **1** having a set of teeth which extends about the inner periphery **22** of the ring. Planet gear assembly **20** includes a sun gear **2** which is driven by a motor **3** and a plurality of planet gears **4**. In the embodiment shown, planet gear assembly **20** is provided with three planet gears **4**. A glass vessel **5** is placed on one of the planet gears **4** and accommodates a suspension **6** comprised of an ethanol suspending medium. a grinding agent **7** of diamond powder, and plastic granulators **8** in the form of beads. Although only one vessel **5** is shown, each of the planet gears **4** can be provided with a vessel **5**.

Figure 7.1: Grinding Agent Carrier

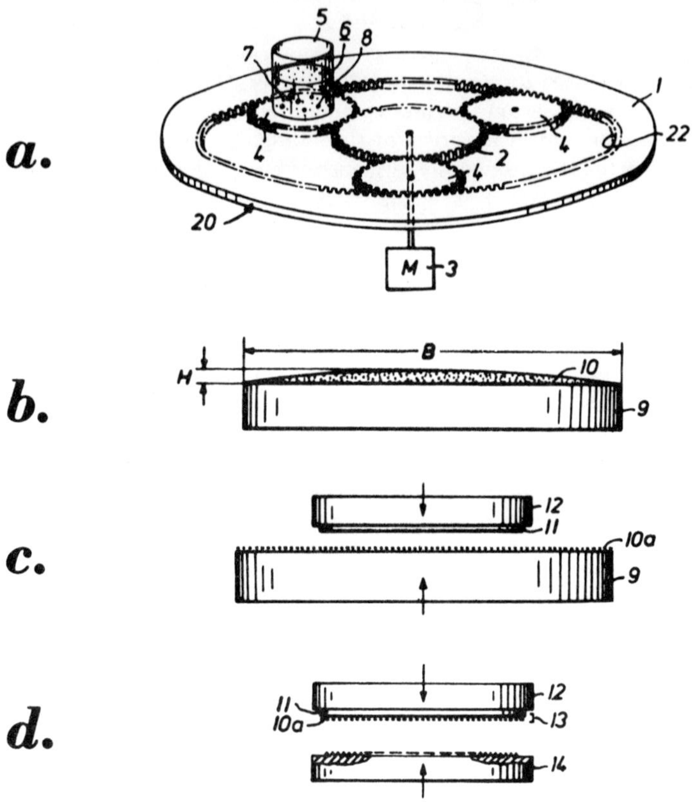

Source: U.S. Patent 4,133,145

After the vessel **5** is placed on planet gear **4**, the planet gears are then rotated to homogeneously distribute the diamond powder in the suspending medium.

Once the diamond powder has been homogeneously distributed in the ethanol, the rotation of the planet gears is stopped and a required partial quantity of the suspension is removed from vessel **5** by means of a pipette and placed onto an auxiliary carrier **9**, shown in Figure 7.1b. Due to its homogeneous distribution, the diamond powder will be uniformly dosaged. The required quantity of liquid is selected, by means of the appropriate mixing ratio of the diamond powder and the suspension medium, so that the entire surface area of auxiliary carrier **9** will be covered with a liquid column **10** of the suspension in the form of a puddle, but the ethanol will not flow over the edges of auxiliary carrier **9** which would result in flow patterns which might interfere with the homogeneous distribution of the grinding agent in the suspension.

Moreover, in this way, it is not necessary to provide an edge limitation for the puddle by means of delimiting walls or dams which can lead to inhomogeneities in the edge zone of the grinding agent layer. Liquid column **10** can be merely constituted by a puddle having a height **H** and a width **B** which is more than 10 times the height **H**. During application and drying of the suspension, the auxiliary carrier **9** must be placed perfectly horizontally. Upon evaporation of the suspending medium and settling of the diamond particles on the surface of auxiliary carrier **9**, a sediment in the form of a grinding agent layer **10a**, Figure 7.1c, is formed which is homogeneously distributed over the entire surface of auxiliary carrier **9** and has a uniform layer thickness.

After grinding agent layer **10a** is formed on auxiliary layer **9**, it is transferred to a carrier body **11**, as shown in Figure 7.1c, by pressing carrier body **11**, which, for example, may be a PVC foil, onto the coated auxiliary carrier **9** with the aid of a pressing member **12** and a pressure of about 20 x 10^3 N/cm^2. The carrier body **11** thus absorbs the diamonds of the grinding agent layer **10a**, but at such a slight penetration depth that there is produced sufficient adhesion for a single-layer coating of diamond and excess diamonds remain on the auxiliary carrier **9**. This effect can be enhanced by effecting the pressing of carrier body **11** without heating it because heating otherwise could effect the unwanted softening of the carrier body **11**. In this way, possibly existing fluctuations in layer thickness on auxiliary carrier **9** are eliminated during the transfer onto carrier body **11**.

Figure 7.1d finally shows how a grinding agent profile can be pressed into a grinding agent carrier **13** comprised of carrier body **11** and the adhering grinding agent layer **10a** with the aid of a hot stamping matrix **14**. Stamping matrix **14** is pressed under the influence of heat against the pressing member **12** carrying grinding agent carrier **13**. Even if no groove profile is desired, it is advisable to press a stamping matrix without a profile against the grinding agent layer **10a** under the influence of heat. The temperature may be chosen in the range of around 150° to 170°C and a suitable pressure range is from about 1,500 to 2,000 N/cm^2. It is advisable to cool down the grinding agent carrier before the pressure is reduced to atmospheric pressure.

The method according to this process, has the advantage that it permits the production of grinding agent carriers having very uniform grinding agent layers which is of importance particularly for single-layer coats. In addition, it results in a saving of grinding agent because the transfer of the grinding agent layer **10a** from the auxiliary carrier **9** to the grinding agent carrier body **11** in the cold state prevents accumulations of grinding agent.

Encapsulated Abrasive Grain

R.N. Howard; U.S. Patent 4,112,631; September 12, 1978; assigned to Minnesota Mining and Manufacturing Company provides an encapsulated abrasive-lubricant slurry which is simple to manufacture and which can utilize a wide variety of abrasive grains and lubricants. The capsules of this process, which provide an efficient means of handling fine grain abrasive, are essentially spherical, making them easily handled and dispersed as abrasive in a carrier liquid for lapping. They also conveniently combine both grinding aid and abrasive intimately mixed in predetermined proportions insuring uniformity of performance. Unbroken capsules are considerably larger than the individual encapsulated abrasive grains, and hence more easily separated from the carrier liquid for salvage.

In coated abrasive constructions the relatively large capsules are substituted for the fine abrasive grains, resulting in a thicker, more score-resistant abrasive coating than could otherwise be obtained. In use, the capsules gradually rupture, and the abrasive not being held in place by the make coat resin provides a loose grain finish with the convenience of a coated abrasive. Where a low melting solid, e.g., wax or pitch, is used as the lubricant, or grinding aid, a liquid can be introduced to the work surface to remove swarf without concurrently removing the abrasive mineral. The capsules also provide a means for using corrosive or toxic grinding aids without danger to workmen and machines, since the grinding aid is not handled as a separate liquid.

The capsules, which contain both fine abrasive grains and liquid or low-melting grinding aids, can be made substantially spheroidal and have moderate diameters, e.g., 10 to 200 microns. The smaller capsules, e.g., 10 to 30 microns, are, of course, suitable for use with the smaller abrasive grains, e.g., 0.5 to 5 microns. Larger capsules, however, may include either fine or coarse abrasive grains. In all cases, however, the capsule will be at least twice the size of the encapsulated abrasive grain, so that a plurality of abrasive grains are generally present in the capsule. The capsules will normally be spherical; this shape is preferred because of ease in handling. Ellipsoidal or irregularly rounded capsules are also satisfactory.

The abrasive grains to be encapsulated may include such well-known abrasive grains as silicon carbide, aluminum oxide, boron carbide, as well as other abrasive grains and mixtures thereof. The abrasive granules of this process are particularly suited for use with expensive, difficult-to-disperse abrasive grains such as diamond or cubic boron nitride.

The abrasive grains to be encapsulated will range in size from submicron, i.e., ½ micron or less, up to 25 microns in diameter. The abrasive grains will generally have a diameter of less than about 25 microns; grains larger than 25 microns can be coated on abrasive articles or used in slurries to form useful materials without the need of encapsulation. In this process, the most preferred abrasive grains are those having a nominal diameter of 15 microns or less, since these finer abrasives are more difficult to use in coated abrasive articles and benefit the most from being encapsulated.

The abrasive grains or grains to be used are mixed or dispersed into a grinding aid, e.g., grinding oil or lubricant. The term "grinding aid" is intended to embrace nonabrasive liquids or low melting solids which reduce the force required to abrade or reduce the frictional force between the workpiece and abrasive. Examples of suitable grinding aids are the Vantrol and other low viscosity, high flash oils; Almag diamond saw lubricant; carnauba wax; oleic acid; water-soluble oils, e.g., Cindol; glycerols; petroleum oil; chlorinated hydrocarbons; paraffin; and pitch. For polishing such materials as zinc, silicon, or aluminum, acidic or basic materials (e.g., $NaHSO_4$ or NaOH) may be employed as grinding aids.

The abrasive grain to grinding aid weight ratios for relatively inexpensive common abrasives such as silicon carbide and aluminum oxide will be generally 40% mineral-60% grinding aid to 60% mineral-40% grinding aid. These ratios represent up to about 10 to 30% by volume of abrasive mineral, the amount of mineral being chosen within this range to maximize the abrasive action and facilitate encapsulation of the abrasive slurries.

For more expensive abrasive grains, e.g., cubic boron nitride or diamond, the abrasive to grinding aid ratio will normally range from one part abrasive:two parts grinding aid to one part abrasive:five parts grinding aid by weight, which corresponds to approximately 5 to 15% by volume of abrasive. The amount of expensive abrasive chosen is lower in order to provide sufficient abrasive action while minimizing the cost to the user.

Example: *Preparation of Lubricant-Abrasive Capsules* – A urea-formaldehyde precondensate was formed by blending 20 grams of urea and 54 grams of 37% formaldehyde in water, adding 0.4 ml tetraethylene amine to render the system basic and slowly stirring the mixture for one hour at 70°C. The system was then diluted with 92 ml of water and the temperature lowered to about 30°C.

An oil-abrasive slurry containing 40 grams of Vantrol 5551-A and 10 grams of 3 micron diamond was added to the system and 50 ml water and 1.8 ml 9% hydrochloric acid added immediately thereafter. The resulting mixture was stirred rapidly to form droplets of the oil-abrasive slurry and the temperature raised to 40°C over a 2 hour period and maintained at 40°C for an additional 2 hours while continuing agitation. In the warm acidic environment the urea-formaldehyde condensed into shell walls enclosing the slurry.

The solution was then neutralized using 5 grams of 8% sodium acetate, after which the capsules were filtered and washed several times in clear water. The diameter of the capsules ranged from 5 to 150 microns. The capsule diameter can be lowered by increasing, or raised by decreasing, the rate of agitation.

Dresser in Friction-Meltable Matrix

According to the process of *P.E. Bonnice; U.S. Patent 4,098,253; July 4, 1978; assigned to General Electric Company* a diamond, cubic boron nitride, silicon carbide, aluminum oxide, or obviously equivalent, grinding wheel is dressed by means of applying an abrasive compound in a normally rigid but friction-meltable matrix against the rotating wheel and causing the dressing composition to melt and form a dressing slurry on the surface of the wheel and then pinching the slurry between the wheel and a workpiece until dressing is substantially complete. Also provided by this process is a wheel dresser which comprises a body of multiplicity of abrasive particles dispersed in a friction-meltable organic polymeric matrix, which in preferred embodiments is in stick form adapted for hand application to the grinding wheel surface.

A grinding wheel dressing stick is made by: (1) melting a quantity of polyethylene glycol (PEG) of molecular weight of 1,300 to 1,600, 43° to 45°C melting point (Baker Type U-220 or equivalent); (2) adding 100 grit aluminum oxide at a mix ratio of 0.6 gram of Al_2O_3 per ml of polyethylene glycol; (3) stirring the mixture until completely mixed and the PEG begins to solidify; (4) pouring mixture into a mold and cooling (a refrigerator can be used to accelerate cooling, and reduce sedimentation); and (5) removing the hardened stick from the mold and storing in a cool dry place until used. Grit size is U.S. Standard Sieve.

Instead of polyethylene glycol of the type described, one having a molecular weight of 1,300 to 1,600 and a melting point of 42° to 46°C can be used. Other organic polymeric matrixes can be used, such as paraffin wax. A cubic boron nitride abrasive wheel or other abrasive-containing wheel can be used, such as a diamond wheel. Instead of aluminum oxide crystals, others can be used such as cubic boron nitride, diamond, silicon carbide, and the like. Instead of 100 grit size, the dressing compound grit can range from 60 to 1,000. Instead of a mix ratio of 0.6 g/ml, other ratios, such as 0.4 to 1.5 g/ml can be used.

Wheel trueing can comprise a preferred preliminary step, using a conventional wheel-forming tool. Washing off the dressing compound is contemplated in many embodiments. If polyethylene glycol or other water-soluble matrixes are used, the dressing compound retained by the wheel is easily removed with flushing or coolant water. Obviously, the wheel can be rotated first and the wheel dresser applied and then the wheel set against the workpiece, or vice versa, or the two operations can be carried out simultaneously.

Solid Triglyceride for Use with Nonferrous Metal

A. Olsin; U.S. Patent 4,060,943; December 6, 1977 provides a metal working compound comprising a glyceride and preferably comprising in predominant amount a triglyceride in which all of the glycerides have alkyl or alkenyl moieties of from about 7 to 17 carbon atoms, the working compound having a melting point within the range of about 100° to 125°F, a softening point within about 10°F of the melting point, an iodine value number no greater than about 10, and a solid fat index at 80°F within the range of about 35 to 60%.

In use, the working compound of this process, which is solid at room temperatures, is merely applied to the metal or the working instrument or to both. The compound has a sufficiently low melting point that under the heat generated by contact between the metal and instrument, the compound is readily reduced from a solid state to a flowable, lubricating liquid. But as soon as the compound leaves the immediate area of contact or heat is otherwise removed, the compound immediately sufficiently congeals, that is, does not splatter or otherwise fly off due to centrifugal force because of its closely associated softening point on the temperature scale.

Further, in solid or liquid form, the working compound alone fills the pores of the working instrument and prevents the buildup in the pores of metal particles created by the operation. Also, this compound does not have the adhesiveness of normal lubricants, such that there is little or no buildup of clusters of particles otherwise held together by a sticky lubricant and which can foul a metal-shaping operation. Still further, this metal working compound successfully counteracts the welding action of nonferrous metals to a forming instrument.

LAPPING COMPOUNDS

Cosmetic Type Holder for Diamond Lapping Compound

In accordance with the process of *F. Rademacher; U.S. Patent 4,218,852; August 26, 1980* there is provided a holder **10**, shown in Figure 7.2, having an elongated inner sleeve guide member **12** connected to a rotatable knob **14**. A receptacle **16** for holding the stick of diamond lapping compound **30** is adapted to be slidable in the inner sleeve guide member **12** in a manner as herein described.

Figure 7.2: Cosmetic Type Holder for Diamond Lapping Compound

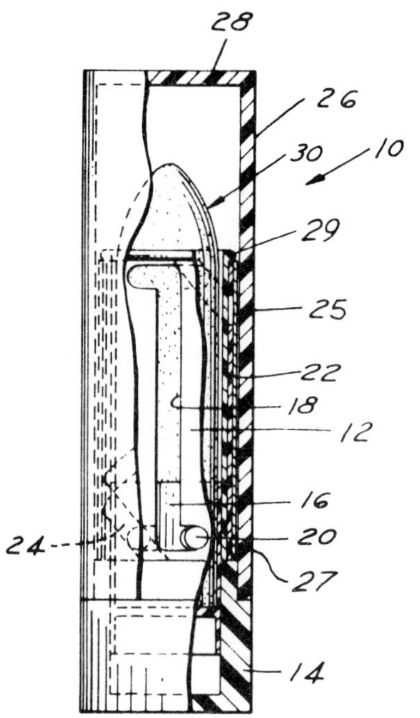

Source: U.S. Patent 4,218,852

The inner sleeve guide member **12** consists of a pair of diametrically opposed elongated slots **18** which allows the receptacle **16** to move longitudinally in the inner sleeve guide member and held therein by means of a pair of projections **20** diametrically opposed to each other on the outer surface of the receptacle **16**. The inner sleeve guide member is encased in an outer sleeve member **22**. The outer sleeve member comprises a slot **24** located longitudinally and diagonally and is entirely encased in a cylinder member **25**. The cylinder member **25** is adjacent the outer sleeve **22** and both members abut the shoulder **27** of the inner sleeve member **12** at one end thereof, and is held from slipping out at the other end by a lip **29** on the opening end of the inner sleeve member **12**.

Thus it is readily apparent that turning the knob **14** and grasping or holding the cylinder member **25**, the receptacle **16** is moved axially or longitudinally within the inner surface of the inner sleeve guide member **12**. This is accomplished by means of the diametrically opposed projections **20** moving longitudinally in the diagonal slot **24**. A cover or outer casing **26** comprising an opening at one end and a closed end **28** provides a seal and cover for the unit.

There is thus provided a stick of diamond lapping compound in an applicator which is compact and easy to use by an operator. That is, an individual may use several different grades of diamond lapping compound before the workpiece is polished to the specification required. In order to do so, an operator will start with coarser grade of diamond lapping compound and end with using an extremely fine or superfinish diamond lapping compound.

Gelled Hydrocarbon

The gelled lapping compositions of the process of *J.W. van Hesden; U.S. Patent 4,046,524; September 6, 1977; assigned to Borg-Warner Corporation* comprise a liquid hydrocarbon gelled by the addition of from about 0.1 to 10% by weight of the liquid hydrocarbon of a pseudo double salt of an alkyl orthophosphoric acid as a gelling agent, and a particulate lapping abrasive grit. The physical characteristics of these gelled compositions distinguish them from merely thickened solutions which have increased viscosities resulting from dissolved solute. Gel consistency will be thinned upon dilution, but the gel structure will not be lost. At very high dilutions, the gel consistency may not be visually apparent and the presence of gel structure will be detected only by measurement of bulk physical properties of the system.

The liquid hydrocarbons useful for the purposes of this process are the liquid petroleum hydrocarbons capable of being gelled by the particular alkyl orthophosphate gelling agents employed, including kerosene, diesel fuel, light mineral oils, spindle oil and the like. While light petroleum fractions including gasoline, naphtha and the like may also be employed, their volatility and flammability make them hazardous and therefore less desirable for most lapping purposes. High viscosity oils and greases may also be gelled, but for most applications these will be too viscous to be of practical value. The preferred liquid hydrocarbons will have a flash point above about 100°F and a viscosity of less than about 200 cp (Brookfield viscometer, No. 2 spindle 50 rpm).

The gelling agents useful for the practice of this process are broadly described as pseudo double salts of alkyl orthophosphoric acids. In general, the pseudo double salt gelling agents are obtained by neutralizing an acidic, nonstoichiometric aluminum salt of an alkyl orthophosphoric acid with a second inorganic base. The acidic nonstoichiometric aluminum salt precursors are the product of the reaction of less than a stoichiometric amount of a basic aluminum compound such as hydrated alumina with an alkyl ester of orthophosphoric acid.

More particularly, an alkyl ester of orthophosphoric acid is first prepared by reacting an equimolar mixture of two or more aliphatic alcohols such as butyl alcohol and decyl alcohol with phosphorus pentoxide to give a mixture of monoalkyl and dialkyl orthophosphoric acids. This mixture is then further reacted with from 20 to 70% of the stoichiometric amount (based on acidic hydrogens in the orthophosphoric acid) of an aluminum compound such as hydrated alumina, aluminum isopropoxide or the like.

The pseudo double salt gelling agent is then formed in situ by dispersing or dissolving the nonstoichiometric product in the liquid hydrocarbon and neutralizing the remaining acidity by adding a second base such as sodium hydroxide, sodium carbonate, trisodium phosphate, calcium oxide or the like to give a gel. The second base may be added in solid form, or where soluble, in aqueous solution. The nonstoichiometric aluminum salts do not give gels until subsequently neutralized by adding the base, and thus are not in themselves gelling agents.

The amount of gelling additive will in part determine the thickness of the gel produced, and at concentrations below about 0.1% by weight the gels are thin and relatively nonviscous. At higher concentrations the gels become increasingly viscous and nonflowing under low shear stresses. The viscosity and gel character also depend in part upon the particular hydrocarbon employed and upon the particular gelling agent as defined both by the alkyl substituents on the orthophosphoric acid, and upon the degree of precision with which neutralization is accomplished. Useful viscosity and suspending characteristics will result only in the range of from about 0.1 to 10.0% by weight of gelling agent and the concentration may be varied within these limits to suit a particular purpose.

The particulate lapping abrasive or grit employed may be any of the abrasives conventionally employed in lapping operations, including silicon carbide, corundum, alumina, garnet, emery, chromic oxide, ferric oxide, cerium oxide, diamond and the like in grit sizes of from No. 20 to No. 3800 and in certain polishing applications, as low as 1.0 micron.

In general, the lapping composition will be prepared by first combining the liquid hydrocarbon with the gelling agent to form the gelled vehicle. As set forth above, the normal procedure will be to combine the liquid hydrocarbon with the requisite amount of acidic, nonstoichiometric aluminum salt, then to neutralize with a second base to form the gel. The acidic, nonstoichiometric aluminum salts disperse readily in hydrocarbons with low shear mixing, however, high speed agitation during the neutralization step is desirable to ensure rapid mixing and uniform gel formation.

It is also possible to prepare concentrates of pseudo double salt gelling agent in a suitable liquid hydrocarbon such as diesel fuel. These concentrates, when added in appropriate amounts to a liquid hydrocarbon and subjected to a high shear mixing operation will be further diluted to form the gelled vehicle.

The desired amount of the lapping abrasive grit is then added to the gelled vehicle with stirring to achieve a uniform dispersion of grit in the vehicle. The resulting composition is then ready for use, or it may be stored in drums for later use. Settling of the grit is a problem with most ungelled vehicles, but with the gelled vehicles of this process little settling occurs even on long storing, and gentle agitation just prior to use is normally sufficient to ensure that the dispersion is uniform. Where large grit sizes are employed, and in compositions containing very high grit loadings, settling is more likely to occur, and it may be desirable to provide more extensive agitation.

Example 1: An alkyl acid orthophosphate was prepared by reacting 53.8 grams of P_2O_5 with a mixture of 30.2 grams of butyl alcohol and 100 grams of a commercial mixture of n-octyl, n-hexyl and n-decyl alcohols in the absence of a solvent. The one hundred grams of butyl octyl acid orthophosphate resulting from this reaction were reacted with 5.2 grams of hydrated Al_2O_3 (30% of theory) by heating the mixture at 110°C for approximately one hour. The water was evaporated off and the resultant partial aluminum salt was cooled. The partial aluminum salt was a viscous oil.

Example 2: Two grams of the nonstoichiometric aluminum salt of Example 1 were added to 250 ml of diesel fuel. The salt dispersed rapidly in the diesel fuel upon stirring at room temperature. No change in viscosity was noted.

Five drops of 30% aqueous sodium hydroxide were added to the mixture with high speed stirring to give a gelled hydrocarbon neutral to methyl red indicator. The gel was highly viscous, and registered a viscosity of 3,200 cp on a Brookfield viscometer at 10 rpm, No. 2 spindle, at room temperature (23°C).

The gelling agent concentration in this example is approximately 1% by weight. While this vehicle may be used without further dilution for many lapping operations, more dilute gels will normally be preferred both by reason of lower viscosity and for economic reasons. Gels having lower concentrations of the gelling agents may be prepared directly by employing lower quantities of partial aluminum salt or 1% by weight gel may be further diluted by addition of a liquid hydrocarbon while stirring.

For Reel Type Lawnmowers

A composition is disclosed by *H.T. Thrower, Jr.; U.S. Patent 4,038,048; July 26, 1977* for lapping or sharpening various cutting edges with primary emphasis on reel type lawnmowers, especially those for cutting golf course greens and other important turf surfaces. The composition is a permanent suspension of a lapping gel and a grinding grit. Contents of the lapping gel include a carboxy vinyl polymer thickener, a neutralizing base, water and a surfactant as essential ingredients, though a concentrated gel may be produced and the water added later. Additionally, other ingredients may be included to improve shelf life, viscosity, stabilize freeze-thaw conditions and the like. Representative lapping compositions according to this process comprise 0.18 to 0.75 weight percent carboxy vinyl polymer; a sufficient amount of a base to neutralize the carboxy vinyl polymer and provide a pH in a range of about 7 to 9; at least about 0.08 weight percent surfactant; 1.2 to 2.4 weight percent of a humectant such as glycerin; a minor amount of a preservative for the surfactant such as formalin; 35 to 70 weight percent of a grinding grit and the balance water.

Company Index

The company names listed below are given exactly as they appear in the patents, despite name changes, mergers and acquisitions which have, at times, resulted in the revision of a company name.

Inventor Index

U.S. Patent Number Index

Copies of U.S. patents are easily obtained from the U.S. Patent Office at 50¢ a copy.

Notice

Other Noyes Publications

REFRACTORY MATERIALS 1980
Developments Since 1977
Edited by J.I. Duffy

Chemical Technology Review No. 178

An up-to-date study of more than 230 processes based on recent technological advances in refractory manufacture. Current processes concerning the conventional silicon oxide refractories are offered. Since there has been accelerated research in the area of silicon carbides and nitrides, these are covered in detail. The unique properties of ceramics based on silicon carbide and silicon nitride make them prime candidates for use in advanced energy conversion systems, vehicular engine design and industrial heat exchangers.

The section on refractory fibers contains pertinent processes for those concerned with high temperature-resistant fibers, e.g., asbestos substitutes for heat shields and protective apparel. Also included are new heat resistant fiber reinforced composites.

Listed below is a partial table of content with **important subtitles,** and in parentheses, the number of processes per topic.

ISBN 0-8155-0827-1

367 pages

Other Noyes Publications

POWDER COATINGS
Recent Developments

Edited by M.T. Gillies

Chemical Technology Review No. 183

The powder coatings industry, buoyed by a favorable environmental climate, could steadily increase their share of the industrial finishes market. The solvent loss and environmental problems associated with conventional coatings figure prominently in the potential of ecologically acceptable powder coatings.

The subject matter of this current book is taken from about 100 recent processes including formulations and additives for powder coatings, their industrial uses and application techniques. Some industrial uses covered are electrical insulation, wire coatings, container coatings, pipe coatings and automotive topcoats. These coatings exhibit excellent uniformity, on particularly difficult areas such as substrate edges and corners, and they can be applied in conventional thicknesses or extremely heavy coats.

Listed below is a summarized table of contents illustrating the **chapter headings and some important subtitles.** In parentheses is the number of processes per topic.

ISBN 0-8155-0836-0 (1981) **326 pages**

Other Noyes Publications

SYNTHETIC GEMS PRODUCTION TECHNIQUES 1980

Edited by L.H. Yaverbaum

Chemical Technology Review No. 149

This book is a new edition of our 1973 volume on the subject which enjoyed universal acceptance. It contains substantial new material. Recent, rapidly rising costs of natural gems have resulted in substantial interest in the manufacture of these artifical gems.

We have retained all previously described processes for making gemstones, also all those processes producing either industrial stones or jewels. In addition we have included all such processes described in the U.S. patent literature since mid-1972.

Most processes for producing synthetic gems have been patented, and technology of any importance in this field can be found primarily in the patent literature.

The partial table of contents below lists **chapter headings, examples of some subtitles** and, in parentheses, the number of processes.

ISBN 0-8155-0788-7 **354 pages**

Other Noyes Publications

TECHNOLOGY OF METAL POWDERS
Recent Developments 1980
Edited by L.H. Yaverbaum

Chemical Technology Review No. 153

In these energy-conscious times powder metallurgical processes have acquired new significance.

The powder metallurgy approach to the manufacture of metal items, when compared with casting and forging, brings with it serendipitous advantages: there is practically no loss by scrap formation; the temperatures applied are much lower and heat is applied for shorter periods. Most sintering reactions take place below the melting point of the metal or metals being shaped.

Research is being hastened trying to satisfy the demand for tailor-made powders with carefully checked granulometric and morphological characteristics.

This book probably represents the most comprehensive data source available for the production of metal powders, ranging from iron and nickel to precious metals and the superalloys.

The partial table of contents below gives **chapter headings, examples of subtitles** and the number of processes per topic in parentheses.

ISBN 0-8155-0794-1 **360 pages**